This book introduces the use of Lie algebra and differential geometry methods to study nonlinear integrable systems of Toda type.

Many challenging problems in theoretical physics are related to the solution of nonlinear systems of partial differential equations. One of the most fruitful approaches in recent years has resulted from a merging of group algebraic and geometric techniques. The book provides a comprehensive introduction to this exciting branch of science. Chapters 1 and 2 review the basic notions of Lie algebras and differential geometry, with an emphasis on further applications to integrable nonlinear systems. Chapter 3 contains a derivation of Toda-type systems and their general solutions based on Lie algebra and differential geometry methods. The final chapter examines the explicit solutions of the corresponding equations. The book is written in an accessible lecture note style, with many examples and exercises given to illustrate key points and to reinforce understanding.

This book will be of interest to graduate students and researchers in theoretical physicics and applied mathematics.

CAMBRIDGE LECTURE NOTES IN PHYSICS 8

General Editors: P. Goddard, J. Yeomans

Lie Algebras, Geometry, and Toda-type Systems

CAMBRIDGE LECTURE NOTES IN PHYSICS

1. Clarke: The Analysis of Space-Time Singularities
2. Dorey: Exact S-Matrices in Two Dimensional Quantum Field Theory
3. Sciama: Modern Cosmology and the Dark Matter Problem
4. Veltman: Diagrammatica – The Path to Feynman Rules
5. Cardy: Scaling and Renormalization in Statistical Physics
6. Heusler: Black Hole Uniqueness Theorems
7. Coles and Ellis: Is the Universe Open or Closed?
8. Razumov and Saveliev: Lie Algebras, Geometry, and Toda-type Systems

Lie Algebras, Geometry, and Toda-type Systems

A. V. RAZUMOV
M. V. SAVELIEV
Institute for High Energy Physics, Protvino, Russia

PUBLISHED BY THE PRESS SYNDICATE OF THE UNIVERSITY OF CAMBRIDGE
The Pitt Building, Trumpington Street, Cambridge CB2 1RP United Kingdom

CAMBRIDGE UNIVERSITY PRESS
The Edinburgh Building, Cambridge CB2 2RU, United Kingdom
40 West 20th Street, New York, NY 10011–4211, USA
10 Stamford Road, Oakleigh, Melbourne 3166, Australia

First published 1997

Typeset by the author

A catalogue record for this book is available from the British Library

Library of Congress Cataloguing in Publication data

Razumov, Alexander V. (Alexander Vitalievich), 1952–
Lie algebras, geometry, and Toda-type systems/Alexander V. Razumov,
Mikhail V. Saveliev.
p. cm. – (Cambridge lecture notes in physics: 8)
Includes bibliographical references and index.
ISBN 0 521 47923 1
1. Lie algebras. 2. Geometry, Differential. 3. Nonlinear theories. 4. Mathematical
physics. I. Saveliev, M. V. (Mikhail Vladimirovich) II. Title. III. Series
QC20.7.L54R39 1997
516.3'62–dc21 96–46118CIP

ISBN 0 521 47923 1 paperback

Transferred to digital printing 2003

To our parents

Vitalii Ivanovich Razumov
Valentina Trofimovna Razumova

and

Vladimir Ivanovich Saveliev
Nonna Mikhailovna Korotkina

with love and gratitude

Contents

Preface

Ce qui fut hier le but est
l'obstacle demain;
Dans les cages les mieux gardées
S'entredévorent les idées
Sans que jamais meure leur faim.
(*Émile Verhaeren: L'impossible*)

Nonlinear integrable systems represent a very important and popular branch of theoretical and mathematical physics, and most of the famous universities and colleges currently include this subject in their educational programmes for students and post-graduate students of physical, mathematical, and even technical specialities. Over the last decade in particular, investigations related to studies of nonlinear phenomena have been in the foreground in an overwhelming majority of areas of modern theoretical and mathematical physics, especially in elementary particle, solid state and plasma physics, nonlinear optics, physics of the Earth, etc. The principal physical properties resulting from the nonlinear nature of the phenomena itself are not in general reproduced here by perturbative methods. This fact leads to the necessity to construct the exact solutions of the corresponding nonlinear differential equations describing the dynamical systems under consideration.

To the present time, physics has placed at our disposal a wide range of nonlinear equations arising repeatedly in its various branches. The methods of their explicit integration began to be efficient in this or that extent, mainly for equations in one and two dimensions, from the end of the 1960s. Some of the principal and important examples given here are Toda systems of various types: abelian and nonabelian finite nonperiodic, periodic and affine Toda systems. For the finite nonperiodic Toda systems the general solution is represented as a finite series made up of nested integrals of arbitrary functions which determine the initial (Cauchy) and boundary (Goursat) value problem; while for affine Toda systems this is given by infinite absolutely convergent series; see, for example, Leznov & Saveliev (1992) and references therein. These systems of second order partial differential equations arise in the investigation of many essentially nonlinear physical phenomena. In particular, in gauge field theories of elementary particle physics, finite nonperiodic systems are relevant to the cylindrically symmetric self-dual configurations of the Yang–Mills–Higgs fields (instantons and dyons), and to the spherically symmetric t'Hooft–Polyakov monopoles; in string and superstring models; etc. At the same time, affine Toda systems possess a set of soliton-type solutions which have entered the lexicon of various branches of physics. For example, such solutions appear in plasma physics, nonlinear optics, superconductivity and aerodynamics. Recently, nonabelian versions of Toda systems have found application in connection with conformal and topological field theories. It is remarkable that the self-dual Yang–Mills system, being a very important tool in gauge theories of elementary particle physics, serves as an origin for a number of integrable nonlinear systems. In particular, it can be dimensionally reduced to the equations of the remarkable Wess–Zumino–Novikov–Witten model, and then to the Toda-type systems.

Looking at the list of integrable, to this or that extent, nonlinear equations, every investigator will feel the need to systematise them, and to work out the general criteria of integrability. The cipher key here is the notion of symmetry. The concept of symmetry, constituting an unalienable part of the mathematical apparatus of modern theoretical physics, is realised by the group-algebraic methods. These represent the relevant language for the formulation of the principal concepts of most branches of contemporary science. The lucidity and relative simplicity of these methods, and especially their application to the problems considered here, which usually incorporate various symmetries related to the general laws of physics, distinguish group-algebraic methods among other methods. Note that there is a deep relation between integrable systems and the representation theory which has been established and used to considerable effect only recently. At the same time, the idea itself goes back to the pioneering work of

S. Lie, who foresaw the leading role of group-theoretical methods as a powerful instrument for the integration of systems of differential equations. According to him, the transformation groups of the equations play the same role as do the Galois groups of algebraic equations. Briefly, the investigation of the action of the group, the widest among those admitted by the system in question, on the variety of the solutions of the system, allows one in principle to describe the general structure of this variety, as well as the special subclasses of the solutions. There are reviews and books concerning the group and algebraic background of the problem; see, for example, Ovsiannikov (1982); Olver (1986); Krasil'shchik, Lychagin & Vinogradov (1986); Ibragimov (1987); Fushchich, Serov & Shtelen (1989); Dickey (1991); Dorfman (1993) and Jimbo & Miwa (1995).

Most of the modern methods for the constructive investigation of nonlinear equations are based on the representation of the equations as the zero curvature condition of some connection on a trivial principal fibre bundle. For some class of equations it takes the form of the Lax representation. The available instruments here fall into the analytic, group-algebraic and differential geometry types. There are excellent monographs on analytical methods for the investigation of integrability problem in low dimensions; see, for example, Ablowitz & Segur (1981); Calogero & Degasperis (1982); Manakov *et al.* (1984); Faddeev & Takhtadjan (1987) and Marchenko (1988).

From the end of the 1970s an approach, based mainly on group-algebraic concepts, has been developed for the study of a wide class of one- and two-dimensional nonlinear integrable systems representable as the zero curvature condition; see Leznov & Saveliev (1989, 1992). For applications to supersymmetric systems, and to the class of multidimensional partial differential and integro-differential equations based on the so-called continuum Lie algebras, see Leites, Saveliev & Serganova (1986) and Saveliev & Vershik (1990). In the framework of the method the integrability criteria are related to such properties of the internal symmetry group of equations under consideration as solvability and finiteness of growth; see Leznov, Smirnov & Shabat (1982).

The main aim of our book is to describe the differential geometry foundation of a wide class of nonlinear integrable systems associated with the zero curvature condition. A large number of interesting results exist concerning the application of the zero curvature representation of nonlinear integrable systems obtained in the framework of differential geometry methods. However, as far as we know, there is no textbook on the subject. Moreover, the intersection of the group-algebraic and the differential geometry background of integrable systems represents an ecological niche in many respects. In a precise sense, the subject is not swallowed up by any traditional section of mathematical physics, while it is characterised by the relative simplicity of the images and proofs. This is why we believe that a monograph on the differential geometry background of integrable systems, especially in lecture note form, with practical examples and exercises, will be quite useful to a wide audience.

Those who are actively working on the problems of modern theoretical and mathematical physics would agree with the statement that there are no mathematics without tears for a physicist. In particular, the theory of integrable systems uses a very large number of concepts from various branches of mathematics. Bearing this in mind, and with the intention of making the book self-contained, we found it reasonable to begin with two introductory chapters containing the necessary notation and definitions, and more or less known information concerning some topics related to Lie algebras and the geometry of complex and real Lie groups. We hope that these chapters will be useful not only for students and newcomers to the theory of integrable systems, but also for those physicists and mathematicians who are not experts in this area. Namely, we use mainly the language of differential geometry, while, as it seems to us, this language is not commonly accepted by the whole audience of possible readers. Moreover, some definitions used in the main body of the book, for example the relation of Lie algebra valued 1-forms with connection forms, principal bundle isomorphisms and gauge transformations, are given in the physical and mathematical literature in different forms and, sometimes, with different meanings; some notions are defined on a different level of generality, for example matrix valued and generic 1-forms taking values in a Lie algebra. This is why we provide some definitions which synthesise and interpolate, in a sense, between them. The need for brevity has meant that we cannot present proofs of the statements given in the introductionary chapters. We have tried to compensate by including some examples to clarify the concepts defined. We hope that these will be more useful for the reader who is beginning work in the field of integrable systems than are proofs of general mathematical results. Nevertheless, we realise that the information given in the first two chapters is not sufficient for a deep understanding of the subject and, hence, we recommend their use as a guideline for further reading.

It is assumed that the reader is acquainted with the basic definitions of analysis, linear algebra and group theory. We believe that Kostrikin & Manin (1989) and Rudin (1964) and the first chapter of Najmark & Stern (1982) provide the necessary background to start reading the book.

Most section include exercises which in part contain the results of original papers on the subject; hopefully these will assist in understanding the material given in our book. The reference list given in the book, especially on the theory of integrable systems, is obviously incomplete; however, the reader can find many missing references in the reference lists of the monographs cited in the book. Being restricted by the standard size of the textbooks in the series, we unfortunately could not include here some remarkable results concerning the geometry of affine Toda systems, their supersymmetric extensions, and some continuous limits associated with continuum Lie algebras.

This book is based in part on lectures given by the authors at the Physical Faculty of the Moscow State University from 1990 to 1995, at the Laboratoire de Physique Théorique de l'École Normale Supérieure de Lyon in 1992, and at the Higher College of Mathematical Physics of the Moscow Independent University in 1993 and 1994.

Acknowledgements

We would like to thank F. E. Burstall, D. Finley, J.-L. Gervais, P. Goddard, A. A. Kirillov, I. M. Krichever, A. N. Leznov, Yu. I. Manin, S. P. Novikov, D. I. Olive, L. O'Raifeartaigh, G. L. Rcheulishvili, P. Sorba, Yu. G. Stroganov, A. M. Vershik and L. M. Woodward for very useful and illuminating discussions. We are also most grateful to our families, without whose permanent help and support the book would never have been completed.

The authors were supported in part by the International Science Foundation, INTAS, and by the Russian Foundation for Basic Research.

1
Introductory data on Lie algebras

In chapters 1 and 2 we review some basic notions of algebra and differential geometry, illustrating various concepts with examples and exercises. These chapters contain no exhaustive treatment of the theory; their purpose is to help the reader in understanding the Lie algebraic and differential geometry formulation of the integrability problem for the dynamical systems considered in the book.

As for the current chapter, those interested in a systematic discussion of the theory of Lie algebras and their representations are urged to consult the books by Bourbaki (1975, 1982); Gorbatsevich, Onishchik & Vinberg (1994); Goto & Grosshans (1978); Helgason (1978); Humphreys (1972); Kac (1990); Kirillov (1976); Najmark & Stern (1982); Serre (1966) and Zhelobenko (1994), which provide an excellent explanation of the subject; we used these books for the preparation of our lectures and in writing the book.

In what follows we suppose that the basic field $\mathbb{K}$ is either the field of real numbers $\mathbb{R}$, or the field of complex numbers $\mathbb{C}$. The transpose and the hermitian conjugation of an arbitrary matrix a are denoted by a^t and $a^\dagger$ respectively. The action of an element α of the dual V^* of a vector space V on an element $v \in V$ is denoted by $\langle \alpha, v \rangle$.

1.1 Basic definitions

1.1.1 Algebras

A vector space A over a field $\mathbb{K}$ is said to be an *algebra* over $\mathbb{K}$ if there is given a product operation $(a, b) \in A \times A \mapsto ab \in A$, which satisfies the conditions

(A1) $(a+b)c = ac + bc, \quad a(b+c) = ab + ac,$

(A2) $k(ab) = (ka)b = a(kb)$,

for all $a, b, c \in A$ and $k \in \mathbb{K}$. An algebra over the field of real numbers $\mathbb{R}$ is called a *real* algebra, while an algebra over the field of complex numbers $\mathbb{C}$ is called a *complex* algebra. An algebra A is called *commutative* or *abelian* if $ab = ba$ for all $a, b \in A$.

An algebra A may possess an *identity* (*unit*) element 1 such that $1a = a1 = a$ for any $a \in A$. Such an algebra is called *unital algebra*. An algebra may possess only one identity element.

An algebra A is called *associative* if $a(bc) = (ab)c$ for all $a, b, c \in A$.

EXAMPLE 1.1 The vector space $\mathrm{Mat}(m, \mathbb{K})$ of all $m \times m$ matrices over the field $\mathbb{K}$ with respect to the matrix multiplication form an associative algebra over the field $\mathbb{K}$. The unit matrix plays the role of the unit of the algebra here.

EXAMPLE 1.2 The set of all linear mappings from a vector space V to a vector space W is denoted by $\mathrm{Hom}(V, W)$. According to standard terminology, a linear operator on a linear space V is called an *endomorphism* of V. The vector space $\mathrm{End}(V) \equiv \mathrm{Hom}(V, V)$ of all endomorphisms of a vector space V with respect to the product of linear operators is an associative algebra.

EXAMPLE 1.3 Let V be a finite-dimensional vector space over a field $\mathbb{K}$. Introduce the notation

$$T^k(V) \equiv \underbrace{V \otimes \cdots \otimes V}_{k}$$

and put $T^0(V) \equiv \mathbb{K}$. The linear space

$$T(V) \equiv \bigoplus_{k=0}^{\infty} T^k(V)$$

with respect to the tensor product operation is an associative algebra with the unit element $1 \in \mathbb{K} = T^0(V)$. This algebra is called the *tensor algebra* on V.

1.1.2 Lie algebras

Before giving the definition of a Lie algebra, note that such algebras are related to Lie groups, which are usually denoted by capital latin letters. It is customary to denote the corresponding

Lie algebra by the corresponding small gothic letter. The product operation of a Lie algebra possesses the properties of the commutator, which can be defined for an arbitrary associative algebra, and denoted by square brackets. We use such a notation for an arbitrary Lie algebra.

An algebra $\mathfrak{g}$ over a field $\mathbb{K}$ with the product satisfying the conditions

(L1) $[x, x] = 0$ for any $x \in \mathfrak{g}$,

(L2) $[[x, y], z] + [[y, z], x] + [[z, x], y] = 0$ for all $x, y, z \in \mathfrak{g}$,

is called a *Lie algebra* over the field $\mathbb{K}$. The condition (L2) is called the *Jacobi identity*. It follows from (L1) that $[x, y] = -[y, x]$ for all $x, y \in \mathfrak{g}$.

EXAMPLE 1.4 Let V be a vector space over a field $\mathbb{K}$. Equip V with a product operation defined by the relation

$$[v, u] = 0$$

for all $v, u \in V$. It is clear that, with respect to this operation, V is a commutative Lie algebra over the field $\mathbb{K}$.

EXAMPLE 1.5 Let A be an associative algebra. Define the *commutator* $[a, b]$ of two elements $a, b \in A$ by

$$[a, b] \equiv ab - ba.$$

It can easily be shown that the vector space A with respect to the commutator is a Lie algebra. This Lie algebra is called the Lie algebra *associated* with the algebra A. The Lie algebra associated with the algebra $\mathrm{End}(V)$ of endomorphisms of the vector space V is denoted by $\mathfrak{gl}(V)$. The Lie algebra associated with the algebra $\mathrm{Mat}(m, \mathbb{K})$ of the square matrices of order n over a field $\mathbb{K}$ is denoted by $\mathfrak{gl}(m, \mathbb{K})$.

1.1.3 Homomorphisms and isomorphisms

Let A and B be algebras over a field $\mathbb{K}$; a linear mapping $\varphi : A \to B$ is called an *(algebra) homomorphism* if

$$\varphi(aa') = \varphi(a)\varphi(a') \tag{1.1}$$

for all $a, a' \in A$. If an algebra homomorphism φ is invertible, then φ is said to be an *(algebra) isomorphism*. One says that algebras A and B are isomorphic if there exists an isomorphism $\varphi : A \to B$.

In this case one writes $A \simeq B$. An isomorphism of an algebra A onto itself is called an (*algebra*) *automorphism*. The set of all automorphisms of an algebra A is a group which is called the *group of automorphisms* of A and which is denoted by $\mathrm{Aut}(A)$.

EXAMPLE 1.6 Let V be an m-dimensional vector space over $\mathbb{K}$, and $\{e_i\}$ a basis of V. For any endomorphism $A \in \mathrm{End}(V)$ and any element $v = \sum_{i=1}^m e_i v_i \in V$ we have

$$Av = \sum_{i=1}^{m} e_i \sum_{j=1}^{m} a_{ij} v_j,$$

where the numbers $a_{ij} \in \mathbb{K}$, $i, j = 1, \ldots, m$, are defined from the relation $Ae_j \equiv \sum_{i=1}^m e_i a_{ij}$. The matrix $a \equiv (a_{ij})$ is called the *matrix of the endomorphism* A with respect to the basis $\{e_i\}$. It can be shown that the mapping $A \in \mathrm{End}(V) \mapsto a \in \mathrm{Mat}(m, \mathbb{K})$ is an algebra isomorphism. Note that this isomorphism depends on the choice of basis of V.

Similarly, the Lie algebra $\mathfrak{gl}(V)$ is isomorphic to the Lie algebra $\mathfrak{gl}(m, \mathbb{K})$.

A linear mapping from an algebra A to an algebra B is said to be an *antihomomorphism* if

$$\varphi(aa') = \varphi(a')\varphi(a) \tag{1.2}$$

for all $a, a' \in A$. An invertible antihomomorphism φ from an algebra A onto itself is called an *antiautomorphism* of A. An antiautomorphism φ of an algebra A satisfying the relation $\varphi \circ \varphi = \mathrm{id}_A$ is called an *involution* of the algebra A.

Let A be an algebra, denote by A' the algebra which coincides with A as a set but has a new product operation $(a, b) \in A' \times A' \mapsto a \cdot b \in A'$ defined as

$$a \cdot b \equiv ba.$$

If e is the unit of A, then e is the unit of A'. If A is an associative algebra (a Lie algebra), then A' is also an associative algebra (a Lie algebra). Let A and B be two algebras, and let φ be an antihomomorphism from A to B. Denote by φ' the mapping φ considered as a mapping from A to B'. The mapping φ' is a homomorphism.

For the case of complex algebras, one also considers antilinear mappings satisfying either (1.1) or (1.2); we call such mappings *antilinear homomorphisms* and *antilinear antihomomorphisms* respectively. Similarly, one defines antilinear automorphisms and antilinear antiautomorphisms. An antilinear antiautomorphism φ of an algebra A satisfying the relation $\varphi \circ \varphi = \mathrm{id}_A$ is called a *hermitian involution* of the algebra A. An antilinear automorphism φ of an algebra A satisfying the relation $\varphi \circ \varphi = \mathrm{id}_A$ is called a *conjugation* of A.

For any complex vector space V we define a complex linear space $\overline{V}$ as follows. The linear space $\overline{V}$ as a set coincides with V. The addition of vectors in $\overline{V}$ is the same as in V. The result of multiplication of a vector $v \in \overline{V}$ by a complex number c coincides with the result of the multiplication of v by $\bar{c}$ in V. Similarly, for an arbitrary complex algebra A we can define the corresponding algebra $\bar{A}$ supposing that the product operations in A and $\bar{A}$ coincide. The unit of A, if it exists, is also the unit of $\bar{A}$. If A is an associative algebra (a Lie algebra), then $\bar{A}$ is also an associative algebra (a Lie algebra). Let A and B be two algebras, and φ be an antilinear homomorphism (an antilinear antihomomorphism) from A to B. Denote by $\bar{\varphi}$ the mapping φ considered as a mapping from A to $\overline{B}$. The mapping $\bar{\varphi}$ is a homomorphism (an antihomomorphism).

EXAMPLE 1.7 The matrix transposition is an antiautomorphism of $\mathrm{Mat}(n, \mathbb{K})$. It is clear that it is an involution of $\mathrm{Mat}(m, \mathbb{K})$. The hermitian conjugation in $\mathrm{Mat}(m, \mathbb{C})$ is an antilinear antiautomorphism which is a hermitian involution.

1.1.4 Subalgebras and ideals

Let A be an algebra, and B, C be subsets of A. We write BC for the subspace of A spanned by the elements of the form bc, where $b \in B$ and $c \in C$. A subspace B of A is said to be a *subalgebra* of A if $BB \subset B$.

EXAMPLE 1.8 Let B be a bilinear form on an m-dimensional vector space V over a field $\mathbb{K}$. Denote by $\mathfrak{gl}_B(V)$ the set formed by the elements $A \in \mathfrak{gl}(V)$ satisfying the relation

$$B(v, Au) + B(Av, u) = 0 \tag{1.3}$$

for all $v, u \in V$. It can easily be shown that $\mathfrak{gl}_B(V)$ is a subalgebra of $\mathfrak{gl}(V)$. Let $\{e_i\}$ be a basis of V. Then for $v = \sum_i v_i e_i$ and $u = \sum_i u_i e_i$ we have

$$B(v, u) = \sum_{i,j} v_i b_{ij} u_j,$$

where $b_{ij} \equiv B(e_i, e_j)$. The matrix $b \equiv (b_{ij})$ is called the *matrix of the bilinear form* B with respect to the basis $\{e_i\}$. Now relation (1.3) can be written in the form

$$\sum_i v_i (b_{ik} a_{kj} + a_{ki} b_{kj}) u_j = 0,$$

which must be valid for all v_i and u_i. This relation is equivalent to the following matrix relation:

$$ba + a^t b = 0, \tag{1.4}$$

where a is the matrix of the endomorphism A with respect to the basis $\{e_i\}$. Thus, the matrix a of an endomorphism $A \in \mathfrak{gl}_B(V)$ with respect to the basis $\{e_i\}$, satisfies (1.4). On the other hand, any matrix a satisfying (1.4) corresponds to some element of $\mathfrak{gl}_B(V)$. Recall that the mapping $A \in \mathfrak{gl}(V) \mapsto a \in \mathfrak{gl}(m, \mathbb{K})$ is an isomorphism. Therefore, relation (1.4) defines a subalgebra of $\mathfrak{gl}(m, \mathbb{K})$ isomorphic to $\mathfrak{gl}_B(V)$. In general, different choices of bases of V lead to different subalgebras of $\mathfrak{gl}(m\mathbb{K})$. Nevertheless, all such subalgebras are isomorphic to $\mathfrak{gl}_B(V)$ and, hence, they are isomorphic to one another.

Let us restrict ourselves to the case of nondegenerate bilinear forms possessing definite symmetry.

Denote by I_m the unit $m \times m$ matrix. Recall that for any symmetric nondegenerate bilinear form B on an m-dimensional real vector space V there is a basis of V such that the matrix of B with respect to this basis coincides with the matrix

$$I_{k,l} \equiv \begin{pmatrix} I_k & 0 \\ 0 & -I_l \end{pmatrix}, \tag{1.5}$$

where $k + l = m$. The corresponding subalgebra of $\mathfrak{gl}(m, \mathbb{R})$ is denoted by $\mathfrak{o}(k, l)$. This Lie algebra is called the *pseudo-orthogonal* algebra. It is clear that the Lie algebras $\mathfrak{o}(k, l)$ and $\mathfrak{o}(l, k)$ are isomorphic. We will use the notation $\mathfrak{o}(m) \equiv \mathfrak{o}(m, 0)$. The Lie algebra $\mathfrak{o}(m)$ is called the *real orthogonal* algebra. The dimension of $\mathfrak{o}(k, l)$ is equal to $(k + l)(k + l - 1)/2$.

For any symmetric nondegenerate bilinear form on an m-dimensional complex vector space V there is a basis of V such that the matrix of B, with respect to this basis, is I_m. The corresponding subalgebra of $\mathfrak{gl}(m, \mathbb{C})$ is called the *complex orthogonal* algebra and is denoted by $\mathfrak{o}(m, \mathbb{C})$. The dimension of $\mathfrak{o}(m, \mathbb{C})$ is equal to $m(m-1)/2$.

A skew-symmetric nondegenerate bilinear form on an m-dimensional vector space V over a field $\mathbb{K}$ may exist only when m is even. In this case there is a basis of V such that the matrix of B, with respect to this basis, coincides with the matrix

$$J_n \equiv \begin{pmatrix} 0 & I_n \\ -I_n & 0 \end{pmatrix}, \tag{1.6}$$

where $n = m/2$. The corresponding subalgebra of $\mathfrak{gl}(2n, \mathbb{K})$ is denoted by $\mathfrak{sp}(n, \mathbb{K})$, and is called the *real* or *complex symplectic* algebra depending on whether $\mathbb{K} = \mathbb{R}$, or $\mathbb{K} = \mathbb{C}$. The dimension of $\mathfrak{sp}(n, \mathbb{K})$ is equal to $2n^2 + n$.

A *left ideal* B of an algebra A is defined as a subalgebra B of A such that $AB \subset B$; a *right ideal* of A is a subalgebra B such that $BA \subset B$. If a left ideal B is also a right ideal of A, then it is called a *two-sided ideal* of A, or simply an *ideal* of A. Any ideal of a Lie algebra is a two-sided ideal. For any algebra A the subspace $\{0\}$ and the whole algebra A are ideals of A. Such ideals are called *trivial* ideals.

EXAMPLE 1.9 The set $\mathfrak{sl}(V)$ of elements of $\mathfrak{gl}(V)$ having zero trace is an ideal of $\mathfrak{gl}(V)$. Similarly, the set $\mathfrak{sl}(m, \mathbb{K})$ of elements of $\mathfrak{gl}(m, \mathbb{K})$ having zero trace is an ideal of $\mathfrak{gl}(m, \mathbb{K})$. The Lie algebra $\mathfrak{sl}(m, \mathbb{R})$ is called the *real special linear* algebra; similarly, $\mathfrak{sl}(m, \mathbb{C})$ is called the *complex special linear* algebra.

If φ is a homomorphism from an algebra A to an algebra B, then $\operatorname{Ker}\varphi$ is an ideal of A, and $\operatorname{Im}\varphi$ is a subalgebra of B.

The intersection of any family of subalgebras (ideals) of an algebra is a subalgebra (an ideal) of the algebra. Let S be a subset of an algebra A. The intersection of all subalgebras (ideals) of A containing S is called the subalgebra (ideal) *generated* by S. Note that it is the minimal ideal (subalgebra) containing S. If A is an associative unital algebra, the ideal generated by S coincides with ASA.

The *centre* $Z(A)$ of an algebra A is defined as

$$Z(A) \equiv \{c \in A \mid [c, a] = 0 \text{ for any } a \in A\}.$$

If A is an associative algebra, the set $Z(A)$ is a commutative subalgebra of A. It is clear that $Z(A) = A$ if and only if A is commutative. For any Lie algebra $\mathfrak{g}$ the centre $Z(\mathfrak{g})$ is a commutative ideal of $\mathfrak{g}$.

Let B be an ideal of an algebra A. The quotient space A/B is an algebra with respect to the product operation defined by

$$(a + B)(a' + B) \equiv aa' + B.$$

This algebra is called the *quotient algebra.* The canonical projection $\pi : A \to A/B$ is a surjective homomorphism and $\operatorname{Ker} \pi = B$. The quotient algebra of an associative algebra (Lie algebra) is an associative algebra (Lie algebra).

EXAMPLE 1.10 Consider the ideal $I(V)$ of the tensor algebra $T(V)$ generated by the tensors of the form $v \otimes w - w \otimes v$ with $v, w \in V$. An associative algebra $S(V) \equiv T(V)/I(V)$ is called the *symmetric algebra* on V.

1.1.5 Derivations

Let A be an algebra; a mapping $D \in \operatorname{End}(A)$ is called a *derivation* of A if

$$D(ab) = (Da)b + a(Db)$$

for all $a, b \in A$. The commutator of any two derivations is a derivation. Therefore, the set $\operatorname{Der}(A)$ of all derivations of A can be considered as a subalgebra of the Lie algebra $\mathfrak{gl}(A)$. If A is an associative algebra or a Lie algebra, then for any $a \in A$ the mapping $D_a \in \operatorname{End}(A)$, defined as

$$D_a b \equiv [a, b],$$

is a derivation of A. Such derivations are called *inner derivations* of A. For any derivation $D \in \operatorname{Der}(A)$ and $a \in A$ we have

$$[D, D_a] = D_{D(a)}.$$

Hence, the set of all inner derivations of A is an ideal of $\operatorname{Der}(A)$ considered as a Lie algebra.

1.1.6 Direct and semidirect products

Let B and C be two algebras. The direct sum $B \oplus C$ of the vector spaces B and C consists of all ordered pairs (b, c), where $b \in B$ and $c \in C$. Supplying $B \oplus C$ with the bilinear operation

$$(b_1, c_1)(b_2, c_2) \equiv (b_1 b_2, c_1 c_2),$$

we obtain an algebra which is called the *direct product* of the algebras B and C, and is denoted $B \times C$. Note that the algebras B and C can be identified with the subalgebras of $B \times C$ formed by the elements $(b, 0)$, $b \in B$, and $(0, c)$, $c \in C$ respectively. Actually, these subalgebras are ideals of $B \times C$. The direct product of associative algebras (Lie algebras) is an associative algebra (Lie algebra).

Now let B and C be ideals of an algebra A such that $A = B \oplus C$. In this case the mapping $(b, c) \in B \times C \mapsto b + c \in A$ is an isomorphism, and we can identify A with $B \times C$.

For Lie algebras there also exists a notion of semidirect product. Let $\mathfrak{h}$ and $\mathfrak{k}$ be Lie algebras. Suppose that there is a homomorphism $\varphi : \mathfrak{h} \to \mathrm{Der}(\mathfrak{k})$. Supplying $\mathfrak{h} \oplus \mathfrak{k}$ with the bilinear operation

$$[(y_1, z_1), (y_2, z_2)] \equiv ([y_1, y_2], [z_1, z_2] + \varphi(y_1)z_2 - \varphi(y_2)z_1],$$

we obtain a Lie algebra which is called the semidirect product of the Lie algebras $\mathfrak{h}$ and $\mathfrak{k}$ and which is denoted by $\mathfrak{h} \ltimes_\varphi \mathfrak{k}$, or simply $\mathfrak{h} \ltimes \mathfrak{k}$. Identifying $\mathfrak{h}$ and $\mathfrak{k}$ with the corresponding subspaces of $\mathfrak{h} \oplus \mathfrak{k}$, we see that $\mathfrak{h}$ is a subalgebra and $\mathfrak{k}$ is an ideal of $\mathfrak{h} \ltimes \mathfrak{k}$.

Let $\mathfrak{h}$ be a subalgebra and let $\mathfrak{k}$ be an ideal of a Lie algebra $\mathfrak{g}$. Suppose that $\mathfrak{g} = \mathfrak{h} \oplus \mathfrak{k}$. Since $\mathfrak{k}$ is an ideal, we can define a homomorphism $\varphi : \mathfrak{h} \to \mathrm{Der}(\mathfrak{k})$ by

$$\varphi(y) \equiv D_y|_\mathfrak{k}.$$

It is now clear that the mapping $(y, z) \in \mathfrak{h} \ltimes_\varphi \mathfrak{k} \to y + z \in \mathfrak{g}$ is an isomorphism, and we can identify $\mathfrak{g}$ with $\mathfrak{h} \ltimes_\varphi \mathfrak{k}$.

1.1.7 Representations and modules

Let A be a real associative algebra, and let V be a real or complex vector space. A homomorphism ρ from the algebra A to the algebra $\mathrm{End}(V)$ is called a *representation* of A in V. One says that ρ is a *real* representation if V is a real vector space, and that it is *complex* when V is a complex vector space. A representation of a complex associative algebra A in a complex vector space is defined as an arbitrary homomorphism from A to the algebra of endomorphisms of V. If A is an associative unital algebra with the unit 1, then it is also required that $\rho(1) = \mathrm{id}_V$. The dimension of the space V is called the dimension of the representation ρ. The space V is called the *representation space* of ρ.

Similarly, if $\mathfrak{g}$ is a Lie algebra and V is a vector space, we define a representation of $\mathfrak{g}$ in V as a homomorphism from $\mathfrak{g}$ to $\mathfrak{gl}(V)$.

The notion of a representation of an associative algebra A is closely related to the notion of an A-module. Let us give a corresponding definition. A vector space V is called a (left) *module* over an associative algebra A, or just an (left) A-module, if there is given a bilinear operation $(a, v) \in A \times V \mapsto av \in V$ such that

$$(ab)v = a(bv)$$

for all $a, b \in A$ and $v \in V$; and

$$1v = v$$

for any $v \in V$, in the case when A is a unital algebra with the identity element 1. For a Lie algebra $\mathfrak{g}$ we define a (left) $\mathfrak{g}$-module as a vector space V endowed with a bilinear operation $(x, v) \in \mathfrak{g} \times V \mapsto xv \in V$ such that

$$[x, y]v = x(yv) - y(xv),$$

for all $x, y \in \mathfrak{g}$ and $v \in V$.

If ρ is a representation of an algebra A in a vector space V, then putting

$$av \equiv \rho(a)v,$$

we endow V with the structure of an A-module. Similarly, if V is an A-module, then reversing the above relation we define a representation of A in V. Therefore, any statement about a representation of an algebra can be reformulated as a statement about the corresponding module and vice versa.

Let V be a module over an algebra A; let B be a subset of A; and let W be a subset of V. Denote by BW the subspace of V spanned by the vectors of the form bw, where $b \in B$ and $w \in W$. A subspace W of the space V is called a *submodule* of the module V, if $AW \subset W$. It is clear that $\{0\}$ and V are submodules of V. Such submodules are called *trivial submodules*.

Let V and W be A-modules, and let $\varphi \in \mathrm{Hom}(V, W)$. The mapping φ is called a (*module*) *homomorphism* if

$$\varphi(av) = a\varphi(v)$$

for any $a \in A$ and $v \in V$. The set of all homomorphisms of A-modules V and W is denoted by $\mathrm{Hom}_A(V, W)$. If $\varphi \in \mathrm{Hom}_A(V, W)$ is an isomorphism of vector spaces V and W, it is called a (*module*) *isomorphism*. If for two A-modules V and W there exists an isomorphism $\varphi \in \mathrm{Hom}_A(V, W)$, the modules V and W are called *isomorphic*. It can easily be shown that if $\varphi \in \mathrm{Hom}_A(V, W)$, then $\mathrm{Ker}\,\varphi$ is a submodule of V, and $\mathrm{Im}\,\varphi$ is a submodule of W. The representations corresponding to isomorphic modules are called *equivalent*. The set $\mathrm{Hom}_A(V, V)$ is denoted by $\mathrm{End}_A(V)$. The elements of $\mathrm{End}_A(V)$ are called (*module*) *endomorphisms*.

An A-module $V \neq \{0\}$ is said to be *simple* if it has only trivial submodules. The corresponding representation of A is called *irreducible*. If a representation of an algebra is not irreducible, we say that it is *reducible*.

Let V and W be two simple A-modules, and let $\varphi \in \mathrm{Hom}_A(V, W)$. In this case $\mathrm{Ker}\,\varphi$ either coincides with V or is equal to $\{0\}$. If $\mathrm{Ker}\,\varphi$ is V, the homomorphism φ is trivial. Suppose that $\mathrm{Ker}\,\varphi$ is $\{0\}$. In this case $\mathrm{Im}\,\varphi$ is a submodule of W which does not coincide with $\{0\}$. Since W is simple, we have the only possibility $\mathrm{Im}\,\varphi = W$; in this case φ is an isomorphism. Thus a homomorphism from one simple A-module to another simple A-module is either trivial or an isomorphism. Furthermore, for any simple finite-dimensional module V over a complex algebra A, the set $\mathrm{End}_A(V)$ is formed by the mappings proportional to the identity mapping id_V. This statement is known as the *Schur lemma*.

Let V be an A-module. For any $v \in V$ the subset Av is a submodule of V. Hence, if V is simple, then for any $v \in V$ either $Av = \{0\}$ or $Av = V$. If A is a unital algebra, then the former is impossible; and in this case the module V is simple if and only if $Av = V$ for any $v \in V$.

Let S be a subset of an A-module V. If $AS = V$, then we say that V is generated by S. If V is generated by its finite subset, it is called *finitely generated*. Finally, if V is generated by a subset consisting of just one element v, then the module V is called a *cyclic module* and v is said to be a *cyclic vector* of V.

An A-module is said to be *semisimple* if it can be represented as a direct sum of simple submodules. The corresponding representation of A is called *completely reducible*. A module is called *indecomposable* if it cannot be represented as a direct sum of simple submodules. Any irreducible module is indecomposable, but an indecomposable module may be reducible.

For any associative algebra A the mapping $\mathrm{ad} : A \to \mathrm{End}(A)$ defined by

$$\mathrm{ad}(a)b \equiv ab$$

is a representation of A in A, called the *adjoint representation*. Similarly, for an arbitrary Lie algebra $\mathfrak{g}$ we define the adjoint representation $\mathrm{ad} : \mathfrak{g} \to \mathfrak{gl}(\mathfrak{g})$ with the help of the relation

$$\mathrm{ad}(x)y \equiv [x, y].$$

Note that the operators $\mathrm{ad}(x)$, $x \in \mathfrak{g}$ are, in fact, the inner derivations of $\mathfrak{g}$.

The adjoint representation equips a Lie algebra $\mathfrak{g}$ with the structure of a $\mathfrak{g}$-module. It is this structure of a $\mathfrak{g}$-module that we will have in mind in saying that we consider a Lie algebra $\mathfrak{g}$ to be a $\mathfrak{g}$-module. A Lie algebra $\mathfrak{g}$ is called *reductive* if it is a semisimple $\mathfrak{g}$-module.

Let V be a module over a Lie algebra $\mathfrak{g}$. The dual vector space V^* becomes a $\mathfrak{g}$-module if we define

$$\langle x\alpha, v\rangle \equiv -\langle \alpha, xv\rangle,$$

for any $x \in \mathfrak{g}$, $\alpha \in V^*$ and $v \in V$. The $\mathfrak{g}$-module V^* is called the *dual module* of the $\mathfrak{g}$-module V. If ρ is the representation of $\mathfrak{g}$ corresponding to the module V, the representation corresponding to the module V^* is called the *dual representation* of ρ and is denoted by ρ^*. The dual representation of the adjoint representation is called the *coadjoint representation*.

EXAMPLE 1.11 The Lie algebra $\mathfrak{sl}(2, \mathbb{C})$ consists of all traceless 2×2 matrices. The matrices

$$x_- = \begin{pmatrix} 0 & 0 \\ 1 & 0 \end{pmatrix}, \qquad h = \begin{pmatrix} 1 & 0 \\ 1 & -1 \end{pmatrix}, \qquad x_+ = \begin{pmatrix} 0 & 1 \\ 0 & 0 \end{pmatrix}$$

form a basis of $\mathfrak{sl}(2, \mathbb{C})$ called the *standard basis*. For these matrices we have

$$[h, x_\pm] = \pm 2x_\pm, \qquad [x_+, x_-] = h.$$

Let n be a nonnegative integer and let $L(n)$ be an $(n+1)$-dimensional complex vector space. Fix a basis $\{v_k\}_{k=0}^{n}$ of $L(n)$ and endow $L(n)$ with the structure of an $\mathfrak{sl}(2,\mathbb{C})$-module, putting

$$\begin{aligned} x_- v_k &\equiv v_{k+1}, \\ h v_k &\equiv (n-2k) v_k, \\ x_+ v_k &\equiv k(n-(k-1)) v_{k-1}, \end{aligned}$$

where $v_{n+1} \equiv 0$. It can be proved that the modules $L(n)$ $n = 0, 1, \ldots$ exhaust all irreducible finite-dimensional $\mathfrak{sl}(2,\mathbb{C})$-modules.

1.1.8 Invariant bilinear forms

A bilinear form B on an A-module V is called *invariant* if

$$B(av, u) + B(v, au) = 0$$

for all $a \in A$ and $v, u \in V$.

Recall that a nondegenerate bilinear form B on a vector space V defines an isomorphism $\nu : V \to V^*$ by

$$\langle \nu(v), u \rangle \equiv B(v, u).$$

If B is an invariant nondegenerate bilinear form on a module V over a Lie algebra $\mathfrak{g}$, then

$$\nu(xv) = x\nu(v)$$

for any $x \in \mathfrak{g}$ and $v \in V$. Therefore, $\nu \in \mathrm{Hom}_{\mathfrak{g}}(V, V^*)$. Since ν is an isomorphism, the $\mathfrak{g}$-modules V and V^* are isomorphic.

Let B_1 and B_2 be two invariant nondegenerate bilinear forms on a simple module V over a complex Lie algebra $\mathfrak{g}$, and let ν_1 and ν_2 be the corresponding mappings from V to V^*. By definition, we have

$$\langle \nu_1(v), u \rangle = B_1(v, u), \quad \langle \nu_2(v), u \rangle = B_2(v, u)$$

for all $v, u \in V$. These relations imply

$$B_2(v, u) = B_1(\nu_1^{-1} \circ \nu_2(v), u).$$

The mapping $\varphi \equiv \nu_1^{-1} \circ \nu_2$ satisfies the relation

$$\varphi(xv) = x\varphi(v)$$

for any $x \in \mathfrak{g}$ and $v \in V$; in other words, $\varphi \in \mathrm{End}_{\mathfrak{g}}(V)$. Since V is a simple module and the mapping φ is nontrivial, then, as follows from the Schur's lemma, φ is proportional to the identity mapping id_V. Thus, $B_2 = cB_1$ for some complex number c, so that an invariant nondegenerate bilinear form on a simple $\mathfrak{g}$-module is unique up to multiplication by complex numbers.

A bilinear form B on a Lie algebra $\mathfrak{g}$ is called *invariant* if it is invariant as a bilinear form on $\mathfrak{g}$ considered as a $\mathfrak{g}$-module defined by the adjoint representation. In other words, a bilinear form B on a Lie algebra $\mathfrak{g}$ is invariant if

$$B([x,y],z) + B(y,[x,z]) = 0$$

for all $x, y, z \in \mathfrak{g}$. Let ρ be a finite-dimensional representation of $\mathfrak{g}$. A symmetric bilinear form B_ρ on $\mathfrak{g}$ defined by

$$B_\rho(x,y) \equiv \operatorname{tr}(\rho(x)\rho(y)),$$

is called associated with ρ. This bilinear form is invariant. The bilinear form associated with the adjoint representation of $\mathfrak{g}$ is called the *Killing form* of the Lie algebra $\mathfrak{g}$. We will denote the Killing form by K. If $\mathfrak{h}$ is an ideal of $\mathfrak{g}$, then the Killing form of $\mathfrak{h}$ coincides with the restriction of the Killing form of $\mathfrak{g}$ to $\mathfrak{h}$. If a Lie algebra $\mathfrak{g}$ can be endowed with a nondegenerate bilinear form then the adjoint and coadjoint representations of $\mathfrak{g}$ are equivalent.

EXAMPLE 1.12 In this example we consider the Killing forms for the Lie algebras $\mathfrak{gl}(m, \mathbb{K})$, $\mathfrak{sl}(m, \mathbb{K})$, $\mathfrak{o}(m, \mathbb{K})$ and $\mathfrak{sp}(n, \mathbb{K})$.

Consider first the general linear algebra $\mathfrak{gl}(m, \mathbb{K})$. It is clear that the $m \times m$ matrices e_{ij}, $i, j = 1, \dots, m$, with the matrix elements

$$(e_{ij})_{rs} = \delta_{ir}\delta_{js},$$

form a basis for $\mathfrak{gl}(m, \mathbb{K})$. It is easy to show that

$$e_{ij}e_{kl} = \delta_{jk}e_{il}.$$

Hence, we obtain

$$[e_{ij}, e_{kl}] = \delta_{jk}e_{il} - \delta_{li}e_{kj}. \tag{1.7}$$

For any element $a = (a_{ij})$ of $\mathfrak{gl}(m, \mathbb{K})$ we have the representation

$$a = \sum_{i,j=1}^{m} a_{ij}e_{ij};$$

which, taking account of (1.7), gives

$$\operatorname{ad}(a)e_{ij} = \sum_{k,l}(a_{ki}\delta_{lj} - a_{jl}\delta_{ki})e_{kl}.$$

Using this equality, we come to the relation

$$\operatorname{ad}(a)\operatorname{ad}(b)e_{ij} = \sum_{k,r,s}(a_{rk}b_{ki}\delta_{sj} + a_{ks}b_{jk}\delta_{ri} - a_{ri}b_{jk}\delta_{sk} - a_{js}b_{ki}\delta_{rk})e_{rs}. \quad (1.8)$$

It is not difficult to show that the Killing form for the Lie algebra $\mathfrak{gl}(m,\mathbb{K})$ can now be written as

$$K(a,b) = 2m\operatorname{tr}(ab) - 2\operatorname{tr}a\operatorname{tr}b.$$

The Lie algebra $\mathfrak{sl}(m,\mathbb{K})$ is an ideal of the Lie algebra $\mathfrak{gl}(m,\mathbb{K})$. Hence, the Killing form of $\mathfrak{sl}(m,\mathbb{K})$ can be written as

$$K(a,b) = 2m\operatorname{tr}(ab). \quad (1.9)$$

Consider now the orthogonal Lie algebra $\mathfrak{o}(m,\mathbb{K})$. Recall that this Lie algebra consists of the $m\times m$ matrices $a = (a_{ij})$ over the field $\mathbb{K}$ satisfying the condition

$$a + a^t = 0.$$

For any such a matrix we have

$$a = \sum_{i,j} a_{ij}e_{ij} = \frac{1}{2}\sum_{i,j} a_{ij}e_{[i,j]} = \sum_{i<j} a_{ij}e_{[i,j]},$$

where

$$e_{[i,j]} \equiv e_{ij} - e_{ji}.$$

The matrices $e_{[i,j]}$, $i<j$, form a basis of $\mathfrak{o}(m,\mathbb{K})$. Using (1.8), one can easily obtain that

$$\begin{aligned}\operatorname{ad}(a)\operatorname{ad}(b)e_{[i,j]} &= \sum_{\substack{k,r,s\\ r<s}}(a_{rk}b_{ki}\delta_{sj} + a_{ks}b_{jk}\delta_{ri} - a_{ri}b_{jk}\delta_{sk} - a_{js}b_{ki}\delta_{rk} \\ &\quad - a_{rk}b_{kj}\delta_{si} - a_{ks}b_{ik}\delta_{rj} + a_{rj}b_{ik}\delta_{sk} + a_{is}b_{kj}\delta_{rk})e_{[r,s]}.\end{aligned}$$

From this equality we obtain

$$K(a,b) = (m-2)\operatorname{tr}(ab) \quad (1.10)$$

for the Lie algebra $\mathfrak{o}(m,\mathbb{K})$.

A similar consideration gives

$$K(a,b) = 2(n+1)\operatorname{tr}(ab) \quad (1.11)$$

for the Lie algebra $\mathfrak{sp}(n,\mathbb{K})$.

1.1.9 Elements of the structural theory of Lie algebras

The fundamental problem in the theory of Lie algebras is the classification of all nonisomorphic Lie algebras. This problem, in a sense, effectively reduces to the classification of two mutually complementary types of algebra, namely, solvable and semisimple. Let us give the relevant definitions.

For any two ideals $\mathfrak{h}$ and $\mathfrak{k}$ of a Lie algebra $\mathfrak{g}$, the subspace $[\mathfrak{h}, \mathfrak{k}]$ is an ideal of $\mathfrak{g}$. From this fact it follows that the subspaces $\mathcal{C}^k\mathfrak{g}$, $k = 0, 1, \ldots$, of $\mathfrak{g}$, defined inductively by

$$\mathcal{C}^{k+1}\mathfrak{g} \equiv [\mathfrak{g}, \mathcal{C}^k\mathfrak{g}],$$

with the initial condition $\mathcal{C}^0\mathfrak{g} \equiv \mathfrak{g}$, are ideals of $\mathfrak{g}$. It is clear that $\mathcal{C}^{k+1}\mathfrak{g} \subset \mathcal{C}^k\mathfrak{g}$. A Lie algebra $\mathfrak{g}$ is called *nilpotent* if there exists an integer n such that $\mathcal{C}^n\mathfrak{g} = \{0\}$. The centre of a nonzero nilpotent Lie algebra is different from zero.

Similarly, the subspaces $\mathcal{D}^k\mathfrak{g}$, $k = 0, 1, \ldots$, of a Lie algebra $\mathfrak{g}$, defined inductively by

$$\mathcal{D}^{k+1}\mathfrak{g} \equiv [\mathcal{D}^k\mathfrak{g}, \mathcal{D}^k\mathfrak{g}],$$

with the initial condition $\mathcal{D}^0\mathfrak{g} \equiv \mathfrak{g}$, are ideals of $\mathfrak{g}$. Here we again have $\mathcal{D}^{k+1}\mathfrak{g} \subset \mathcal{D}^k\mathfrak{g}$. A Lie algebra $\mathfrak{g}$ is called *solvable* if there exists an integer n such that $\mathcal{D}^n\mathfrak{g} = \{0\}$. Any nilpotent Lie algebra is a solvable Lie algebra. On the other hand, it can be shown that a Lie algebra $\mathfrak{g}$ is solvable if and only if $[\mathfrak{g}, \mathfrak{g}]$ is a nilpotent Lie algebra.

EXAMPLE 1.13 An $m \times m$ matrix (a_{ij}) is called *upper (lower) triangular* if $a_{ij} = 0$ for $i > j$ ($i < j$). The set $\mathfrak{t}_+(m, \mathbb{K})$ ($\mathfrak{t}_-(m, \mathbb{K})$) of $m \times m$ upper (lower) triangular matrices is a subalgebra of $\mathfrak{gl}(m, \mathbb{K})$. The Lie algebras $\mathfrak{t}_\pm(m, \mathbb{C})$ are solvable.

An $m \times m$ matrix (a_{ij}) is called *strictly upper (lower) triangular* if $a_{ij} = 0$ for $i \geq j$ ($i \leq j$). The set $\mathfrak{n}_+(m, \mathbb{K})$ ($\mathfrak{n}_-(m, \mathbb{K})$) of $m \times m$ strictly upper (lower) triangular matrices is a subalgebra of $\mathfrak{gl}(m, \mathbb{K})$. The Lie algebras $\mathfrak{n}_\pm(m, \mathbb{C})$ are nilpotent.

Denote by $\mathfrak{d}(m, \mathbb{K})$ the set of all $m \times m$ diagonal matrices. It is clear that $\mathfrak{t}_\pm(m, \mathbb{K}) = \mathfrak{d}(m, \mathbb{K}) \oplus \mathfrak{n}_\pm(m, \mathbb{K})$.

Let $\mathfrak{h}$ and $\mathfrak{k}$ be solvable ideals of a Lie algebra $\mathfrak{g}$. It can be shown that $\mathfrak{h} + \mathfrak{k}$ is also a solvable ideal of $\mathfrak{g}$. From this fact it follows that, among all solvable ideals of an arbitrary Lie algebra $\mathfrak{g}$, there is the ideal which contains all other solvable ideals.

Such an ideal is called the *radical* of $\mathfrak{g}$. A Lie algebra $\mathfrak{g}$ is called *semisimple* if it has no nonzero solvable ideals. In other words, a Lie algebra is called semisimple if its radical is zero. It can be shown that a Lie algebra is semisimple if and only if it has no nonzero commutative ideals. There is a complete classification of the semisimple Lie algebras. This classification is based on the fact that a Lie algebra is semisimple if and only if its Killing form is nondegenerate. In particular, as follows from example 1.12, the Lie algebra $\mathfrak{gl}(m, \mathbb{K})$ is not semisimple, while $\mathfrak{sl}(m, \mathbb{K})$ $(m \geq 2)$, $\mathfrak{o}(m, \mathbb{K})$ $(m \geq 3)$ and $\mathfrak{sp}(n, \mathbb{K})$ $(n \geq 1)$ are semisimple Lie algebras. There is an important property of the representations of the semisimple Lie algebras, namely, according to the *Weyl theorem*, any finite-dimensional module over a finite-dimensional semisimple Lie algebra is semisimple.

A Lie algebra is called *simple* if it is noncommutative and has no nontrivial ideals. The ideals of a Lie algebra $\mathfrak{g}$ are exactly the submodules of $\mathfrak{g}$ considered as a $\mathfrak{g}$-module. From this point of view, a Lie algebra $\mathfrak{g}$ is simple if and only if it is noncommutative and simple as a $\mathfrak{g}$-module.

A Lie algebra is semisimple if and only if it is the direct product of simple Lie algebras. Thus, to classify all semisimple Lie algebras, it suffices to classify all simple Lie algebras. The classification of simple Lie algebras will be considered in the next section.

The above statement that the classification of Lie algebras reduces to the classification of solvable and semisimple Lie algebras is based on *Levi theorem* which states that any Lie algebra $\mathfrak{g}$ can be represented as the semidirect product

$$\mathfrak{g} = \mathfrak{l} \ltimes \mathfrak{r},$$

where $\mathfrak{r}$ is the radical of $\mathfrak{g}$ and $\mathfrak{l}$ is a semisimple subalgebra of $\mathfrak{g}$, called a *Levi subalgebra* of $\mathfrak{g}$.

1.1.10 Universal enveloping algebra

Let $\mathfrak{g}$ be a Lie algebra, and let $T(\mathfrak{g})$ be the tensor algebra on $\mathfrak{g}$. Denote by $J(\mathfrak{g})$ the two-sided ideal of $T(\mathfrak{g})$ generated by tensors of the form $x \otimes y - y \otimes x - [x, y]$, where $x, y \in \mathfrak{g}$. The associative algebra $U(\mathfrak{g}) \equiv T(\mathfrak{g})/J(\mathfrak{g})$ is called the *universal enveloping algebra* of $\mathfrak{g}$. For the case when $\mathfrak{g}$ is an abelian algebra, the ideal $J(\mathfrak{g})$

coincides with the ideal $I(\mathfrak{g})$ introduced in example 1.8. Therefore, the universal enveloping algebra in this case is nothing but the symmetric algebra $S(\mathfrak{g})$.

Denote by π the canonical projection from $T(\mathfrak{g})$ to $U(\mathfrak{g})$, and by ι the embedding of $\mathfrak{g}$ into $T(\mathfrak{g})$. The composition $\sigma \equiv \pi \circ \tau$ is called the canonical mapping of $\mathfrak{g}$ into $U(\mathfrak{g})$. For any $x, y \in \mathfrak{g}$ we have

$$\sigma(x)\sigma(y) - \sigma(y)\sigma(x) = \sigma([x, y]).$$

Let $\{e_i\}_{i=1}^m$ be a basis of $\mathfrak{g}$, and

$$f_i \equiv \sigma(e_i).$$

The *Poincaré–Birkhoff–Witt theorem* states that the monomials $f_1^{k_1} f_2^{k_2} \cdots f_m^{k_m}$, where $k_1, \ldots, k_m$ are nonnegative integers, form a basis of $U(\mathfrak{g})$. From this theorem it follows, in particular, that the canonical mapping σ is injective. This fact allows one identify the Lie algebra $\mathfrak{g}$ with its image $\sigma(\mathfrak{g})$.

Let A be a unital algebra, and let φ be a linear mapping from a Lie algebra $\mathfrak{g}$ to A, such that $\varphi(x)\varphi(y) - \varphi(y)\varphi(x) = \varphi([x, y])$ for all $x, y \in \mathfrak{g}$. It can be shown that the mapping φ can be uniquely extended to a homomorphism from $U(\mathfrak{g})$ to A. From this it follows that any representation of a Lie algebra can be uniquely extended to the representation of the universal enveloping algebra. In other words, any $\mathfrak{g}$-module has the natural structure of a $U(\mathfrak{g})$-module.

Let $\mathfrak{h}$ be a subalgebra of a Lie algebra $\mathfrak{g}$. The inclusion mapping of $\mathfrak{h}$ into $\mathfrak{g}$ is a homomorphism from $\mathfrak{h}$ to $\mathfrak{g}$. This homomorphism can be uniquely extended to an injective homomorphism from $U(\mathfrak{h})$ to $U(\mathfrak{g})$. Taking this into account, we identify $U(\mathfrak{h})$ with the corresponding subalgebra of $U(\mathfrak{g})$. Further, let $\mathfrak{h}$ and $\mathfrak{k}$ be subalgebras of $\mathfrak{g}$, such that $\mathfrak{g} = \mathfrak{h} \oplus \mathfrak{k}$. It can easily be shown that $U(\mathfrak{g}) = U(\mathfrak{h})U(\mathfrak{k})$.

1.1.11 Contraction of Lie algebras

Let φ be a linear operator acting on a Lie algebra $\mathfrak{g}$. If φ is invertible, we can define in $\mathfrak{g}$ a new Lie algebra operation by

$$[x, y]' \equiv \varphi^{-1}([\varphi(x), \varphi(y)]).$$

As a result we obtain the Lie algebra $\mathfrak{g}'$ which coincides with $\mathfrak{g}$ as a vector space but which has the new Lie algebra operation. Actually, this new Lie algebra is isomorphic to the initial Lie algebra $\mathfrak{g}$.

Suppose now that we have a family φ_λ of linear operators acting in $\mathfrak{g}$, parametrised by a real parameter λ. Suppose that for all values of the parameter λ, except $\lambda = 0$, the mappings φ_λ are invertible, while for $\lambda = 0$ the mapping φ_λ has no inverse. If, nevertheless, there exists the limit

$$[x, y]' \equiv \lim_{\lambda \to 0} \varphi_\lambda^{-1}([\varphi_\lambda(x), \varphi_\lambda(y)]),$$

we again obtain a new Lie algebra $\mathfrak{g}'$ which is now not isomorphic to the Lie algebra $\mathfrak{g}$. Such a procedure is called a *contraction* of $\mathfrak{g}$. A contraction of a semisimple Lie algebra yields in general a nonsemisimple Lie algebra.

The most famous example here is the *Inönü–Wigner contraction*, see Inönü & Wigner (1953). This contraction is performed as follows. Let a Lie algebra $\mathfrak{g}$ be represented as a direct sum of its vector subspaces $\mathfrak{k}_0$ and $\mathfrak{k}_1$; then any element $x \in \mathfrak{g}$ can be uniquely represented as $x = x_0 + x_1$, where $x_0 \in \mathfrak{k}_0$ and $x_1 \in \mathfrak{k}_1$. Define the family of linear operators φ_λ by

$$\varphi_\lambda(x) \equiv x_0 + \lambda x_1.$$

It is evident that for $\lambda \neq 0$ one has

$$\varphi_\lambda^{-1}(x) = x_0 + \lambda^{-1} x_1,$$

therefore,

$$\varphi_\lambda^{-1}([\varphi(x), \varphi(y)]) = \varphi_\lambda^{-1}([x_0, y_0] + \lambda[x_0, y_1] + \lambda[x_1, y_0] + \lambda^2[x_1, y_1]).$$

From this equality one sees that the corresponding limit exists if and only if $[x_0, y_0] \in \mathfrak{k}_0$. In other words, $\mathfrak{k}_0$ must be a subalgebra of $\mathfrak{g}$. Note that, after the contraction, the subspace $\mathfrak{k}_1$ becomes an abelian ideal of the new Lie algebra $\mathfrak{g}'$, while $\mathfrak{k}_0$ remains a subalgebra. A more general contraction procedure was considered in Saletan (1961).

1.1.12 Realification and complexification

Let V be a complex vector space. We can multiply the elements of V by complex numbers, in particular by real numbers. Hence, we can consider V as a real vector space. This vector space is called the *realification* of V and is denoted by $V_\mathbb{R}$. If $\{e_i\}$ is a basis for a complex vector space V, then $\{e_i, \sqrt{-1}e_i\}$ is a basis for the real vector space $V_{\mathbb{R}}$. It follows from this fact that the dimensions of the spaces V and $V_{\mathbb{R}}$ are connected by the relation

$$\dim V_{\mathbb{R}} = 2\dim V.$$

Let $\mathfrak{g}$ be a complex Lie algebra. The real vector space $\mathfrak{g}_{\mathbb{R}}$ has the natural structure of a real Lie algebra.

EXAMPLE 1.14 For an arbitrary complex vector space V, a mapping from $V \times V$ to $\mathbb{C}$ which is antilinear with respect to the first argument and linear with respect to the second one is called a *sesquilinear form* on V. A sesquilinear form B on V satisfying the condition

$$B(u, v) = \overline{B(v, u)}$$

for all $v, u \in V$, is called a *hermitian* form on V.

Let V be an m-dimensional complex vector space, and let B be a sesquilinear form on V. It is clear that relation (1.3) defines a subalgebra $\mathfrak{gl}_B(V)$ of the Lie algebra $\mathfrak{gl}(V)_{\mathbb{R}}$. Let $\{e_i\}$ be a basis of V and let b be the matrix of B with respect to $\{e_i\}$. It can be shown that, in the case under consideration, relation (1.3) is equivalent to the matrix relation

$$ba + a^{\dagger}b = 0, \tag{1.12}$$

where a is the matrix of the endomorphism A with respect to the basis $\{e_i\}$. This relation defines a subalgebra of $\mathfrak{gl}(m, \mathbb{C})_{\mathbb{R}}$ which is isomorphic to $\mathfrak{gl}_B(V)$.

Suppose that B is a positive definite hermitian form. It is known that there exists a basis $\{e_i\}$ of V such that the matrix of B with respect to $\{e_i\}$ coincides with the unit matrix I_m. Hence, in this case, relation (1.12) takes the form

$$a + a^{\dagger} = 0.$$

The corresponding Lie subalgebra of $\mathfrak{gl}(m, \mathbb{C})_{\mathbb{R}}$ is called the *unitary* algebra and is denoted by $\mathfrak{u}(m)$. The intersection $\mathfrak{sl}(m, \mathbb{C})_{\mathbb{R}} \cap \mathfrak{u}(m)$ is called the *special unitary* algebra. This algebra is denoted by $\mathfrak{su}(m)$.

Let V and W be two complex vector spaces, and let $\varphi \in \mathrm{Hom}(V, W)$. We can consider φ as an element of $\mathrm{Hom}(V_{\mathbb{R}}, W_{\mathbb{R}})$. This homomorphism is called the realification of φ and is denoted by $\varphi_{\mathbb{R}}$. For example, let V be an arbitrary complex vector space.

The endomorphism J of $V_{\mathbb{R}}$, corresponding to the multiplication by $\sqrt{-1}$ in V, satisfies the relation

$$J^2 = -1.$$

Let us try to find for a given real vector space V a complex vector space $\widetilde{V}$ such that V is the realification of $\widetilde{V}$. Suppose that this space does exist, then the operator of the multiplication by $\sqrt{-1}$ in $\widetilde{V}$ induces the operator J in V, satisfying the relation $J^2 = -1$. These reasonings lead to the following definition. An endomorphism J of a real vector space V satisfying the relation $J^2 = -1$ is called a *complex structure* on V. Note that a complex structure exists only for even-dimensional real vector spaces. Having such a structure, define the operation of the multiplication of the elements of V by complex numbers as

$$(a + \sqrt{-1}b)u \equiv au + bJu.$$

It can be shown that this definition endows V with the structure of a complex vector space denoted by $\widetilde{V}$. Here we have $\widetilde{V}_{\mathbb{R}} = V$. Hence, in particular,

$$\dim V = 2 \dim \widetilde{V}.$$

A complex structure J on a real Lie algebra $\mathfrak{g}$ is said to be a *Lie complex structure* if

$$[x, J(y)] = J([x, y])$$

for all $x, y \in \mathfrak{g}$. If a real Lie algebra $\mathfrak{g}$ is endowed with a Lie complex structure J, then the complex vector space $\widetilde{\mathfrak{g}}$ has the natural structure of a complex Lie algebra inherited from $\mathfrak{g}$. It is clear that $\widetilde{\mathfrak{g}}_{\mathbb{R}} = \mathfrak{g}$.

There is another way to construct a complex vector space, proceeding from a given real vector space. Let V be a real vector space. Define on the space $V \oplus V$ the operator J_c acting on an element $(u, v) \in V \oplus V$ in accordance with the rule

$$J_c(v, u) = (-u, v).$$

It is clear that $J_c^2 = -1$; hence, J_c is a complex structure called the *canonical complex structure* on $V \oplus V$. The corresponding complex linear space is called the *complexification* of the real vector space V and is denoted by $V^{\mathbb{C}}$. The initial real vector space V can be identified with the subset of $V^{\mathbb{C}}$, formed by the vectors of the form

$(v, 0)$. Note that

$$\sqrt{-1}(v,0) = J_c(v,0) = (0,v);$$

hence, an arbitrary vector w of $V^{\mathbb{C}}$ can be represented as

$$w = v + \sqrt{-1}u,$$

where $v, u \in V$; and such a representation is unique.

Again let $\mathfrak{g}$ be a real Lie algebra. The complex vector space $\mathfrak{g}^{\mathbb{C}}$ has the natural structure of a complex Lie algebra given by

$$[x + \sqrt{-1}y, z + \sqrt{-1}t] \equiv [x,z] - [y,t] + \sqrt{-1}([y,z] + [x,t]),$$

where $x, y, z, t \in \mathfrak{g}$. The complex Lie algebra $\mathfrak{g}^{\mathbb{C}}$ is called the *complexification* of the real Lie algebra $\mathfrak{g}$.

Exercises

1.1 Let A be an algebra over $\mathbb{K}$, and let $\{e_i\}$ be a basis of A. It is clear that

$$e_i e_j = \sum_k e_k f^k{}_{ij}$$

for some $f^k{}_{ij} \in \mathbb{K}$. The coefficients $f^k{}_{ij}$ are called the *structure constants* of the algebra A. Prove that if A is an associative algebra, then

$$f^s{}_{ir} f^r{}_{jk} = f^s{}_{rk} f^r{}_{ij}.$$

Show that the structure constants of a Lie algebra satisfy the conditions

$$f^s{}_{ij} = -f^s{}_{ji},$$

$$f^s{}_{rk} f^r{}_{ij} + f^s{}_{ri} f^r{}_{jk} + f^s{}_{rj} f^r{}_{ki} = 0.$$

1.2 Prove that the Jacobi identity for a Lie algebra $\mathfrak{g}$ can be written as

$$[x,[y,z]] + [y,[z,x]] + [z,[x,y]] = 0$$

for all $x, y, z \in \mathfrak{g}$.

1.3 Show that any bilinear form B on a vector space V may be written uniquely as $B = B_s + B_a$, where B_s is symmetric and B_a is skew-symmetric. Prove that

$$\mathfrak{gl}_B(V) = \mathfrak{gl}_{B_s}(V) \cap \mathfrak{gl}_{B_a}(V).$$

1.4 Let B be a bilinear form on $\mathbb{K}^m$, given by $B(a,b) \equiv \sum_{i=0}^{n} a^i b^i$, where $n \leq m$. Describe the Lie algebra $\mathfrak{gl}_B(m,\mathbb{K})$. Show that

$$\mathfrak{gl}_B(m,\mathbb{K}) \simeq (\mathfrak{so}(n,\mathbb{K}) \times \mathfrak{gl}(m-n,\mathbb{K})) \ltimes \mathbb{K}^{(m-n)n},$$

where $\mathbb{K}^{(m-n)n}$ is treated as a commutative Lie algebra.

1.5 Show that for any vector v of $\mathfrak{sl}(2,\mathbb{C})$-module $L(n)$ defined in example 1.11, we have

$$2(x_+x_-v + x_-x_+v) + h^2v = n(n+2)v.$$

1.6 Prove that the Killing forms of the Lie algebras $\mathfrak{sl}(m,\mathbb{K})$ $(m \geq 2)$, $\mathfrak{o}(m,\mathbb{K})$ $(m \geq 3)$ and $\mathfrak{sp}(n,\mathbb{K})$ $(n \geq 1)$ are non-degenerate.

1.7 Let a Lie algebra $\mathfrak{g}$ be represented as the direct sum of vector subspaces $\mathfrak{k}_i$, $i = 0, 1, \ldots, s$. Represent an arbitrary element $x \in \mathfrak{g}$ as $x = \sum_{i=1}^{s} x_i$, where $x_i \in \mathfrak{k}_i$; and define a family of the linear operators φ_λ in $\mathfrak{g}$ by

$$\varphi_\lambda(x) \equiv \sum_{i=1}^{s} \lambda^i x_i.$$

Investigate, under which conditions the family φ_λ provides a contraction of $\mathfrak{g}$.

1.2 Semisimple Lie algebras

We start this section with a consideration of abstract root systems. This notion is crucial for the classification of complex semisimple Lie algebras.

1.2.1 Root systems

Let V be a vector space over a field $\mathbb{K}$, and let $\alpha \in V^*$. A linear operator $s : V^* \to V^*$ is called a *reflection* in V^* if $\dim \operatorname{Im}(1-s) = 1$ and $s^2 = 1$. Let α be an arbitrary nonzero element of $\operatorname{Im}(1-s)$; then for any $\beta \in V^*$ we have $\beta - s(\beta) = \varphi(\beta)\alpha$, where $\varphi(\beta) \in \mathbb{K}$. It is clear that φ is a linear mapping. Hence, there exists an element $\alpha^\vee \in V$, such that $\varphi(\beta) = \langle \beta, \alpha^\vee \rangle$ for any $\beta \in V^*$. The requirement $s^2 = 1$ gives $\langle \alpha, \alpha^\vee \rangle = 2$.

On the other hand, let α be an arbitrary element of V^*, and let $\alpha^\vee$ be an element of V satisfying the condition $\langle \alpha, \alpha^\vee \rangle = 2$; then the linear operator $s_{\alpha,\alpha^\vee}$ defined by

$$s_{\alpha,\alpha^\vee}(\beta) \equiv \beta - \langle \beta, \alpha^\vee \rangle \alpha,$$

is a reflection in V^*. Thus, any reflection in V^* has the form $s_{\alpha,\alpha^\vee}$ for some $\alpha \in V^*$ and $\alpha^\vee$ satisfying the condition $\langle \alpha, \alpha^\vee \rangle = 2$. Note that the reflection $s_{\alpha,\alpha^\vee}$ leaves the hyperplane

$$H_{\alpha^\vee} \equiv \{\beta \in V^* \mid \langle \beta, \alpha^\vee \rangle = 0\}$$

pointwise fixed, and sends the element α to $-\alpha$.

By definition of the transpose of a mapping, we have

$$\langle \beta, s^t_{\alpha,\alpha^\vee}(v) \rangle = \langle s_{\alpha,\alpha^\vee}(\beta), v \rangle.$$

From this relation it follows that

$$s^t_{\alpha,\alpha^\vee}(v) = v - \alpha^\vee \langle \alpha, v \rangle.$$

Hence, for any reflection s in V^*, the mapping s^t is a reflection in V, and $s^t_{\alpha,\alpha^\vee} = s_{\alpha^\vee,\alpha}$.

Let V be a vector space over a field $\mathbb{K}$. A subset Δ of V^* is called a *root system* in V^* if

(RS1) Δ is finite, spans V^*, and does not contain 0;

(RS2) for any $\alpha \in \Delta$ there exists a vector $\alpha^\vee \in V$ such that $\langle \alpha, \alpha^\vee \rangle = 2$, and the reflection $s_{\alpha,\alpha^\vee}$ leaves Δ invariant;

(RS3) $\langle \alpha, \beta^\vee \rangle \in \mathbb{Z}$ for all $\alpha, \beta \in \Delta$.

The dimension of the vector space V is called the *rank* of Δ, and the elements of Δ are called the *roots*. It can be shown that the vector $\alpha^\vee$ is uniquely determined by the vector α. Thus, it is natural to denote $s_{\alpha,\alpha^\vee}$ just by s_α.

It is clear that if $\alpha \in \Delta$, then $-\alpha \in \Delta$. A root system Δ is called *reduced* if for any $\alpha \in \Delta$ the element $-\alpha$ is the only root proportional to α.

Let Δ be a root system in V^*. An automorphism of V^* leaving Δ invariant is called an *automorphism* of Δ. The automorphisms of Δ form the group $\mathrm{Aut}(\Delta)$, which is called the *group of automorphisms* of Δ. Since the set Δ is finite and spans V^*, the group $\mathrm{Aut}(\Delta)$ can be identified with a subgroup of the symmetric group of Δ. From this fact it follows, in particular, that $\mathrm{Aut}(\Delta)$ is finite. The subgroup $W(\Delta)$ of $\mathrm{Aut}(\Delta)$, generated by the reflections s_α, $\alpha \in \Delta$, is called the *Weyl group* of Δ. The group $W(\Delta)$ is a normal subgroup of the group $\mathrm{Aut}(\Delta)$.

The set $\Delta^\vee$ formed by the vectors $\alpha^\vee$, $\alpha \in \Delta$, is a root system in V. This root system is called the *dual*, or *inverse* of Δ. The mapping $a \mapsto (a^t)^{-1}$ defines an isomorphism of the groups $\mathrm{Aut}(\Delta)$ and $\mathrm{Aut}(\Delta^\vee)$.

Let Δ be a root system in V^* of rank r. Since Δ spans V^*, we can choose a set of r roots $\{\beta_i\}_{i=1}^r$ forming a basis of V^*. Hence, for any root $\alpha \in \Delta$, we have a unique decomposition $\alpha = \sum_{i=1}^r k_i\beta_i$, which implies the relations

$$\langle \alpha, \beta_j^\vee \rangle = \sum_{j=1}^r k_i \langle \beta_i, \beta_j^\vee \rangle, \quad j = 1, \ldots, r.$$

We can consider these relations as a system of equations for the coefficients k_i. The matrix $(\langle \beta_i, \beta_j^\vee \rangle)$ is nondegenerate and has integer matrix elements; $\langle \alpha, \beta_j^\vee \rangle$ are also integers, and thus the coefficients k_i are rational numbers. Actually, it is important that they are real numbers.

Consider a root system Δ in a complex vector space V^*. The subset $V(\mathbb{R})$ of V, composed by the vectors $v \in V$ satisfying the condition $\langle \alpha, v \rangle \in \mathbb{R}$ for all $\alpha \in \Delta$, is a real vector space. From the above discussion we conclude that $V(\mathbb{R})$ is r-dimensional. The vector space $V(\mathbb{R})$ can be characterised as a subset of V formed by all linear combinations of the vectors $\alpha^\vee$, $\alpha \in \Delta$, with real coefficients. The dual space $V(\mathbb{R})^*$ can be identified with the set of all real linear combinations of roots $\alpha \in \Delta$. In particular, Δ is a subset of $V(\mathbb{R})^*$ and $\Delta^\vee$ is a subset of $V(\mathbb{R})$. It is quite clear that Δ is a root system in $V(\mathbb{R})^*$, while $\Delta^\vee$ is a root system in $V(\mathbb{R})$. Thus, any root system in a complex vector space generates a root system in a real vector space. On the other hand, taking the complexification, we see that a root system in a real vector space defines a root system in the corresponding complex vector space. In fact, we have here a bijective correspondence between root systems in complex and real vector spaces.

Recall that a finite-dimensional real vector space endowed with a positive definite nondegenerate bilinear form is called a *Euclidean space.*

For any root system Δ in V^* we can define a bilinear form $(\cdot\,,\cdot)$ in V by

$$(v, u) \equiv \sum_{\alpha\in\Delta} \langle \alpha, v \rangle \langle \alpha, u \rangle.$$

This bilinear form is symmetric and nondegenerate. Moreover, it is invariant under the group $\mathrm{Aut}(\Delta^\vee)$. If V is a real vector space, this bilinear form is positive definite. If V is a complex vector space, then the restriction of $(\cdot,\cdot)$ to $V(\mathbb{R})$ is also a positive definite bilinear form.

Denote by ν the isomorphism from the space V to V^* induced by the bilinear form $(\cdot,\cdot)$. Recall that this isomorphism is defined by

$$\langle \nu(v), u\rangle \equiv (v,u).$$

The isomorphism ν allows one to define a bilinear form on V^* by

$$(\alpha,\beta) \equiv (\nu^{-1}(\alpha), \nu^{-1}(\beta))$$

The bilinear form $(\cdot,\cdot)$ on V^* is symmetric, nondegenerate and invariant with respect to the group $\mathrm{Aut}(\Delta)$. In particular, for any reflection s_α we have $(s_\alpha(\beta), s_\alpha(\gamma)) = (\beta,\gamma)$ for any $\beta,\gamma \in V^*$. Taking $\gamma = \alpha$, we obtain $(s_\alpha(\beta),\alpha) = -(\beta,\alpha)$. From this equality it follows that

$$\langle \beta, \alpha^\vee\rangle = \frac{2(\beta,\alpha)}{(\alpha,\alpha)}.$$

For any $\beta,\alpha \in V^*$ we have

$$(\beta,\alpha) = \langle \beta, \nu^{-1}(\alpha)\rangle.$$

Thus, for any $\alpha \in \Delta$ we obtain

$$\alpha^\vee = \frac{2\nu^{-1}(\alpha)}{(\alpha,\alpha)} = \frac{2\nu^{-1}(\alpha)}{(\nu^{-1}(\alpha),\nu^{-1}(\alpha))}.$$

The above-defined bilinear form $(\cdot,\cdot)$ on V^* is called the *canonical bilinear form.*

Suppose that V is a real vector space. Denote by $|\alpha|$ the *length* of the element $\alpha \in V^*$; in other words $|\alpha| \equiv (\alpha,\alpha)^{1/2}$. The angle between two elements $\alpha,\beta \in V^*$ is, by definition, a unique angle θ, $0 \le \theta \le \pi$, such that

$$|\alpha||\beta|\cos\theta = (\alpha,\beta).$$

Hence, we can write

$$\langle \alpha, \beta^\vee\rangle = 2\frac{|\alpha|}{|\beta|}\cos\theta.$$

Therefore,

$$\langle \alpha,\beta^\vee\rangle\langle \beta,\alpha^\vee\rangle = 4\cos^2\theta.$$

From condition (RS3) entering the definition of a root system, we see that $4\cos^2\theta$ can take the values 0, 1, 2, 3, or 4. Here the value

Table 1.1.

$\langle\alpha,\beta^\vee\rangle$	$\langle\beta,\alpha^\vee\rangle$	θ	$\lvert\beta\rvert^2/\lvert\alpha\rvert^2$
0	0	$\pi/2$	-
1	1	$\pi/3$	1
-1	-1	$2\pi/3$	1
1	2	$\pi/4$	2
-1	-2	$3\pi/4$	2
1	3	$\pi/6$	3
-1	-3	$5\pi/6$	3

4 corresponds to the case of proportional roots. Considering only nonproportional roots, and supposing that $|\beta| \geq |\alpha|$, we arrive at the possibilities described in table 1.1. In the case where V is a complex vector space, we define the length of a root and the angle between two roots, treating Δ as a root system in a real vector space $V(\mathbb{R})^*$.

A subset Π of a root system Δ in V^* is called a *base* of Δ if

(B1) Π is a basis of V^*;

(B2) for any $\beta \in \Delta$ the coefficients m_α of the expansion $\beta = \sum_{\alpha\in\Pi} m_\alpha\alpha$ are integers, and either all m_α are nonnegative or all they are nonpositive.

The elements of Π are called *simple roots*. A base of a root system is often called a *system of simple roots*. The *height*, $\operatorname{ht}\beta$, of the root $\beta = \sum_{\alpha\in\Pi} m_\alpha\alpha$ with respect to Π is defined as $\operatorname{ht}\beta \equiv \sum_{\alpha\in\Pi} m_\alpha$. A root $\beta = \sum_{\alpha\in\Pi} m_\alpha\alpha$ is called *positive* (*negative*) if all $m_\alpha \geq 0$ (all $m_\alpha \leq 0$).

A relation $\prec$ on a set S is said to be a *partial order* on S if

(P1) $x \prec x$ for each $x \in S$;

(P2) $x \prec y$ and $y \prec x$ imply that $x = y$;

(P3) $x \prec y$ and $y \prec z$ imply that $x \prec z$.

If $x \prec y$, we also write $y \succ x$. If $\prec$ is a partial order on S, then an element $y \in S$ is called a *maximal element* if $y \prec x$ implies that $x = y$. A *minimal element* is defined similarly.

A base Π of the root system Δ in V^* specifies a partial order on V^* defined as follows. Let $\alpha, \beta \in V^*$, define $\beta \prec \alpha$ if and only if either $\alpha - \beta$ is a positive root or $\beta = \alpha$. In particular, for any positive (negative) root β we have $\beta \succ 0$ ($\beta \prec 0$).

It can be shown that any root system has a base. Actually, it is not difficult to describe a procedure for constructing all possible bases which looks as follows. Suppose, first, that V is a real vector space. Call a vector $v \in V$ *regular* if $v \in V - \bigcup_{\alpha \in \Delta} H_\alpha$, where the hyperplanes H_α are defined by

$$H_\alpha \equiv \{v \in V \mid \langle \alpha, v \rangle = 0\}.$$

In other words, v is regular if $\langle \alpha, v \rangle \neq 0$ for all $\alpha \in \Delta$. Let v be a regular element of V. Introduce the notations

$$\Delta^+(v) \equiv \{\alpha \in \Delta \mid \langle \alpha, v \rangle > 0\}, \quad \Delta^-(v) \equiv -\Delta^+(v).$$

Since v is regular, we have $\Delta = \Delta^+(v) \bigcup \Delta^-(v)$. A root $\alpha \in \Delta^+(v)$ is called *decomposable* if $\alpha = \beta + \gamma$ for some $\beta, \gamma \in \Delta^+(v)$. Otherwise we say that the root α is *indecomposable*. It can be shown that the set $\Pi(v)$ of all indecomposable roots is a base of Δ. Moreover, any base of Δ can be obtained in such a way. If V is a complex vector space, we can restrict ourselves to the space $V(\mathbb{R})^*$ and construct a base of Δ by the above procedure. Any such base is a base of Δ considered as a root system in V^*.

Suppose again that V is a real vector space. A connected component of the set $V - \bigcup_{\alpha \in \Delta} H_\alpha$ is called a *Weyl chamber*. It is easy to see that $\Pi(v) = \Pi(v')$ if and only if the vectors v and v' belong to the same Weyl chamber. Thus, there is a bijective correspondence between Weyl chambers and bases. Denote by $C(\Pi)$ the Weyl chamber corresponding to the base Π, and call this chamber the *fundamental Weyl chamber* with respect to Π. For a given base Π, the fundamental Weyl chamber $C(\Pi)$ can be described as the set of all $v \in V$ such that $\langle \alpha, v \rangle > 0$ for all $\alpha \in \Pi$. The union of all Weyl chambers is called the *Tits cone*. In the case of a complex vector space V, the Weyl chambers are defined as subsets of $V(\mathbb{R})$.

Let Δ be a reduced root system in V^*. For any bases Π and Π' there exists a unique element $w \in W(\Delta)$ such that $w(\Pi) = \Pi'$. It can be also shown that, for any base Π of Δ, the Weyl group $W(\Delta)$ is generated by the reflections s_α, $\alpha \in \Pi$. Furthermore, any root $\alpha \in \Delta$ can be represented as $w(\pi_i)$, where $w \in W(\Delta)$ and π_i is a simple root.

1.2.2 Irreducible root systems

Let Δ_1 and Δ_2 be root systems in V_1^* and V_2^* respectively. Using the canonical injection of V_1^* and V_2^* into $(V_1 \oplus V_2)^* \simeq V_1^* \oplus V_2^*$, we can identify Δ_1 and Δ_2 with subsets of V^*, where $V \equiv V_1 \oplus V_2$. It can easily be shown that $\Delta \equiv \Delta_1 \cup \Delta_2$ is a root system in V^*. This root system is called the direct sum of the root systems Δ_1 and Δ_2. On the other hand, let Δ be a root system in V^*, and let $V = V_1 \oplus V_2$, where V_1 and V_2 are such subspaces of V that $\Delta \subset V_1^* \cup V_2^*$. Denote $\Delta_1 \equiv \Delta \cap V_1^*$ and $\Delta_2 \equiv \Delta \cap V_2^*$. Now we can easily see that Δ_1 and Δ_2 are root systems in V_1^* and V_2^* respectively. Moreover, the subspaces V_1^* and V_2^* are orthogonal. In such a situation we say that the root system Δ is the direct sum of its subsystems Δ_1 and Δ_2.

A root system Δ in V^* is called *irreducible* if Δ cannot be represented as the direct sum of its subsystems. Any root system is the sum of irreducible root subsystems.

Let Δ be an irreducible root system, and let Π be a base of Δ. With respect to the partial ordering $\prec$ specified by the base Π, there is the unique *maximal root* $\tilde{\alpha}$. It is clear that $-\tilde{\alpha}$ is the unique *minimal root* of Δ. Here, if $\beta \neq \tilde{\alpha}$, then $\operatorname{ht}\beta < \operatorname{ht}\tilde{\alpha}$, and all the coefficients of the expansion of $\tilde{\alpha}$ over the base Π are positive. The sum of these coefficients plus 1 is called the *Coxeter number* of Δ.

For any irreducible root system Δ there are at most two root lengths. If Δ has two distinct root lengths, we speak of *long* and *short* roots. If all the roots have the same length, the Lie algebra is called *simply laced*. In this case we call all the roots long. The maximal root is always long.

Now let Π be a base of a root system Δ of rank r. Choosing some ordering of the elements of Δ, we write them as $\alpha_1, \ldots, \alpha_r$. The $r \times r$ matrix $k = (k_{ij})$, where

$$k_{ij} \equiv \langle \alpha_i, \alpha_j^\vee \rangle = \frac{2(\alpha_i, \alpha_j)}{(\alpha_j, \alpha_j)},$$

is called the Cartan matrix of the root system Δ. The Cartan matrix depends on the ordering of the simple roots entering the base Π. Here the Cartan matrices corresponding to different orderings can be reduced one to another by a simultaneous interchange of rows and columns. Up to this freedom in ordering, the Cartan

Table 1.2.

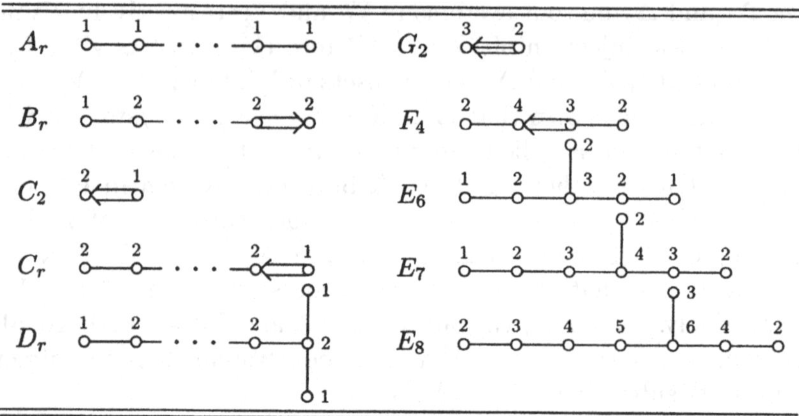

matrix is independent on the choice of Π. Since Π is a basis of V^*, the Cartan matrix is nondegenerate. It appears that any root system is determined by its Cartan matrix up to isomorphism.

It is convenient to describe Cartan matrices with the help of Dynkin diagrams. The *Dynkin diagram* corresponding to an $r \times r$ Cartan matrix (k_{ij}) has r vertices. For any $i, j = 1, \ldots, r$, such that $i \neq j$, the ith vertex is connected with the jth vertex by $k_{ij}k_{ji}$ edges with an arrow pointing to the ith vertex if $|k_{ij}| < |k_{ji}|$. It is clear that the Cartan matrix is uniquely determined by its Dynkin diagram. A root system is irreducible if and only if the Dynkin diagram, corresponding to its Cartan matrix, is connected.

We restrict ourselves to the case of reduced root systems because only such root systems arise for the semisimple Lie algebras. It can be shown that, up to an isomorphism, there are four classical series of irreducible reduced root systems: A_r $(r \geq 1)$, B_r $(r \geq 3)$, C_r $(r \geq 2)$, D_r $(r \geq 4)$, and five exceptional ones: E_6, E_7, E_8, F_4, G_2. Here the lower index means the rank of the root system. The corresponding Dynkin diagrams are given in table 1.2. The labels of the vertices are the expansion coefficients of the maximal root over the simple roots.

The explicit form of the Cartan matrices for the classical series A_r, B_r, C_r and D_r is presented in chapter 4.

1.2.3 Cartan subalgebras

Let $\mathfrak{h}$ be a vector subspace of a Lie algebra $\mathfrak{g}$. The set

$$N_{\mathfrak{g}}(\mathfrak{h}) \equiv \{x \in \mathfrak{g} \mid [x, \mathfrak{h}] \subset \mathfrak{h}\}$$

is called the *normaliser* of $\mathfrak{h}$. For any Lie algebra $\mathfrak{g}$ and any vector subspace $\mathfrak{h}$ of $\mathfrak{g}$, the normaliser $N_{\mathfrak{g}}(\mathfrak{h})$ is a subalgebra of $\mathfrak{g}$. If $\mathfrak{h}$ is a subalgebra of a Lie algebra $\mathfrak{g}$, then $\mathfrak{h}$ is an ideal of $N_{\mathfrak{g}}(\mathfrak{h})$. Actually, $N_{\mathfrak{g}}(\mathfrak{h})$ is the largest subalgebra of $\mathfrak{g}$ which includes $\mathfrak{h}$ as an ideal.

A subalgebra $\mathfrak{h}$ of a Lie algebra $\mathfrak{g}$ is called a *Cartan subalgebra* if $\mathfrak{h}$ is nilpotent and coincides with its normaliser $N_{\mathfrak{g}}(\mathfrak{h})$. It can be proved that any Lie algebra has a Cartan subalgebra. Moreover, in the case where $\mathfrak{g}$ is a complex Lie algebra, the group $\mathrm{Int}(\mathfrak{g})$ of inner automorphisms of $\mathfrak{g}$ acts transitively on the set of all Cartan subalgebras of $\mathfrak{g}$. This is not, in general, true for real Lie algebras. Nevertheless, all Cartan subalgebras of a given Lie algebra have the same dimension. Hence, we can define the *rank* of a Lie algebra $\mathfrak{g}$ as the dimension of any of its Cartan subalgebra.

A linear operator $A \in \mathrm{End}(V)$ is said to be *semisimple* if any invariant subspace of A has an invariant complement. A linear operator in a complex vector space is semisimple if and only if it is diagonalisable. A linear operator $A \in \mathrm{End}(V)$ is called *nilpotent* if $A^k = 0$ for some integer $k > 0$. For any $A \in \mathrm{End}(V)$ there exist unique operators $A_s, A_n \in \mathrm{End}(V)$ such that $A = A_s + A_n$; A_s is semisimple, A_n is nilpotent, and A_s and A_n commute. Furthermore, the operators A_s and A_n are polynomials in A. The decomposition $A = A_s + A_n$ is called the *Jordan decomposition* of A; A_s and A_n are called the *semisimple part* and the *nilpotent part* of A respectively.

An element x of a Lie algebra $\mathfrak{g}$ is called *semisimple* (*nilpotent*) if the linear operator $\mathrm{ad}(x)$ is semisimple (nilpotent). If $\mathfrak{g}$ is a semisimple complex Lie algebra and ρ is a representation of $\mathfrak{g}$, then for any semisimple (nilpotent) element $x \in \mathfrak{g}$ the linear operator $\rho(x)$ is semisimple (nilpotent).

Let S be a subset of a Lie algebra $\mathfrak{g}$. The set

$$C_{\mathfrak{g}}(S) \equiv \{x \in \mathfrak{g} \mid [x, S] = 0\}$$

is called the *centraliser* of S. It is clear that $C_{\mathfrak{g}}(\mathfrak{g}) = Z(\mathfrak{g})$. For any subset S of a Lie algebra $\mathfrak{g}$, the centraliser $C_{\mathfrak{g}}(S)$ is a subalgebra of $\mathfrak{g}$.

If $\mathfrak{g}$ is a semisimple Lie algebra, then any Cartan subalgebra $\mathfrak{h}$ of $\mathfrak{g}$ is commutative, all elements of $\mathfrak{h}$ are semisimple, $\mathfrak{h}$ coincides with its centraliser $C_{\mathfrak{g}}(\mathfrak{h})$, and the restriction of the Killing form of $\mathfrak{g}$ to $\mathfrak{h} \times \mathfrak{h}$ is a nondegenerate bilinear form. Actually, in this case we can define a Cartan subalgebra as a maximal commutative subalgebra consisting entirely of semisimple elements.

Let $\mathfrak{h}$ be a Cartan subalgebra of a complex semisimple Lie algebra $\mathfrak{g}$. Since $\mathfrak{h}$ is commutative, the linear operators $\mathrm{ad}_{\mathfrak{g}}(h)$, $h \in \mathfrak{h}$, form a commuting set of semisimple linear operators in $\mathfrak{g}$. Therefore, these operators are simultaneously diagonalisable. This means that there exists a basis for $\mathfrak{g}$ consisting of common eigenvectors of the operators $\mathrm{ad}_{\mathfrak{g}}(h)$, $h \in \mathfrak{h}$. Let x be such a vector, then for any $h \in \mathfrak{h}$ we can write

$$[h, x] = \alpha(h)x,$$

where $\alpha(h)$ is a complex number. It is clear that α is a linear mapping from $\mathfrak{h}$ to $\mathbb{C}$. In other words, α is an element of $\mathfrak{h}^*$.

Let $\alpha \in \mathfrak{h}^*$, denote by $\mathfrak{g}^\alpha$ the linear subspace of $\mathfrak{g}$ given by

$$\mathfrak{g}^\alpha \equiv \{x \in \mathfrak{g} \mid [h, x] = \alpha(h)x \text{ for all } h \in \mathfrak{h}\}. \tag{1.13}$$

From the Jacobi identity it follows that

$$[\mathfrak{g}^\alpha, \mathfrak{g}^\beta] \subset \mathfrak{g}^{\alpha+\beta} \tag{1.14}$$

for any $\alpha, \beta \in \mathfrak{h}^*$. Since $\mathfrak{h}$ coincides with its centraliser $C_{\mathfrak{g}}(\mathfrak{h})$, we have $\mathfrak{g}^0 = \mathfrak{h}$. An element $\alpha \in \mathfrak{h}$, such that $\alpha \neq 0$ and $\mathfrak{g}^\alpha \neq \{0\}$ is called a *root* of $\mathfrak{g}$ with respect to $\mathfrak{h}$. Here the subspace $\mathfrak{g}^\alpha$ is said to be a *root subspace.* Denote the set of roots of $\mathfrak{g}$ with respect to $\mathfrak{h}$ by Δ. It is clear that

$$\mathfrak{g} = \mathfrak{h} \oplus \bigoplus_{\alpha \in \Delta} \mathfrak{g}^\alpha. \tag{1.15}$$

Note here that for any $h, h' \in \mathfrak{h}$ we have

$$K(h, h') = \sum_{\alpha \in \Delta} \langle \alpha, h \rangle \langle \alpha, h' \rangle.$$

Hence, the restriction of the Killing form of $\mathfrak{g}$ to $\mathfrak{h}$ induces the canonical bilinear form on $\mathfrak{h}^*$.

For any $\alpha \in \Delta$ the subspace $\mathfrak{g}^\alpha \subset \mathfrak{g}$ is one-dimensional. Then for any $\alpha, \beta \in \Delta$, such that $\alpha + \beta \in \Delta$ and $\alpha + \beta \neq 0$, the subspace $[\mathfrak{g}^\alpha, \mathfrak{g}^\beta] \subset \mathfrak{g}^{\alpha+\beta}$ is either one-dimensional, or trivial; actually,

$$[\mathfrak{g}^\alpha, \mathfrak{g}^\beta] = \mathfrak{g}^{\alpha+\beta}.$$

It can be shown that if $\alpha \in \Delta$, then $-\alpha \in \Delta$. The subspace $[\mathfrak{g}^\alpha, \mathfrak{g}^{-\alpha}] \subset \mathfrak{h}$ is again either one-dimensional, or trivial. It appears that all such subspaces are one-dimensional; moreover

$$[\mathfrak{g}^\alpha, \mathfrak{g}^{-\alpha}] = \mathbb{C}\nu^{-1}(\alpha) = \mathbb{C}\alpha^\vee,$$

where ν is the canonical isomorphism from $\mathfrak{h}$ to $\mathfrak{h}^*$ induced by the restriction of the Killing form of $\mathfrak{g}$ to $\mathfrak{h}$.

1.2.4 Defining relations of complex semisimple Lie algebras

A Lie algebra $\mathfrak{g}$ is said to be *free* if there exists a subset $S \subset \mathfrak{g}$ such that $\mathfrak{g}$ is generated by S, and any mapping from S to an arbitrary Lie algebra $\mathfrak{k}$ can be extended to a homomorphism from $\mathfrak{g}$ to $\mathfrak{k}$, which is actually unique. We also say in such a situation that $\mathfrak{g}$ is *free on* S and that $\mathfrak{g}$ is *freely generated by* S. Taking into account the connection of antihomomorphisms and antilinear (anti)antihomomorphisms with the usual homomorphisms, we see that if a Lie algebra $\mathfrak{g}$ is free on a subset S, then any mapping from S to a Lie algebra $\mathfrak{k}$ can be uniquely extended either to an antihomomorphism or to antilinear (anti)homomorphism from $\mathfrak{g}$ to $\mathfrak{h}$. The existence of free Lie algebras is demonstrated by the following construction.

Let S be an arbitrary set. Consider the set V of all formal finite linear combinations of elements of S with the coefficients from a field $\mathbb{K}$. In other words, an element $v \in V$ has the form $v = \sum_{s \in S} v_s s$, where only a finite number of the coefficients k_s differ from zero. Introduce in V the operation of addition and multiplication by elements of $\mathbb{K}$, defined as

$$v + w \equiv \sum_{s \in S} (v_s + w_s) s, \qquad kv \equiv \sum_{s \in S} (k v_s) s,$$

where $v = \sum_{s \in S} v_s s$, $w = \sum_{s \in S} w_s s$, and $k \in \mathbb{K}$. With respect to these operations, V is a vector space over the field $\mathbb{K}$. This vector space is often denoted by $\mathbb{K}^S$. Consider the Lie algebra associated with the tensor algebra $T(\mathbb{K}^S)$. The subalgebra of this Lie algebra generated by S is free on S. This Lie algebra is called the *free Lie algebra* over S and is denoted by $\mathfrak{g}\langle S \rangle$. It is evident that any Lie algebra which is free on some set S is isomorphic to $\mathfrak{g}\langle S \rangle$.

Any Lie algebra is isomorphic to a quotient algebra of some free Lie algebra. Indeed, let a Lie algebra $\mathfrak{g}$ be generated by a subset $S \subset \mathfrak{g}$. Consider a mapping $\sigma : S \to \mathfrak{g}$ which sends any element of S, as an element of $\mathfrak{g}\langle S\rangle$, to the same element, but as an element of $\mathfrak{g}$. Since $\mathfrak{g}\langle S\rangle$ is free on S, this mapping can be extended to a unique homomorphism from $\mathfrak{g}\langle S\rangle$ to $\mathfrak{g}$, which we denote by π. The mapping π is surjective and, therefore, $\mathfrak{g} \simeq \mathfrak{g}\langle S\rangle/\mathfrak{i}$, where $\mathfrak{i} \equiv \ker\pi$. The mapping π will be called the canonical projection from $\mathfrak{g}\langle S\rangle$ onto $\mathfrak{g}$.

Thus, any set S, generating a Lie algebra $\mathfrak{g}$, gives the corresponding ideal $\mathfrak{i}$ of the free Lie algebra $\mathfrak{g}\langle S\rangle$. Usually the ideal $\mathfrak{i}$ is specified by pointing out some subset $R \subset \mathfrak{i}$ which generates $\mathfrak{i}$. Any such subset is called a system of *defining relations* of $\mathfrak{g}$ with respect to the system of generators S. Since S generates $\mathfrak{g}\langle S\rangle$, the elements of R can always be expressed in terms of elements of S. For any relation r we have $\pi(r) = 0$, where π is the canonical projection from $\mathfrak{g}\langle S\rangle$ to $\mathfrak{g}$ defined in the previous paragraph. Therefore, instead of writing $r \in R$, it is customary to write formally $r = 0$. A relation of the form $r - r' = 0$ is also written as $r = r'$.

Return now to the case of complex semisimple Lie algebras. Let $\mathfrak{g}$ be such an algebra, let $\mathfrak{h}$ be its Cartan subalgebra, and let Δ be the root system of $\mathfrak{g}$ with respect to $\mathfrak{h}$. Choose a base $\Pi = \{\alpha_1, \ldots, \alpha_r\}$ of Δ, and introduce the notation $h_i \equiv \alpha_i^\vee$. It is clear that

$$[h_i, h_j] = 0. \tag{1.16}$$

For any choice of the elements $x_{+i} \in \mathfrak{g}^{\alpha_i}$, $i = 1, \ldots, r$, we can choose elements $x_{-i} \in \mathfrak{g}^{-\alpha_i}$, $i = 1, \ldots, r$, in such a way that

$$[x_{+i}, x_{-j}] = \delta_{ij} h_i. \tag{1.17}$$

Furthermore, we have

$$[h_i, x_{+j}] = k_{ji} x_{+j}, \qquad [h_i, x_{-j}] = -k_{ji} x_{-j}. \tag{1.18}$$

The elements h_i and $x_{\pm i}$, $i = 1, \ldots, r$, are called *Cartan generators* and *Chevalley generators* respectively. The Cartan generators form a basis of $\mathfrak{h}$ and, together with the Chevalley generators, generate the whole Lie algebra $\mathfrak{g}$. The following equalities are also valid:

$$\operatorname{ad}(x_{+i})^{-k_{ij}+1} x_{+j} = 0, \quad \operatorname{ad}(x_{-i})^{-k_{ij}+1} x_{-j} = 0, \quad i \neq j. \tag{1.19}$$

These equalities are called the *Serre relations*. Let S be the set formed by Cartan and Chevalley generators. It appears that equalities (1.16)–(1.19), where h_i and $x_{\pm i}$ are treated as elements of the free Lie algebra $\mathfrak{g}\langle S\rangle$, describe a system of defining relations of the Lie algebra $\mathfrak{g}$ with respect to the system of generators h_i and $x_{\pm i}$.

Let a Lie algebra $\mathfrak{g}$ be generated by a subset S, let $R \subset \mathfrak{g}\langle S\rangle$ be a system of defining relations of $\mathfrak{g}$ with respect to S, and let φ be a mapping from S to $\mathfrak{g}$. Since S generates both $\mathfrak{g}$ and $\mathfrak{g}\langle S\rangle$, we can define the mapping $\widetilde{\varphi} : S \to \mathfrak{g}\langle S\rangle$ in such a way that $\varphi = \pi \circ \widetilde{\varphi}$, where π is the canonical projection from $\mathfrak{g}\langle S\rangle$ to $\mathfrak{g}$. The mapping $\widetilde{\varphi}$ is, in general, not unique. Since $\mathfrak{g}\langle S\rangle$ is a free Lie algebra, the mapping $\widetilde{\varphi}$ can be uniquely extended to a homomorphism $\widetilde{\psi}$ from $\mathfrak{g}\langle S\rangle$ to $\mathfrak{g}\langle S\rangle$. Suppose that $\widetilde{\varphi}(R) \subset \mathfrak{i}$, where $\mathfrak{i}$ is the ideal generated by R, which, by definition, coincides with $\ker \pi$. Note that the condition $\widetilde{\varphi}(R) \subset \mathfrak{i}$ is valid for any mapping $\widetilde{\varphi}$ satisfying the relation $\pi \circ \widetilde{\varphi} = \varphi$ if and only if it is valid for some mapping satisfying this relation. Hence, this condition is actually a restriction on the mapping φ. Further, the requirement $\widetilde{\varphi}(R) \subset \mathfrak{i}$ is equivalent to the condition $\widetilde{\psi}(\mathfrak{i}) \subset \mathfrak{i}$. Therefore, in the case under consideration, there exists a unique homomorphism $\psi : \mathfrak{g} \to \mathfrak{g}$, such that $\pi \circ \widetilde{\psi} = \psi \circ \pi$. It can be proved that the mapping ψ does not depend on the choice of the mapping $\widetilde{\varphi}$. It is evident that $\psi(s) = \varphi(s)$ for any $s \in S$; in other words, the mapping ψ is an extension of the mapping φ from S to the whole Lie algebra $\mathfrak{g}$. Taking this into account, one usually uses for the mapping ψ the notation of the original mapping φ.

The above discussion demonstrates a method of defining homomorphisms of a Lie algebra $\mathfrak{g}$ into itself by specifying their action on some system of generators of $\mathfrak{g}$. It is clear that a similar method can be used to construct antihomomorphisms, antilinear homomorphisms and antilinear antihomomorphisms of $\mathfrak{g}$ into itself. For example, consider a mapping φ acting on the Cartan and Chevalley generators of a complex semisimple Lie algebra $\mathfrak{g}$ as

$$\sigma(h_i) = -h_i, \qquad \sigma(x_{\pm i}) = -x_{\mp i}.$$

Using the defining relations (1.16)–(1.19), it is not difficult to show that this mapping can be extended to an automorphism of $\mathfrak{g}$ satisfying the condition $\sigma \circ \sigma = \mathrm{id}_{\mathfrak{g}}$. The mapping $x \in \mathfrak{g} \mapsto x' \equiv -\sigma(x)$ is an involution of $\mathfrak{g}$, which is called the *Chevalley involution*.

It can be shown that the set Δ of roots of a complex semisimple Lie algebra is a reduced root system in $\mathfrak{h}^*$. Different Cartan subalgebras lead to isomorphic root systems. Moreover, two complex semisimple Lie algebras having isomorphic root systems are isomorphic as Lie algebras; and for any reduced root system Δ there exists a complex semisimple Lie $\mathfrak{g}$ whose root system is isomorphic to Δ. Furthermore, a complex semisimple Lie algebra $\mathfrak{g}$ is simple if and only if its root system is irreducible. Therefore, to classify complex simple Lie algebras it is enough to classify irreducible reduced root systems. The results of the classification are summarised in table 1.2 in terms of Dynkin diagrams.

1.2.5 Kac–Moody algebras

In this book we do not consider the integrable systems associated with infinite-dimensional Lie algebras. Nevertheless, in this section we discuss a class of infinite-dimensional Lie algebras known as Kac–Moody algebras. We believe that this discussion will provide the reader with a good starting point for a study of the original papers. The presentation of the subject is based mainly on a remarkable book by Kac (1990), which deserves special attention in the theory of infinite-dimensional Lie algebras of finite growth, but also contains much useful information on finite-dimensional Lie algebras.

An $r\times r$ matrix k is called a *generalised Cartan matrix* if it obeys the following conditions:

(CM1) $k_{ii} = 2$, $i = 1, \ldots, r$;

(CM2) $-k_{ij} \in \mathbb{Z}_+$ for all $i \neq j$;

(CM3) if $k_{ij} = 0$, then $k_{ji} = 0$.

Note that if k is a generalised Cartan matrix, its transpose k^t is also such a matrix.

In the case where all the components of the real vector $u = (u_1, \ldots, u_r)$ are positive (nonnegative), we write $u > 0$ ($u \geq 0$). Let k be an $r\times r$ generalised Cartan matrix. The matrix k is said to be of *finite type* if $\det k \neq 0$, there exists a vector $u > 0$, such that $ku > 0$, and from the inequality $kv \geq 0$ it follows that either $v > 0$ or $v = 0$. We say that k is of *affine type* if $\operatorname{rank} k = r - 1$; there exists a vector $u > 0$ such that $ku = 0$, and the inequality $kv \geq 0$ implies $kv = 0$. Finally, the matrix k is said to be of *indefinite type* if there exists a vector $u > 0$ such that $ku < 0$, and the inequalities $kv \geq 0$, $v \geq 0$ imply that $v = 0$. It can be shown that any generalised Cartan matrix belong to one of the above three types, and the generalised Cartan matrices k and k^t are of the same type.

The matrix k is called *symmetrisable* if there exists a nondegenerate diagonal matrix $v = \mathrm{diag}(v_1, ..., v_r)$ such that the matrix vk is symmetric, i. e., $v_i k_{ij} = v_j k_{ji}$. Any generalised Cartan matrix of finite or affine type is symmetrisable.

A generalised Cartan matrix k is called *decomposable* if there is a simultaneous permutation of its rows and columns which brings k into a block-diagonal form. Correspondingly, k is said to be *indecomposable* if there are no simultaneous permutations of its rows and columns bringing k into a block-diagonal form.

Let S be a subset of the set $\{1, \ldots, r\}$, and let k be an $r \times r$ matrix. The matrix $k_S \equiv (k_{i,j})_{i,j \in S}$ is called a *principal submatrix* of k. The determinant of a principal submatrix of k is called a *principal minor* of k.

Let k be an indecomposable generalised Cartan matrix. It can be shown that k is of *finite type* if and only if all its principal minors are positive. Further, the matrix k is of *affine type* if and only if all its proper principal minors are positive and $\det k = 0$.

It is clear that we can use Dynkin diagrams to describe generalised Cartan matrices. In this case we have a bijective correspondence between Dynkin diagrams and Cartan matrices. Note that the Dynkin diagram associated with the Cartan matrix k^t can be obtained from the Dynkin diagram corresponding to the Cartan matrix k by reversing the direction of the arrows.

It can be shown that the Cartan matrices corresponding to the Dynkin diagrams given in table 1.2 exhaust all Cartan matrices of finite type. On the other hand, table 1.2 describes all Cartan matrices corresponding to irreducible reduced root systems. Such root systems are in bijective correspondence with complex finite-dimensional simple Lie algebras. Hence, we can say that the conditions (C1)–(C3), together with the finiteness condition, single out the matrices corresponding to complex finite-dimensional simple Lie algebras. Recall that such an algebra can be constructed with the help of the corresponding generators h_i, $x_{\pm i}$ and relations (1.16)–(1.19). It is natural to consider the Lie algebras defined by the same generators and relations, but with k being an arbitrary generalised Cartan matrix, not necessarily of finite type. Such algebras are called Kac–Moody algebras. They are, in general, infinite-dimensional but, nevertheless, possess many properties of complex finite-dimensional Lie algebras. Strictly speaking, the definition of Kac–Moody algebras, which is usually used nowadays, is slightly different and looks as follows.

Let k be a generalised $r \times r$ Cartan matrix of rank s, and let $\mathfrak{h}$ be a complex vector space. The triple $(\mathfrak{h}, \Pi, \Pi^\vee)$, where $\Pi = \{\alpha_1, \ldots, \alpha_r\}$ and $\Pi^\vee = \{\alpha_1^\vee, \ldots, \alpha_r^\vee\}$ are ordered subsets of $\mathfrak{h}$ and $\mathfrak{h}^*$, respectively, is called a *realisation* of k if

(R1) Π and $\Pi^\vee$ are linearly independent sets;

(R2) $\langle \alpha_i, \alpha_j^\vee \rangle = k_{ij}$, $i, j = 1, \ldots, r$;

(R3) $\dim \mathfrak{h} = 2r - s$.

Two realisations $(\mathfrak{h}_1, \Pi_1, \Pi_1^\vee)$ and $(\mathfrak{h}_2, \Pi_2, \Pi_2^\vee)$ are said to be *isomorphic* if there exists an isomorphism $\varphi : \mathfrak{h}_1 \to \mathfrak{h}_2$ such that $\varphi(\Pi_1^\vee) = \Pi_2^\vee$ and $\varphi^*(\Pi_2) = \Pi_1$. It can be proved that for any generalised Cartan matrix there exists a realisation, and any two realisations of the same matrix are isomorphic. Furthermore, realisations of two generalised Cartan matrices k_1 and k_2 are isomorphic if and only if k_2 can be obtained from k_1 by a permutation of the rows and columns.

Now let k be an $r \times r$ generalised Cartan matrix of rank s, and let $(\mathfrak{h}, \Pi, \Pi^\vee)$ be a realisation of k. Denote $\alpha_i^\vee \equiv h_i$, and supplement the set $\{h_i\}_{i=1,\ldots,r}$ to be a basis $\{h_a\}_{a=1,\ldots,2r-s}$ of $\mathfrak{h}$. Consider the Lie algebra $\widetilde{\mathfrak{g}}(k)$ defined by the generators $x_{\pm i}$, $i = 1, \ldots, r$, h_a, $a = 1, \ldots, 2r - s$, and the relations

$$[h_a, h_b] = 0, \tag{1.20}$$

$$[x_{+i}, x_{-j}] = \delta_{ij} h_i, \tag{1.21}$$

$$[h_a, x_{+j}] = k_{ja} x_{+j}, \qquad [h_a, x_{-j}] = -k_{ja} x_{-j}, \tag{1.22}$$

where $k_{ja} \equiv \langle \alpha_j, h_a \rangle$. Identify $\mathfrak{h}$ with the linear span of the elements h_a. There exists a unique maximal ideal $\mathfrak{j}(k)$ of $\widetilde{\mathfrak{g}}(k)$ trivially intersecting $\mathfrak{h}$. The quotient algebra $\mathfrak{g}(k) \equiv \widetilde{\mathfrak{g}}(k)/\mathfrak{j}(k)$ is called a *Kac–Moody algebra*. It appears that in the case of a symmetrisable matrix k, the ideal $\mathfrak{j}(k)$ is generated by the Serre relations (1.19). Thus, in this case we can define the Kac–Moody algebra $\mathfrak{g}(k)$ as the Lie algebra generated by the generators $x_{\pm i}$, h_a and relations (1.20)–(1.22) and (1.19). Sometimes it is useful to consider the algebra

$$\mathfrak{g}'(k) \equiv [\mathfrak{g}(k), \mathfrak{g}(k)],$$

which is obtained by using the generators $x_{\pm i}$, h_i, $i = 1, \ldots, r$ and relations (1.16)–(1.19).

It can be shown that

$$\mathfrak{g}(k) = \mathfrak{g}'(k) + \mathfrak{h}.$$

Denote $\mathfrak{h}' \equiv \bigoplus_{i=1}^{r} \mathbb{C}h_i$, then $\mathfrak{g}'(k) \cap \mathfrak{h} = \mathfrak{h}'$. The centre of the Lie algebra $\mathfrak{g}(k)$ is

$$Z(\mathfrak{g}(k)) = \{h \in \mathfrak{h} \mid \langle \alpha_i, h \rangle = 0 \text{ for all } i = 1, \ldots, r\}.$$

Here we have $\dim Z(\mathfrak{g}(k)) = r - s$ and $Z(\mathfrak{g}'(k)) = Z(\mathfrak{g}(k))$.

Let k be a symmetrisable generalised Cartan matrix, and let $(\mathfrak{h}, \Pi, \Pi^\vee)$ be its realisation. Choose a subspace $\mathfrak{h}'' \subset \mathfrak{h}$ complementary to the subspace $\mathfrak{h}'$, and define a symmetric bilinear form B on $\mathfrak{h}$ by the relations

$$B(h_i, h) = v_i \langle \alpha_i, h \rangle, \quad h \in \mathfrak{h},$$
$$B(h', h'') = 0, \quad h', h'' \in \mathfrak{h}''.$$

This form is nondegenerate. It can be proved that there exists an extension of this form to $\mathfrak{g}(k)$, which is unique if we require that the resulting form be nondegenerate and invariant. It is clear that for k of finite type we obtain the form proportional to the Killing form.

Exercises

1.8 Prove that for any root system Δ in V^*, the set $\Delta^\vee$ is a root system in V.

1.9 Let Δ be a root system in V^*. Show that for any $\alpha \in \Delta$ and $a \in \mathrm{Aut}(\Delta)$

$$a^{-1t}\alpha^\vee = (a\alpha)^\vee.$$

Using this fact, prove that $W(\Delta)$ is a normal subgroup of $\mathrm{Aut}(\Delta)$; and show that $a s_\alpha a^{-1} = s_{a\alpha}$.

1.10 Find all 2×2 and 3×3 generalised Cartan matrices of finite and affine types.

1.3 Classical complex simple Lie algebras

1.3.1 Series A_r

Consider the complex special linear algebra $\mathfrak{sl}(r+1,\mathbb{C})$. It is not difficult to see that $\mathfrak{h} \equiv \mathfrak{sl}(r+1,\mathbb{C}) \cap \mathfrak{d}(r+1,\mathbb{C})$ is a maximal commutative subalgebra of $\mathfrak{sl}(r+1,\mathbb{C})$. Any element of $h \in \mathfrak{h}$ can be written as $h = \sum_{i=1}^{r+1} a_i e_{ii}$, where $\sum_{i=1}^{r+1} a_i = 0$, and the matrices e_{ij} are introduced in example 1.12. Denote by ϵ_i, $i = 1, \ldots, r+1$, the elements of $\mathfrak{h}^*$ defined by

$$\langle \epsilon_i, h \rangle \equiv a_i.$$

Note that the elements ϵ_i satisfy the relation $\sum_{i=1}^{r+1} \epsilon_i = 0$.

The elements e_{ij} with $1 \leq i, j \leq r+1$ and $i \neq j$, form a basis of $\mathfrak{sl}(r+1,\mathbb{C})$ modulo $\mathfrak{h}$. For any $i \neq j$ we have

$$[h, e_{ij}] = (a_i - a_j) e_{ij}.$$

Hence, the elements of $\mathfrak{h}$ are semisimple and, since $\mathfrak{sl}(r+1,\mathbb{C})$ is semisimple, $\mathfrak{h}$ is a Cartan subalgebra of $\mathfrak{sl}(r+1,\mathbb{C})$. The root system Δ is formed by the elements $\epsilon_i - \epsilon_j$, $i \neq j$, and $\mathbb{C}e_{ij}$ are the corresponding root subspaces. Denote

$$\alpha_i \equiv \epsilon_i - \epsilon_{i+1}, \quad i = 1, \ldots, r.$$

From the relation

$$\pm(\epsilon_i - \epsilon_j) = \pm \sum_{k=i}^{j-1} \alpha_k, \quad i < j,$$

we conclude that the set Π formed by the elements α_i, $i = 1, \ldots, r$, is a base of Δ.

Using relation (1.9), we immediately obtain

$$\nu^{-1}(\epsilon_i) = \frac{1}{2(r+1)} \left(e_{ii} - \frac{1}{r+1} \sum_{j=1}^{r+1} e_{jj} \right);$$

therefore

$$(\epsilon_i, \epsilon_j) = \frac{1}{2(r+1)} \left(\delta_{ij} - \frac{1}{r+1} \right).$$

Hence, for the simple roots we have

$$(\alpha_i, \alpha_j) = \frac{1}{2(r+1)} (2\delta_{ij} - \delta_{i+1,j} - \delta_{i,j+1}).$$

This equality implies that the nonzero nondiagonal matrix elements of the Cartan matrix are

$$k_{i,i+1} = k_{i+1,i} = -1, \quad 1 \leq i \leq r.$$

Thus, the Lie algebra $\mathfrak{sl}(r+1, \mathbb{C})$ is a simple Lie algebra of type A_r. The Cartan and Chevalley generators in this case have the form

$$h_i = e_{ii} - e_{i+1,i+1}, \quad x_{+i} = e_{i,i+1}, \quad x_{-i} = e_{i+1,i}.$$

1.3.2 Series B_r

Consider a subalgebra of the Lie algebra $\mathfrak{gl}(m, \mathbb{C})$ formed by the matrices $a \in \mathfrak{gl}(m, \mathbb{C})$ satisfying the condition

$$ba + a^t b = 0, \tag{1.23}$$

where b is some fixed $m \times m$ matrix. Let s be a nondegenerate $m \times m$ matrix. If a matrix a satisfies the condition (1.23), then the matrix $\widetilde{a} \equiv s^{-1} a s$ satisfies the condition

$$\widetilde{b}\widetilde{a} + \widetilde{a}^t \widetilde{b} = 0, \tag{1.24}$$

where $\widetilde{b} \equiv s^t b s$. On the other hand, the matrices $\widetilde{a}$, satisfying the condition (1.24), form a subalgebra of $\mathfrak{gl}(m, \mathbb{C})$, which is clearly isomorphic to the subalgebra specified by (1.23).

Now we are going to show that the Lie algebra $\mathfrak{o}(2r+1, \mathbb{C})$ is of type B_r. Recall that this Lie algebra is defined by relation (1.23) with $b = I_{2r+1}$. It is convenient for our purposes to consider another subalgebra of $\mathfrak{gl}(2r+1, \mathbb{C})$, which is isomorphic to $\mathfrak{o}(2r+1, \mathbb{C})$. To this end, define the matrix s by

$$s \equiv \frac{1}{\sqrt{2}} \begin{pmatrix} I_r & 0 & \widetilde{I}_r \\ 0 & 1 & 0 \\ \sqrt{-1}\widetilde{I}_r & 0 & -\sqrt{-1} I_r \end{pmatrix}.$$

Here and in what follows $\widetilde{I}_m$ denotes the antidiagonal unit $m \times m$ matrix. The matrix s satisfies the relation

$$s^t s = \widetilde{I}_{2r+1}.$$

Thus, the Lie algebra $\mathfrak{o}(2r+1, \mathbb{C})$ is isomorphic to the subalgebra of $\mathfrak{gl}(2r+1, \mathbb{C})$ formed by the matrices $\widetilde{a}$ satisfying the relation

$$\widetilde{I}_{2r+1} \widetilde{a} + \widetilde{a}^t \widetilde{I}_{2r+1} = 0. \tag{1.25}$$

We denote this Lie algebra by $\widetilde{\mathfrak{o}}(2r+1, \mathbb{C})$.

For any $m \times n$ matrix a define the $n \times m$ matrix a^T by

$$a^T \equiv \widetilde{I}_n a^t \widetilde{I}_m.$$

Using this definition, one can write the relation (1.25) as

$$\tilde{a} + \tilde{a}^T = 0.$$

Any matrix $\tilde{a} \in \tilde{\mathfrak{o}}(2r+1, \mathbb{C})$ has the following block form:

$$\tilde{a} = \begin{pmatrix} X & x & Y \\ y & 0 & -x^T \\ Z & -y^T & -X^T \end{pmatrix},$$

where X, Y, Z are $r \times r$ matrices, x is an $r \times 1$ matrix, and y is a $1 \times r$ matrix such that

$$Y^T = -Y, \qquad Z^T = -Z.$$

The subalgebra $\mathfrak{h} \equiv \tilde{\mathfrak{o}}(2r+1, \mathbb{C}) \cap \mathfrak{d}(2r+1, \mathbb{C})$ is a maximal commutative subalgebra of $\tilde{\mathfrak{o}}(2r+1, \mathbb{C})$. The elements $e_{ii} - e_{2r+2-i, 2r+2-i}$, $i = 1, \ldots, r$, form a basis of $\mathfrak{h}$ and, therefore, any element h of $\mathfrak{h}$ can be written as

$$h = \sum_{i=1}^{r} a_i (e_{ii} - e_{2r+2-i, 2r+2-i}).$$

Define the elements $\epsilon_i \in \mathfrak{h}^*$, $i = 1, \ldots, r$, by

$$\epsilon_i(h) \equiv a_i.$$

It can be shown that $\mathfrak{h}$ is a Cartan subalgebra of $\tilde{\mathfrak{o}}(2r+1, \mathbb{C})$. The corresponding roots and the elements generating the root subspaces are given in table 1.3. From this table we conclude that the root system Δ consists of the elements $\pm(\epsilon_i - \epsilon_j)$, $\pm(\epsilon_i + \epsilon_j)$ with $1 \le i < j \le r$, and of the elements $\pm\epsilon_i$ with $1 \le i \le r$. Introduce the notation

$$\alpha_i \equiv \epsilon_i - \epsilon_{i+1}, \quad i = 1, \ldots, r-1, \quad \alpha_r \equiv \epsilon_r.$$

It follows from the relations

$$\pm(\epsilon_i - \epsilon_j) = \pm \sum_{k=i}^{j-1} \alpha_k, \quad i < j, \quad \pm\epsilon_i = \pm \sum_{k=i}^{r} \alpha_k,$$

$$\pm(\epsilon_i + \epsilon_j) = \pm \left(\sum_{k=i}^{j-1} \alpha_k + 2 \sum_{k=j}^{r} \alpha_k \right), \quad i < j,$$

that the set Π formed by the elements α_i, $i = 1, \ldots, r$, is a base of the root system Δ.

For any two elements a and a' of the Lie algebra $\mathfrak{o}(2r+1, \mathbb{C})$ we have

$$\operatorname{tr}(aa') = \operatorname{tr}(\tilde{a}\tilde{a}'),$$

Table 1.3. *Root vectors and roots*

$\mathfrak{sl}(r+1,\mathbb{C}),\ 1 \le i,j \le r+1$		
e_{ij}	$i<j$	$\epsilon_i - \epsilon_j$
e_{ji}	$i<j$	$-\epsilon_i + \epsilon_j$
$\tilde{\mathfrak{o}}(2r+1,\mathbb{C}),\ 1 \le i,j \le r$		
$e_{ij} - e_{2r+2-j,2r+2-i}$	$i<j$	$\epsilon_i - \epsilon_j$
$e_{ji} - e_{2r+2-i,2r+2-j}$	$i<j$	$-\epsilon_i + \epsilon_j$
$e_{i,r+1} - e_{r+1,2r+2-i}$		ϵ_i
$e_{r+1,i} - e_{2r+2-i,r+1}$		$-\epsilon_i$
$e_{i,2r+2-j} - e_{j,2r+2-i}$	$i<j$	$\epsilon_i + \epsilon_j$
$e_{2r+2-j,i} - e_{2r+2-i,j}$	$i<j$	$-\epsilon_i - \epsilon_j$
$\tilde{\mathfrak{sp}}(r,\mathbb{C}),\ 1 \le i,j \le r$		
$e_{ij} - e_{2r+1-j,2r+1-i}$	$i<j$	$\epsilon_i - \epsilon_j$
$e_{ji} - e_{2r+1-i,2r+1-j}$	$i<j$	$-\epsilon_i + \epsilon_j$
$e_{i,2r+1-j} + e_{j,2r+1-i}$	$i \le j$	$\epsilon_i + \epsilon_j$
$e_{2r+1-j,i} + e_{2r+1-i,j}$	$i \le j$	$-\epsilon_i - \epsilon_j$
$\tilde{\mathfrak{o}}(2r,\mathbb{C}),\ 1 \le i,j \le r$		
$e_{ij} - e_{2r+1-j,2r+1-i}$	$i<j$	$\epsilon_i - \epsilon_j$
$e_{ji} - e_{2r+1-i,2r+1-j}$	$i<j$	$-\epsilon_i + \epsilon_j$
$e_{i,2r+1-j} - e_{j,2r+1-i}$	$i<j$	$\epsilon_i + \epsilon_j$
$e_{2r+1-j,i} - e_{2r+1-i,j}$	$i<j$	$-\epsilon_i - \epsilon_j$

where $\tilde{a} \equiv s^{-1}as$ and $\tilde{a}' \equiv s^{-1}a's$ are the corresponding elements of the Lie algebra $\tilde{\mathfrak{o}}(2r+1,\mathbb{C})$. Using expression (1.10), we see that the Killing form of $\tilde{\mathfrak{o}}(2r+1,\mathbb{C})$ can be written as

$$K(a,a') = (2r-1)\,\mathrm{tr}(aa');$$

therefore

$$\nu^{-1}(\epsilon_i) = \frac{1}{2(2r-1)}(e_{ii} - e_{2r+2-i,2r+2-i}).$$

Now it is not difficult to obtain the following relations:

$$(\alpha_i, \alpha_j) = \frac{1}{2(2r-1)}(2\delta_{ij} - \delta_{i+1,j} - \delta_{i,j+1}), \quad 1 \le i, j < r,$$

$$(\alpha_i, \alpha_r) = (\alpha_r, \alpha_i) = -\frac{1}{2(2r-1)}\delta_{i+1,r}, \quad 1 \le i < r,$$

$$(\alpha_r, \alpha_r) = \frac{1}{2(2r-1)}.$$

Hence, the nonzero nondiagonal matrix elements of the Cartan matrix are

$$k_{i,i+1} = k_{i+1,i} = -1, \quad 1 \le i < r-1,$$
$$k_{r-1,r} = -2, \quad k_{r,r-1} = -1;$$

and the Lie algebra $\widetilde{\mathfrak{o}}(2r+1, \mathbb{C})$ is a simple Lie algebra of type B_r. The same is true for the Lie algebra $\mathfrak{o}(2r+1, \mathbb{C})$. Due to this fact, the Lie algebra $\mathfrak{o}(2r+1, \mathbb{C})$ is often denoted by B_r in the literature. The Cartan and Chevalley generators for the Lie algebra $\widetilde{\mathfrak{o}}(2r+1, \mathbb{C})$ are given in table 1.4. Note that the normalisation of the root vectors corresponding to the Chevalley generators does not always coincide with the normalisation of the roots given in the table 1.3.

1.3.3 Series C_r

The corresponding Lie algebra is the complex symplectic algebra $\mathfrak{sp}(r, \mathbb{C})$. Recall that this subalgebra is defined as a subalgebra of $\mathfrak{gl}(2r, \mathbb{C})$ formed by the matrices a satisfying the relation

$$J_r a + a^t J_r = 0. \tag{1.26}$$

It is more convenient to introduce another Lie subalgebra of $\mathfrak{gl}(2r, \mathbb{C})$. Define a $2r \times 2r$ matrix s by

$$s \equiv \frac{1}{\sqrt{2}} \begin{pmatrix} I_r & \widetilde{I}_r \\ -I_r & \widetilde{I}_r \end{pmatrix}.$$

This matrix has the following property

$$s^t J_r s = \widetilde{J}_r,$$

Table 1.4. *Cartan and Chevalley generators*

	$\mathfrak{sl}(r+1,\mathbb{C})$	
h_i	$e_{ii} - e_{i+1,i+1}$	$1 \leq i \leq r$
x_{+i}	$e_{i,i+1}$	$1 \leq i \leq r$
x_{-i}	$e_{i+1,i}$	$1 \leq i \leq r$
	$\widetilde{\mathfrak{o}}(2r+1,\mathbb{C})$	
h_i	$e_{ii} - e_{i+1,i+1} + e_{2r+1-i,2r+1-i} - e_{2r+2-i,2r+2-i}$	$1 \leq i < r$
h_r	$2(e_{rr} - e_{r+2,r+2})$	
x_{+i}	$e_{i,i+1} - e_{2r+1-i,2r+2-i}$	$1 \leq i < r$
x_{+r}	$\sqrt{2}(e_{r,r+1} - e_{r+1,r+2})$	
x_{-i}	$e_{i+1,i} - e_{2r+2-i,2r+1-i}$	$1 \leq i < r$
x_{-r}	$\sqrt{2}(e_{r+1,r} - e_{r+2,r+1})$	
	$\widetilde{\mathfrak{sp}}(r,\mathbb{C})$	
h_i	$e_{ii} - e_{i+1,i+1} + e_{2r-i,2r-i} - e_{2r+1-i,2r+1-i}$	$1 \leq i < r$
h_r	$e_{rr} - e_{r+1,r+1}$	
x_{+i}	$e_{i,i+1} - e_{2r-i,2r+1-i}$	$1 \leq i < r$
x_{+r}	$e_{r,r+1}$	
x_{-i}	$e_{i+1,i} - e_{2r+1-i,2r-i}$	$1 \leq i < r$
x_{-r}	$e_{r+1,r}$	
	$\widetilde{\mathfrak{o}}(2r,\mathbb{C})$	
h_i	$e_{ii} - e_{i+1,i+1} + e_{2r-i,2r-i} - e_{2r+1-i,2r+1-i}$	$1 \leq i < r$
h_r	$e_{r-1,r-1} + e_{rr} - e_{r+1,r+1} - e_{r+2,r+2}$	
x_{+i}	$e_{i,i+1} - e_{2r-i,2r+1-i}$	$1 \leq i < r$
x_{+r}	$e_{r-1,r+1} - e_{r,r+2}$	
x_{-i}	$e_{i+1,i} - e_{2r+1-i,2r-i}$	$1 \leq i < r$
x_{-r}	$e_{r+1,r-1} - e_{r+2,r}$	

where the matrix $\widetilde{J}_r$ is given by

$$\widetilde{J}_r \equiv \begin{pmatrix} 0 & \widetilde{I}_r \\ -\widetilde{I}_r & 0 \end{pmatrix}.$$

Thus, if a matrix a satisfies (1.26), then the matrix $\widetilde{a} \equiv s^{-1}as$ obeys the relation

$$\widetilde{J}_r\widetilde{a} + \widetilde{a}^t\widetilde{J}_r = 0, \tag{1.27}$$

and the set $\widetilde{\mathfrak{sp}}(r,\mathbb{C})$ of the matrices $\widetilde{a}$ satisfying (1.27) is a subalgebra of $\mathfrak{gl}(2r,\mathbb{C})$ isomorphic to $\mathfrak{sp}(r,\mathbb{C})$. Every matrix of $\widetilde{\mathfrak{sp}}(r,\mathbb{C})$ has the following block form:

$$\widetilde{a} = \begin{pmatrix} X & Y \\ Z & -X^T \end{pmatrix},$$

where X, Y, Z are $r\times r$ matrices submitted to the conditions

$$Y^T = Y, \qquad Z^T = Z.$$

It can be shown that the subalgebra $\mathfrak{h} \equiv \widetilde{\mathfrak{sp}}(r,\mathbb{C}) \cap \mathfrak{d}(2r,\mathbb{C})$ is a maximal commutative subalgebra of $\widetilde{\mathfrak{sp}}(r,\mathbb{C})$. The elements $e_{ii}-e_{2r+1-i,2r+1-i}$, $i = 1,\dots,r$, form a basis of $\mathfrak{h}$, and every element h of $\mathfrak{h}$ can be written as

$$h = \sum_{i=1}^{r} a_i(e_{ii} - e_{2r+1-i,2r+1-i}).$$

Define the elements $\epsilon_i \in \mathfrak{h}^*$, $i = 1,\dots,r$, by

$$\epsilon_i(h) \equiv a_i.$$

One sees that $\mathfrak{h}$ is a Cartan subalgebra of $\widetilde{\mathfrak{sp}}(r,\mathbb{C})$. The corresponding roots and the elements generating the root subspaces are given in table 1.3. The root system Δ consists of the elements $\pm(\epsilon_i - \epsilon_j)$ with $1 \le i < j \le r$, and of the elements $\pm(\epsilon_i + \epsilon_j)$ with $1 \le i \le j \le r$. Let

$$\alpha_i \equiv \epsilon_i - \epsilon_{i+1}, \quad i = 1,\dots,r-1, \quad \alpha_r \equiv 2\epsilon_r.$$

Taking into account the relations

$$\pm(\epsilon_i - \epsilon_j) = \pm\sum_{k=i}^{j-1}\alpha_k, \quad i < j,$$

$$\pm(\epsilon_i + \epsilon_j) = \pm\left(\sum_{k=i}^{r-1}\alpha_k + \sum_{k=j}^{r-1}\alpha_k + \alpha_r\right), \quad i \le j,$$

we conclude that the set Π, formed by the elements α_i, $i = 1, \ldots, r$, is a base of the root system Δ.

For any two elements a and a' of the Lie algebra $\mathfrak{sp}(r, \mathbb{C})$ we have

$$\operatorname{tr}(aa') = \operatorname{tr}(\tilde{a}\tilde{a}'),$$

where $\tilde{a} \equiv s^{-1}as$ and $\tilde{a}' \equiv s^{-1}a's$ are the corresponding elements of the Lie algebra $\widetilde{\mathfrak{sp}}(r, \mathbb{C})$. From (1.11) and from the definition of the Lie algebra $\widetilde{\mathfrak{sp}}(r, \mathbb{C})$, we obtain that the Killing form of $\widetilde{\mathfrak{sp}}(r, \mathbb{C})$ has the form

$$K(\tilde{a}, \tilde{a}') = 2(r+1)\operatorname{tr}(\tilde{a}\tilde{a}').$$

This relation implies

$$\nu^{-1}(\epsilon_i) = \frac{1}{4(r+1)}(e_{ii} - e_{2r+1-i,2r+1-i});$$

and for the simple roots we have

$$(\alpha_i, \alpha_j) = \frac{1}{4(r+1)}(2\delta_{ij} - \delta_{i+1,j} - \delta_{i,j+1}), \quad 1 \le i, j < r,$$

$$(\alpha_i, \alpha_r) = (\alpha_r, \alpha_i) = -\frac{1}{2(r+1)}\delta_{i+1,r}, \quad 1 \le i < r,$$

$$(\alpha_r, \alpha_r) = \frac{1}{(r+1)}.$$

Therefore, the nonzero nondiagonal matrix elements of the Cartan matrix are

$$k_{i,i+1} = k_{i+1,i} = -1, \quad 1 \le i < r-1,$$
$$k_{r-1,r} = -1, \quad k_{r,r-1} = -2.$$

Thus, the Lie algebra $\widetilde{\mathfrak{sp}}(r, \mathbb{C})$ is a simple Lie algebra of type C_r. The isomorphic Lie algebra $\mathfrak{sp}(r, \mathbb{C})$ is also simple and of type C_r. Due to this fact, the Lie algebra $\mathfrak{sp}(r, \mathbb{C})$ is often denoted C_r. For the Cartan and Chevalley generators for the Lie algebra $\widetilde{\mathfrak{sp}}(r, \mathbb{C})$ we refer to table 1.4.

1.3.4 Series D_r

Let us prove that the Lie algebra $\mathfrak{o}(2r, \mathbb{C})$ is of type D_r. As it is now customary for us, consider another Lie algebra which is isomorphic to $\mathfrak{o}(2r, \mathbb{C})$. Define the matrix s by

$$s \equiv \frac{1}{\sqrt{2}} \begin{pmatrix} I_r & \widetilde{I}_r \\ \sqrt{-1}\widetilde{I}_r & -\sqrt{-1}I_r \end{pmatrix}.$$

The matrix s satisfies the relation

$$s^t s = \widetilde{I}_{2r}.$$

From this relation it follows that the Lie algebra $\mathfrak{o}(2r, \mathbb{C})$ is isomorphic to the subalgebra of $\mathfrak{gl}(2r, \mathbb{C})$ formed by the matrices $\widetilde{a}$ obeying the relation

$$\widetilde{a} + \widetilde{a}^T = 0.$$

Denote this Lie algebra by $\widetilde{\mathfrak{o}}(2r, \mathbb{C})$. Any matrix $\widetilde{a} \in \widetilde{\mathfrak{o}}(2r, \mathbb{C})$ can be written in the following block form:

$$\widetilde{a} = \begin{pmatrix} X & Y \\ Z & -X^T \end{pmatrix},$$

where X, Y, Z are $r \times r$ matrices which obey the relations

$$Y^T = -Y, \qquad Z^T = -Z.$$

The subalgebra $\mathfrak{h} \equiv \widetilde{\mathfrak{o}}(2r, \mathbb{C}) \bigcap \mathfrak{d}(2r+1, \mathbb{C})$ is a maximal commutative subalgebra of $\widetilde{\mathfrak{o}}(2r, \mathbb{C})$. The elements $e_{ii} - e_{2r+1-i,2r+1-i}$, $i = 1, \ldots, r$, form a basis of $\mathfrak{h}$, and any element h of $\mathfrak{h}$ can be written uniquely in the form

$$h = \sum_{i=1}^{r} a_i (e_{ii} - e_{2r+1-i,2r+1-i}).$$

One can verify that $\mathfrak{h}$ is a Cartan subalgebra of $\widetilde{\mathfrak{o}}(2r, \mathbb{C})$. The corresponding roots and the elements generating the root subspaces are given in table 1.3. As above, the elements $\epsilon_i \in \mathfrak{h}^*$, $i = 1, \ldots, r$, are defined by

$$\epsilon_i(h) \equiv a_i.$$

Thus, the root system Δ consists in this case of the elements $\pm(\epsilon_i - \epsilon_j)$, $\pm(\epsilon_i + \epsilon_j)$ with $1 \leq i < j \leq r$. Introduce the notation

$$\alpha_i \equiv \epsilon_i - \epsilon_{i+1}, \quad i = 1, \ldots, r-1,$$

$$\alpha_r \equiv \epsilon_{r-1} + \epsilon_r.$$

It follows from the relations

$$\pm(\epsilon_i - \epsilon_j) = \pm \sum_{k=1}^{j-1} \alpha_k, \quad i < j,$$

$$\pm(\epsilon_i + \epsilon_j) = \pm\left(\sum_{k=i}^{j-1} \alpha_k + 2\sum_{k=j}^{r-2} \alpha_k + \alpha_{r-1} + \alpha_r\right), \quad i < j,$$

that the set Π formed by the elements α_i, $i = 1, \ldots, r$, is a base of the root system Δ.

In accordance with (1.10), the Killing form for $\widetilde{\mathfrak{o}}(2r, \mathbb{C})$ is given by

$$K(\widetilde{a}, \widetilde{a}') = 2(r-1)\operatorname{tr}(\widetilde{a}\widetilde{a}');$$

therefore one can easily obtain

$$\nu^{-1}(\epsilon_i) = \frac{1}{4(r-1)}(e_{ii} - e_{2r+1-i,2r+1-i}).$$

Hence, for the simple roots we have

$$(\alpha_i, \alpha_j) = \frac{1}{4(r-1)}(2\delta_{ij} - \delta_{i+1,j} - \delta_{i,j+1}), \quad 1 \le i, j < r,$$

$$(\alpha_i, \alpha_r) = (\alpha_r, \alpha_i) = -\frac{1}{4(r-1)}\delta_{i+1,r-1}, \quad 1 \le i < r,$$

$$(\alpha_r, \alpha_r) = \frac{1}{2(r-1)}.$$

Therefore, the nonzero nondiagonal matrix elements of the Cartan matrix are

$$k_{i,i+1} = k_{i+1,i} = -1, \quad 1 \le i < r-1,$$
$$k_{r-2,r} = k_{r,r-2} = -1.$$

Thus, the Lie algebra $\widetilde{\mathfrak{o}}(2r, \mathbb{C})$ is a simple Lie algebra of type D_r. The same properties has the Lie algebra $\mathfrak{o}(2r, \mathbb{C})$. Due to this fact, in the literature the Lie algebra $\mathfrak{o}(2r, \mathbb{C})$ is often denoted D_r. The Cartan and Chevalley generators for the Lie algebra $\widetilde{\mathfrak{o}}(2r, \mathbb{C})$ are given in table 1.4.

1.3.5 Real forms

Let $\mathfrak{g}$ be a complex Lie algebra. Denote by K and $K_{\mathbb{R}}$ the Killing forms of $\mathfrak{g}$ and $\mathfrak{g}_{\mathbb{R}}$ respectively. It can be shown that

$$K_{\mathbb{R}}(x, y) = 2\operatorname{Re} K(x, y)$$

for all $x, y \in \mathfrak{g}_{\mathbb{R}}$. From this relation it follows that the Lie algebra $\mathfrak{g}_{\mathbb{R}}$ is (semi)simple if and only if the Lie algebra $\mathfrak{g}$ is (semi)simple.

Now let $\mathfrak{g}$ be a real Lie algebra endowed with a Lie complex structure. The discussion given above shows that the Lie algebra $\widetilde{\mathfrak{g}}$ is (semi)simple if and only if the Lie algebra $\mathfrak{g}$ is (semi)simple. Note that the Killing form $\widetilde{K}$ of the complex Lie algebra $\widetilde{\mathfrak{g}}$ is connected to the Killing form K of the real Lie algebra $\mathfrak{g}$ by

$$\widetilde{K}(x,y) = \frac{1}{2}(K(x,y) - \sqrt{-1}K(x,Jy)).$$

Let B be a bilinear form on a real vector space V. There is the unique bilinear form $B^{\mathbb{C}}$ on $V^{\mathbb{C}}$ such that

$$B^{\mathbb{C}}(v,u) = B(v,u)$$

for all $v, u \in V$. The bilinear form $B^{\mathbb{C}}$ is called the *complexification* of the bilinear form B. In fact, it is given by

$$\begin{aligned} &B^{\mathbb{C}}(v + \sqrt{-1}u, w + \sqrt{-1}t) \\ &\qquad = B(v,w) - B(u,t) + \sqrt{-1}(B(v,t) + B(u,w)). \end{aligned}$$

If a bilinear form B is (skew-)symmetric, then the bilinear form $B^{\mathbb{C}}$ is also (skew-)symmetric. If B is nondegenerate, then $B^{\mathbb{C}}$ is also nondegenerate.

A bilinear form B on $V^{\mathbb{C}}$ is said to be real if for any $v, u \in V^{\mathbb{C}}$ one has

$$\overline{B(v,u)} = B(\bar{v},\bar{u}).$$

It is clear that for any bilinear form B, the bilinear form $B^{\mathbb{C}}$ is real.

Again let $\mathfrak{g}$ be a real Lie algebra. The Killing form of the Lie algebra $\mathfrak{g}^{\mathbb{C}}$ is the complexification $K^{\mathbb{C}}$ of the Killing form K of the Lie algebra $\mathfrak{g}$. Hence, the complexification $\mathfrak{g}^{\mathbb{C}}$ of the real Lie algebra $\mathfrak{g}$ is semisimple if and only if $\mathfrak{g}$ is semisimple.

On the other hand, if a real Lie algebra $\mathfrak{g}$ is simple, the Lie algebra $\mathfrak{g}^{\mathbb{C}}$ may not be simple. To show this, let us recall some more facts from linear algebra.

Let V be a real vector space. Any vector $v \in V^{\mathbb{C}}$ has the unique representation of the form $v = u + \sqrt{-1}w$, where $u, w \in V$. The *complex conjugation* in $V^{\mathbb{C}}$ is defined as the mapping sending a vector $w = v + \sqrt{-1}u$ to the vector $\overline{w} = v - \sqrt{-1}u$. A vector $v \in V^{\mathbb{C}}$ is said to be *real* if $\bar{v} = v$, and it is said to be *imaginary* if $\bar{v} = -v$. The set of real vectors coincides with the space V considered as a subset of $V^{\mathbb{C}}$.

Let V, W be two real vector spaces, and $\varphi : V^{\mathbb{C}} \to W^{\mathbb{C}}$. The linear mapping $\bar{\varphi} : V^{\mathbb{C}} \to W^{\mathbb{C}}$, defined by

$$\bar{\varphi}(v) \equiv \overline{\varphi(\bar{v})},$$

is called the *complex conjugate* of the mapping φ. The mapping φ is called *real* if $\bar{\varphi} = \varphi$.

Let V, W be two real vector spaces, and let $\varphi : V \to W$ be a linear mapping. There is the unique linear mapping $\varphi^{\mathbb{C}} : V^{\mathbb{C}} \to W^{\mathbb{C}}$ such that $\varphi^{\mathbb{C}}(v) = \varphi(v)$ for all $v \in V$. The mapping $\varphi^{\mathbb{C}}$ is called the *complexification* of the mapping φ. It acts on a vector $w = v + \sqrt{-1}u \in V^{\mathbb{C}}$ as

$$\varphi^{\mathbb{C}}(w) = \varphi(v) + \sqrt{-1}\varphi(u).$$

Note that the mapping $\varphi^{\mathbb{C}}$ is real.

Now let V be a complex vector space. Denote by J the complex structure on the real vector space $V_{\mathbb{R}}$ induced by the multiplication by $\sqrt{-1}$ in V. Recall that the operator J can be uniquely extended to the operator $J^{\mathbb{C}}$ in $(V_{\mathbb{R}})^{\mathbb{C}}$. It is clear that

$$(J^{\mathbb{C}})^2 = -1.$$

For any $v \in V^{\mathbb{C}}$ we can write

$$v = v^{(1,0)} + v^{(0,1)},$$

where

$$v^{(1,0)} \equiv \frac{1}{2}(v - \sqrt{-1}J^{\mathbb{C}}v), \quad v^{(0,1)} \equiv \frac{1}{2}(v + \sqrt{-1}J^{\mathbb{C}}v).$$

The set of the vectors $v \in (V_{\mathbb{R}})^{\mathbb{C}}$, for which $v^{(0,1)} = 0$, is a linear subspace of $(V_{\mathbb{R}})^{\mathbb{C}}$; denote it by $V^{(1,0)}$. Similarly, the set of the vectors $v \in (V_{\mathbb{R}})^{\mathbb{C}}$ for which $v^{(1,0)} = 0$ is a linear subspace of $(V_{\mathbb{R}})^{\mathbb{C}}$, denoted by $V^{(1,0)}$. Moreover, we have the following decomposition:

$$(V_{\mathbb{R}})^{\mathbb{C}} = V^{(1,0)} \oplus V^{(0,1)}. \tag{1.28}$$

The subspaces $V^{(1,0)}$ and $V^{(0,1)}$ can be characterised as the eigenspaces of the operator $J^{\mathbb{C}}$ corresponding to the eigenvalues $+\sqrt{-1}$ and $-\sqrt{-1}$ respectively.

Introduce the mapping $\varphi : V \to V^{(1,0)}$ defined as follows. First consider a vector $v \in V$ as an element of $V_{\mathbb{R}}$, then as a real vector of $(V_{\mathbb{R}})^{\mathbb{C}}$, and take its $(1,0)$-component. One can easily show that the mapping φ is an isomorphism. We can also define the mapping $\bar{\varphi} : V \to V^{(0,1)}$ taking the $(0,1)$-component of a vector $v \in V$ considered as an element of $(V_{\mathbb{R}})^{\mathbb{C}}$. It appears that the mapping $\bar{\varphi}$

is antilinear in the sense that $\overline{\varphi}(av) = \bar{a}\overline{\varphi}(v)$ for any $a \in \mathbb{C}$ and $v \in V$. Hence, the mapping $\overline{\varphi}$ can be treated as an isomorphism from $\overline{V}$ to $V^{(0,1)}$, where the vector space $\overline{V}$ is defined in section 1.1.3. Thus, for any complex linear space V, we have a natural isomorphism

$$(V_{\mathbb{R}})^{\mathbb{C}} \simeq V \oplus \overline{V}.$$

From this relation it follows immediately that if V is a real vector space endowed with a complex structure J, then there is a natural isomorphism

$$V^{\mathbb{C}} \simeq \widetilde{V} \oplus \overline{\widetilde{V}}.$$

Suppose that a real Lie algebra $\mathfrak{g}$ has a Lie complex structure J. From the discussion given above, it follows that there is an isomorphism of the vector spaces

$$\mathfrak{g}^{\mathbb{C}} \simeq \widetilde{\mathfrak{g}} \oplus \overline{\widetilde{\mathfrak{g}}};$$

moreover, $\mathfrak{g}^{\mathbb{C}} \simeq \widetilde{\mathfrak{g}} \times \overline{\widetilde{\mathfrak{g}}}$. Here $\overline{\widetilde{\mathfrak{g}}}$ is a complex Lie algebra obtained from the real Lie algebra $\mathfrak{g}$ with the help of the Lie complex structure $-J$. Thus, if a real simple Lie algebra $\mathfrak{g}$ has a Lie complex structure, then $\mathfrak{g}^{\mathbb{C}}$ is not simple. Actually, it can be shown that if $\mathfrak{g}$ is a real simple Lie algebra, and $\mathfrak{g}^{\mathbb{C}}$ is not simple, then $\mathfrak{g}$ has a Lie complex structure.

Given a complex Lie algebra $\mathfrak{g}$, a subalgebra $\mathfrak{u}$ of $\mathfrak{g}_{\mathbb{R}}$ is called a *real form* of $\mathfrak{g}$ if

$$\mathfrak{g}_{\mathbb{R}} = \mathfrak{u} \oplus J\mathfrak{u}.$$

Here the linear operator J is a complex structure on $\mathfrak{g}_{\mathbb{R}}$ induced by multiplication by $\sqrt{-1}$ in $\mathfrak{g}$. It is clear that any real Lie algebra $\mathfrak{g}$ can be naturally considered as a real form of $\mathfrak{g}^{\mathbb{C}}$.

Now let $\mathfrak{g}$ be a real simple Lie algebra. If $\mathfrak{g}^{\mathbb{C}}$ is not simple, then $\mathfrak{g}$ has a Lie complex structure, and $\mathfrak{g}$ coincides with $\widetilde{\mathfrak{g}}_{\mathbb{R}}$. On the other hand, if $\mathfrak{g}^{\mathbb{C}}$ is simple, then $\mathfrak{g}$ can be considered as a real form of $\mathfrak{g}^{\mathbb{C}}$. Hence, any real simple Lie algebra is either a realification of a simple complex Lie algebra, or a real form of a simple complex Lie algebra. Thus, the problem of the classification of real simple Lie algebras is equivalent to the problem of the classification of real forms of complex simple Lie algebras.

Let $\mathfrak{u}$ be a real form of a complex Lie algebra $\mathfrak{g}$. Any element $x \in \mathfrak{g}$ has the unique representation of the form $x = y + \sqrt{-1}z$, where $y, z \in \mathfrak{u}$. Hence, we can define the mapping $\sigma_{\mathfrak{u}} : \mathfrak{g} \to \mathfrak{g}$ by

$$\sigma_{\mathfrak{u}}(y + \sqrt{-1}z) \equiv y - \sqrt{-1}z.$$

This mapping is antilinear and has the following properties:

$$\sigma_{\mathfrak{u}} \circ \sigma_{\mathfrak{u}} = \mathrm{id}_{\mathfrak{g}}, \qquad \sigma_{\mathfrak{u}}([x, y]) = [\sigma_{\mathfrak{u}}(x), \sigma_{\mathfrak{u}}(y)].$$

So, the mapping σ is a conjugation of $\mathfrak{g}$, which is called the *conjugation with respect to* $\mathfrak{u}$. One can easily show that any conjugation of $\mathfrak{g}$ defines a real form $\mathfrak{g}_\sigma$ of $\mathfrak{g}$ by

$$\mathfrak{g}_\sigma \equiv \{x \in \mathfrak{g} \mid \sigma(x) = x\}.$$

For any conjugation σ of $\mathfrak{g}$ the mapping $x \in \mathfrak{g} \mapsto x^\dagger \equiv -\sigma(x)$ is a hermitian involution of $\mathfrak{g}$.

All nonisomorphic real forms of complex simple Lie algebras were found by É. Cartan; hardly had a quarter of a century passed before the classification of complex simple Lie algebras was established.

A real Lie algebra $\mathfrak{g}$ is called *compact* if there exists an invariant nondegenerate bilinear form on $\mathfrak{g}$. It appears that a real semisimple Lie algebra is compact if and only if its Killing form is negatively definite.

Let $\mathfrak{g}$ be a complex semisimple Lie algebra. Using Cartan and Chevalley generators of $\mathfrak{g}$, we can define a conjugation σ of $\mathfrak{g}$ by

$$\sigma(h_i) \equiv -h_i, \qquad \sigma(x_{\pm i}) \equiv -x_{\mp i}. \tag{1.29}$$

In this case the corresponding real form $\mathfrak{g}_\sigma$ of $\mathfrak{g}$ is compact. Thus, any complex semisimple Lie algebra has a compact real form. The mapping $x \to x^\dagger \equiv -\sigma(x)$ for the mapping σ defined by (1.29) is usually called the *hermitian Chevalley involution.*

The second real form which any complex semisimple Lie algebra $\mathfrak{g}$ possesses is the *normal* real form. This real form is determined by the conjugation σ of $\mathfrak{g}$ given by

$$\sigma(h_i) \equiv h_i, \qquad \sigma(x_{\pm i}) \equiv x_{\pm i}.$$

Let us give as an illustration the list of all nonisomorphic real forms of the classical simple Lie algebras. Here, to describe a real form, we point out the corresponding conjugation.

The Lie algebra $\mathfrak{sl}(r, \mathbb{C})$ has the real forms $\mathfrak{sl}(r, \mathbb{R})$ $(\sigma(x) = \bar{x})$, $\mathfrak{su}(p, q)$ $(p + q = r,\ \sigma(x) = -I_{p,q} x^\dagger I_{p,q})$, and for an even $r = 2p$ it also has the real form $\mathfrak{su}^*(r)$ $(\sigma(x) = -J_p x^\dagger J_p)$.

The Lie algebra $\mathfrak{o}(r, \mathbb{C})$ has the real forms $\mathfrak{o}(p, q)$ $(p + q = r$, $\sigma(x) = -I_{p,q}x^\dagger I_{p,q})$, and for an even $r = 2p$ it also has the real form $\mathfrak{o}^*(r)$ $(\sigma(x) = -J_p x^\dagger J_p)$.

The Lie algebra $\mathfrak{sp}(r, \mathbb{C})$ has the real forms $\mathfrak{sp}(r, \mathbb{R})$ $(\sigma(x) = \bar{x})$, $\mathfrak{sp}(p, q)$ $(p + q = r$, $\sigma(x) = -K_{p,q}x^\dagger K_{p,q})$.

Here the matrices $I_{p,q}$ and J_p are defined by (1.5) and (1.6), while the matrix $K_{p,q}$ is given by

$$K_{p,q} \equiv \begin{pmatrix} I_{p,q} & 0 \\ 0 & I_{p,q} \end{pmatrix}.$$

It is clear that

$$\mathfrak{su}(p, q) \simeq \mathfrak{su}(q, p), \quad \mathfrak{o}(p, q) \simeq \mathfrak{o}(q, p), \quad \mathfrak{sp}(p, q) \simeq \mathfrak{sp}(q, p).$$

The compact real forms are $\mathfrak{su}(r) \equiv \mathfrak{su}(r, 0)$, $\mathfrak{o}(n) \equiv \mathfrak{o}(n, 0)$, and $\mathfrak{sp}(n) \equiv \mathfrak{sp}(n, 0)$.

There are the following isomorphisms between classical real Lie algebras:

$$\mathfrak{o}(3) \simeq \mathfrak{su}(2) = \mathfrak{sp}(1), \quad \mathfrak{o}(4) \simeq \mathfrak{su}(2) \times \mathfrak{su}(2),$$
$$\mathfrak{o}(5) \simeq \mathfrak{sp}(2), \quad \mathfrak{o}(6) \simeq \mathfrak{su}(4),$$
$$\mathfrak{o}(1, 2) \simeq \mathfrak{su}(1, 1) \simeq \mathfrak{sl}(2, \mathbb{R}) = \mathfrak{sp}(1, \mathbb{R}), \quad \mathfrak{o}(1, 3) \simeq \mathfrak{sl}(2, \mathbb{C})_{\mathbb{R}},$$
$$\mathfrak{o}(1, 4) \simeq \mathfrak{sp}(1, 1), \quad \mathfrak{o}(1, 5) \simeq \mathfrak{su}^*(4),$$
$$\mathfrak{o}(2, 2) \simeq \mathfrak{sl}(2, \mathbb{R}) \times \mathfrak{sl}(2, \mathbb{R}), \quad \mathfrak{o}(2, 3) \simeq \mathfrak{sp}(2, \mathbb{R}),$$
$$\mathfrak{o}(2, 4) \simeq \mathfrak{su}(2, 2), \quad \mathfrak{o}(2, 6) \simeq \mathfrak{o}^*(8),$$
$$\mathfrak{o}(3, 3) \simeq \mathfrak{sl}(4, \mathbb{R}),$$
$$\mathfrak{o}^*(4) \simeq \mathfrak{su}(2) \times \mathfrak{sl}(2, \mathbb{R}), \quad \mathfrak{o}^*(6) \simeq \mathfrak{su}(1, 3).$$

Exercises

1.11 Prove the following isomorphisms of complex Lie algebras:

$$\mathfrak{o}(3, \mathbb{C}) \simeq \mathfrak{sl}(2, \mathbb{C}), \qquad \mathfrak{o}(4, \mathbb{C}) \simeq \mathfrak{o}(3, \mathbb{C}) \times \mathfrak{o}(3, \mathbb{C}),$$
$$\mathfrak{sp}(1, \mathbb{C}) \simeq \mathfrak{sl}(2, \mathbb{C}), \qquad \mathfrak{sp}(2, \mathbb{C}) \simeq \mathfrak{o}(5, \mathbb{C}).$$

1.12 Show that the mapping $a \in \mathfrak{gl}(m, \mathbb{K}) \mapsto a^T \in \mathfrak{gl}(m, \mathbb{K})$ is an involution of the Lie algebra $\mathfrak{gl}(m, \mathbb{K})$.

2
Basic notions of differential geometry

As a general reading on differential geometry we recommend the books by Dubrovin, Fomenko & Novikov (1992, 1985); Helgason (1978); Kirillov (1976); Kobayashi & Nomizu (1963, 1969); Narasimhan (1968) and Warner (1983). As a simple introduction to topology, the reader can use the book by Kosniowski (1980); more advanced problems are treated, for example, in Kelley (1957). In this chapter, unlike the previous one, we distinguish between contravariant and covariant indices; repeated indices imply summation.

2.1 Topological spaces

2.1.1 Definition of a topological space

A collection $\mathcal{U}$ of subsets of a set X is called a *topology* on X if

(T1) $\emptyset \in \mathcal{U}$, $X \in \mathcal{U}$;

(T2) the intersection of any two members of $\mathcal{U}$ belongs to $\mathcal{U}$;

(T3) the union of any collection of members of $\mathcal{U}$ belongs to $\mathcal{U}$.

The members of $\mathcal{U}$ are called *open sets*, and their complements in X are called *closed sets*. A pair $(X,\mathcal{U})$, where X is a set and $\mathcal{U}$ is a topology on X, is called a *topological space*. It is customary to denote a topological space $(X,\mathcal{U})$ simply by X.

EXAMPLE 2.1 Let $\mathcal{U}$ be a collection of all subsets of a set X. It is evident that $\mathcal{U}$ is a topology on X. This topology is called the *discrete topology*. A set endowed with the discrete topology is called a *discrete space*.

The collection $\mathcal{U} \equiv \{\emptyset, X\}$ is a topology on the set X called the *indiscrete topology*. Sometimes such a topology is called the *trivial topology*.

On a set X let there be given two topologies $\mathcal{U}$ and $\mathcal{V}$. If $\mathcal{U} \subset \mathcal{V}$ we say that the topology $\mathcal{V}$ is *stronger* than the topology $\mathcal{U}$, and $\mathcal{U}$ is *weaker* than $\mathcal{V}$. It is clear that the discrete topology is the strongest topology that can be defined on a set, while the indiscrete topology is the weakest one.

EXAMPLE 2.2 The space $\mathbb{R}^m$ can be provided with a topology in the following way. The *Euclidean metric* on $\mathbb{R}^m$ is the mapping $d : \mathbb{R}^m \times \mathbb{R}^m \to \mathbb{R}$, defined by

$$d(a, b) \equiv \sqrt{\sum_{i=1}^{m} (a^i - b^i)^2}.$$

For any real number $r \geq 0$ and $a \in \mathbb{R}^m$ the set

$$B_r^m(a) \equiv \{b \in \mathbb{R}^m \mid d(a, b) < r\}$$

is called an *open ball* in $\mathbb{R}^m$. Here the point a and the real number r are called the *centre* and the *radius* of the open ball $B_r^m(a)$ respectively. A subset U of $\mathbb{R}^m$ is said to be open if for any point $a \in U$ there exists a number $r > 0$ such that $B_r^m(a) \subset U$. The empty set $\emptyset$ is considered as an open set by definition. It can be shown that a collection of open sets, defined as above, is a topology on $\mathbb{R}^m$. This topology is called the *standard topology* on $\mathbb{R}^m$. Below, the space $\mathbb{R}^m$, considered as a topological space, is understood as having the standard topology.

Let V be an m-dimensional real vector space, and let $\{e_i\}_{i=1}^m$ be a basis of V. Any vector of V has a unique representation of the form $v = v^i e_i$. The mapping $\varphi : v \in V \mapsto (v^1, \ldots, v^m) \in \mathbb{R}^m$ is bijective. We call the subset $U \subset V$ open if the subset $\varphi(U) \subset \mathbb{R}^m$ is open. This definition endows V with a topology which does not depend on the choice of the basis $\{e_i\}$. This topology is called the *standard topology* on the real vector space V.

EXAMPLE 2.3 The space $\mathbb{C}^m$ can be identified with the space $\mathbb{R}^{2m}$ in the following way. Let $c = (c^1, \ldots, c^m) \in \mathbb{C}^m$, represent each complex number c^i, $i = 1, \ldots, m$, as $c^i = a^i + \sqrt{-1}b^i$, where $a^i, b^i \in \mathbb{R}$, and associate with the point c the point $(a^1, b^1, \ldots, a^m, b^m) \in \mathbb{R}^{2m}$. Then a subset of $\mathbb{C}^m$ is considered as open if the corresponding subset of $\mathbb{R}^{2m}$ is open. Such a topology on $\mathbb{C}^m$ is called the *standard topology* on $\mathbb{C}^m$. Using the procedure similar to one discussed in the preceding example, we can introduce the *standard topology* on an arbitrary complex vector space.

Any open set containing a point p of a topological space X is called a *neighbourhood* of the point p.

Let S be a subset of a topological space X. The intersection of all closed sets containing S is called the *closure* of S and is denoted by $\overline{S}$. The subset S is called *dense* in X if $\overline{S} = X$.

A mapping φ from a topological space X to a topological space Y is called *continuous* if for any open set $U \subset Y$ the set $\varphi^{-1}(U)$ is open. A mapping $\varphi : X \to Y$ is continuous if and only if for any closed set $V \subset Y$ the set $\varphi^{-1}(V)$ is closed. The composition of continuous mappings is a continuous mapping.

A mapping φ from a topological space X to a topological space Y is said to be *open* (*closed*) if for any open set $U \subset X$ the set $\varphi(U)$ is open (closed). An open (closed) mapping need not be continuous.

A bijective mapping φ from a topological space X to a topological space Y is called a *homeomorphism* if the mappings φ and φ^{-1} are continuous simultaneously. Two topological spaces X and Y are said to be *homeomorphic*, or *topologically equivalent* if there exists a homeomorphism from X to Y.

Let $(X, \mathcal{U})$ be a topological space. A subcollection $\mathcal{B}$ of $\mathcal{U}$ is called a *base of the topology* $\mathcal{U}$ if any element of $\mathcal{U}$ can be represented as a union of elements of $\mathcal{B}$.

A collection $\mathcal{B}$ of sets is a base of some topology on $X \equiv \bigcup_{B \in \mathcal{B}} B$, if and only if for any two elements U and V of $\mathcal{B}$ and any point $p \in U \cap V$ there exists an element W of $\mathcal{B}$, such that $p \in W$ and $W \subset U \cap V$.

A topological space whose topology has a countable base is called *second countable*.

EXAMPLE 2.4 The set of open balls in the space $\mathbb{R}^m$ is a base of the standard topology on $\mathbb{R}^m$. We call the point $a = (a^1, \dots, a^m) \in \mathbb{R}^m$ rational if all a^i, $i = 1, \dots, m$, are rational numbers. The set of open balls with rational radii and centres is also a base of the standard topology on $\mathbb{R}^m$. Hence, $\mathbb{R}^m$ is a second countable space.

Let $(X, \mathcal{U})$ be a topological space and let Y be a subset of X. Denote by $\mathcal{V}$ the system of subsets of Y consisting of all intersections of the elements of $\mathcal{U}$ with Y. In other words, the set $V \subset Y$ belongs to $\mathcal{V}$ if and only if there exists an element $U \in \mathcal{U}$ such that $V = U \cap Y$. It can be shown that $\mathcal{V}$ is a topology on Y. This topology is called the topology induced by the topology $\mathcal{U}$, or simply by the *induced topology*. The topological space $(Y, \mathcal{V})$ is called a *subspace* of the topological space $(X, \mathcal{U})$.

Let a topological space $(Y, \mathcal{V})$ be a subspace of a topological space $(X, \mathcal{U})$. Suppose that Y is an open (closed) subset of X. In this case an open (closed) subset of Y is simultaneously an open (closed) subset of X.

EXAMPLE 2.5 The subset $S_r^m(a)$ of the space $\mathbb{R}^{m+1}$, defined as

$$S_r^m(a) \equiv \left\{ b \in \mathbb{R}^{m+1} \;\middle|\; \sum_{i=1}^{m+1} (b^i - a^i)^2 = r^2 \right\},$$

and considered as a topological space with the topology induced by the standard topology on $\mathbb{R}^n$, is called the n-dimensional *sphere* of radius r and with the centre at a. The sphere $S^m \equiv S_1^m(0)$ is called the *standard* n-dimensional sphere.

Let X and Y be topological spaces, and let $\varphi : X \to Y$ be a continuous mapping. The restriction $\varphi|_S$ of the mapping φ to the subset $S \subset X$ is continuous with respect to the induced topology on S.

2.1.2 Product topology and quotient topology

Let X and Y be two topological spaces. Recall that the direct product $X \times Y$ of the sets X and Y is the set of all ordered pairs (p, q), where $p \in X$ and $q \in Y$. The *product topology* on $X \times Y$ consists of all subsets of $X \times Y$ which are unions of sets of the form $U \times V$, where U is open in X and V is open in Y. The set $X \times Y$ endowed with the product topology is called the *topological product* of X and Y.

Similarly, we can define the topological product of an arbitrary number of topological spaces.

EXAMPLE 2.6 The topological product T^n of n copies of the standard one-dimensional sphere is called the *standard n-dimensional torus*.

There are the *canonical projections* $\pi_X : X \times Y \to X$, and $\pi_Y : X \times Y \to Y$, defined as

$$\pi_X(p,q) \equiv p, \qquad \pi_Y(p,q) \equiv q.$$

The canonical projections π_X and π_Y are continuous mappings. For any $q \in Y$, the subspace $X \times \{q\} \subset X \times Y$ is homeomorphic to X. Similarly, for any $p \in X$ the subspace $\{p\} \times Y \subset X \times Y$ is homeomorphic to Y.

Let $(X, \mathcal{U})$ be a topological space and let Y be a set. Consider an arbitrary surjective mapping from X to Y. Define a topology $\mathcal{V}$ on X by

$$\mathcal{V} \equiv \{V \subset Y \mid \varphi^{-1}(V) \in \mathcal{U}\}.$$

The topology $\mathcal{V}$ is called the *quotient topology* on Y with respect to φ. It is clear that if the set Y is endowed with the quotient topology with respect to φ, then the mapping φ is continuous.

In fact, if we have a surjective mapping φ from a set X to a set Y, we can identify the set Y with the quotient set with respect to the following equivalence relation. Let us call two points p and q of the set X equivalent if $\varphi(p) = \varphi(q)$. It is clear that we do have an equivalence relation, and the set of the corresponding equivalence classes can be put into the bijective correspondence with the points of Y. Thus, the set Y in the definition of the quotient topology is, in a sense, irrelevant. We can start, in general, with some equivalence relation established for the points of the initial topological space.

2.1.3 Some types of topological space

A topological space is called *connected* if it cannot be represented as a union of two disjoint open subsets. A topological space is said to be *disconnected* if it is not connected. A subset of a topological space is called *connected* if it is connected as a topological subspace.

Let X be a topological space and let φ be a surjective continuous mapping from X to a topological space Y. If X is connected, then Y is also connected.

Any maximal connected subset of a topological space X is called a *component* of X. If a topological space is connected, it has only one component which coincides with X. If a topological space is discrete, then each of its components consists of just one point.

A continuous mapping λ from the interval $[0, 1] \subset \mathbb{R}$ to a topological space X, such that $\lambda(0) = p$ and $\lambda(1) = q$, is called a *path* in X from p to q. A topological space X is said to be *path-connected* if for any two points $p, q \in X$ there is a path from p to q. Any path-connected topological space is connected, but not any connected topological space is path-connected.

Let p be a point of a topological space X. A path in X from p to p is called a *loop* in X based at p. The loop in X, which takes all of $[0, 1]$ to p, is said to be the *constant loop* based at p. Two loops λ_0 and λ_1 based at p are called *homotopy equivalent* if there is a continuous mapping $\varphi : [0, 1] \times [0, 1] \to X$ such that

$$\varphi(t, 0) = \varphi(t, 1) = p,$$
$$\varphi(0, s) = \lambda_0(s), \qquad \varphi(1, s) = \lambda_1(s),$$

for all $t, s \in \mathbb{R}$. A topological space X is said to be *simply connected* if it is path-connected and, for any point $p \in X$, each loop based at p is homotopy equivalent to the constant loop based at p. Actually, it is sufficient if the requirement of the above definition is satisfied for some point of X.

Let I be some set of indices. A collection $\mathcal{C} = \{U_\alpha\}_{\alpha \in I}$ of subsets of a topological space X is called a *cover* of X if $X = \bigcup_{\alpha \in I} U_\alpha$. A cover is said to be an *open cover* if it is formed by open sets. A subcollection $\mathcal{D}$ of a cover $\mathcal{C}$ is called a *subcover* if $\mathcal{D}$ is a cover by itself.

A topological space X is called *compact* if every open cover of X has a finite subcover.

Let X be a compact topological space and let φ be a continuous surjective mapping from X to a topological space Y; then Y is a compact topological space. A topological space is called *Hausdorff* if its distinct points always have disjoint neighbourhoods. A subspace of a Hausdorff topological space is Hausdorff, and a finite product of Hausdorff spaces is also Hausdorff.

EXAMPLE 2.7 Let a and a' be two distinct points of the space $\mathbb{R}^m$. Let us show that if $r + r' < d(a, a')$, then $B_r(a) \cap B_{r'}(a') = \emptyset$. Indeed, suppose that $b \in B_r(a) \cap B_{r'}(a')$, then $d(a, b) < r$ and $d(a', b) < r'$; therefore,

$$d(a, a') \leq d(a, b) + d(a', b) < r + r'.$$

Thus, the space $\mathbb{R}^m$, provided with the standard topology, is a Hausdorff topological space.

2.2 Differentiable manifolds

2.2.1 Definition of a manifold

Let U be an open subset of the space $\mathbb{R}^m$. A continuous mapping φ from U to $\mathbb{R}^n$ is called a mapping of class C^0. A mapping $\varphi : U \to \mathbb{R}^n$ is said to be of class C^1 if the coordinates of the point $\varphi(a)$ have continuous partial derivatives over the coordinates of the point $a \in U$. For an arbitrary $r > 1$ we define inductively the notion of a *mapping of class* C^r: a mapping $\varphi : U \to \mathbb{R}^n$ is of class C^r if the partial derivatives of the coordinates of the point $\varphi(a)$ over the coordinates of the point $a \in U$ are of class C^{r-1}. If a mapping $\varphi : U \to \mathbb{R}^n$ is of class C^r for any $r \geq 1$, we say that φ is *of class* C^∞. In this case we also say that the mapping φ is *smooth*. A mapping $\varphi : U \to \mathbb{R}^n$ is called (*real*) *analytic* on U if the coordinates of the point $\varphi(a) \in \mathbb{R}^n$ are real analytic functions of the coordinates of the point $a \in U$. A real analytic mapping is also called a mapping *of class* C^ω.

Let U and V be open subsets of the spaces $\mathbb{R}^m$ and $\mathbb{R}^n$ respectively. A bijective mapping $\varphi : U \to V$ is called a *diffeomorphism* of class C^r, $r = 0, 1, \ldots, \infty, \omega$, or a C^r-*diffeomorphism*, if the mappings φ and φ^{-1} are of class C^r. It is clear that a C^0-diffeomorphism is just a homeomorphism.

Let M be a topological space. A pair (U, φ) where U is an open subset of M and φ is a homeomorphism from U to an open subset of the space $\mathbb{R}^m$ is called a *chart* on M. Here the nonnegative integer m is called the *dimension* of the chart (U, φ). For any $p \in U$ we can write

$$\varphi(p) = (x^1(p), \ldots, x^m(p)).$$

This representation gives m continuous functions $x^i : U \to \mathbb{R}$, which are called the *coordinate functions* corresponding to the chart (U, φ). On the other hand, the functions x^i unambiguously define the mapping φ. It is customary to denote a chart (U, φ) with the coordinate functions x^i by $(U, x^1, \ldots, x^m)$. The numbers $x^i(p)$ are called the *coordinates* of the point p with respect to the chart (U, φ).

Two charts (U, φ) and (V, ψ) are called C^r-*compatible* if either $U \cap V = \emptyset$ or if $\varphi(U \cap V)$ and $\psi(U \cap V)$ are open sets and the mapping

$$\varphi \circ \psi^{-1} : \psi(U \cap V) \to \varphi(U \cap V)$$

is a C^r-diffeomorphism.

Let $\mathcal{A}$ be some set of indices. A family of charts $\{(U_\alpha, \varphi_\alpha)\}_{\alpha \in \mathcal{A}}$ on a topological space M is called an *atlas* of class C^r on M, or a C^r-atlas on M if any two charts in this family are C^r-compatible and

$$\bigcup_{\alpha \in \mathcal{A}} U_\alpha = M.$$

The set of all charts, C^r-compatible with any chart of a given C^r-atlas, is also a C^r-atlas. Such an atlas cannot be a proper subset of any other C^r-atlas, and, by this reason, it is called a *maximal* C^r*-atlas*. Thus, any C^r-atlas generates the corresponding maximal C^r-atlas. Two C^r-atlases on a set M are called *equivalent* if their union is a C^r-atlas. Different C^r-atlases generate one and the same maximal C^r-atlas if and only if they are equivalent. A maximal C^r-atlas on a topological space M is called a *differentiable structure* of class C^r on M, or a C^r*-differentiable structure* on M. It is clear that to specify a C^r-differentiable structure it suffices to show an arbitrary C^r-atlas which is a subset of the corresponding maximal atlas.

A set M, endowed with a C^r-differentiable structure, is called a *differentiable manifold* of class C^r, or a C^r*-manifold*. A chart belonging to the differentiable structure of a differentiable manifold is called an *admissible chart*. A differentiable manifold of class C^0 is usually called a *topological manifold*.

The definition of a manifold we have just given is slightly too wide for our purposes. It is useful to impose some restrictions which will allow us to apply some powerful tools of analysis in consideration of various structures defined on a differentiable manifold. First suppose that all charts forming an atlas have the dimension equal to some fixed positive number, which is called the *dimension* of the manifold. Further it is convenient to consider only manifolds being second countable and Hausdorff as topological spaces. It can be shown that a differentiable manifold is a second countable topological space if and only if it has an atlas with a countable family of charts.

In this book we consider only manifolds of class C^∞. It is customary to use the term 'smooth' as a synonym of the term 'differentiable of class C^∞'. In particular, a *smooth manifold* is a differentiable manifold of class C^∞; a *smooth differentiable structure* is a differentiable structure of class C^∞. Moreover, we usually write 'manifold' instead of 'smooth manifold'. In fact, the majority of the manifolds considered in this book are of class C^ω. Such manifolds are called (*real*) *analytic* manifolds. It is evident that any real analytic manifold is a smooth manifold.

EXAMPLE 2.8 The *standard differentiable structure* on the space $\mathbb{R}^m$ is defined by an atlas which consists of just one chart $(\mathbb{R}^m, \mathrm{id}_{\mathbb{R}^m})$. For the corresponding coordinate functions x^i, $i = 1, \dots, m$, we have

$$x^i(a) = a^i,$$

where $a = (a^1, \dots, a^m)$. These coordinate functions are called the *standard coordinate functions on* $\mathbb{R}^m$.

The identification of the space $\mathbb{C}^m$ with the space $\mathbb{R}^{2m}$ given in example 2.3 allows us to consider the space $\mathbb{C}^m$ as a $2m$-dimensional differentiable manifold. Here the standard differentiable structure on $\mathbb{R}^{2m}$ is called the *standard differentiable structure* on the space $\mathbb{C}^m$. Let $c = (c^1, \dots, c^m)$ be an arbitrary point of $\mathbb{C}^m$; represent each c^i in the form $c^i = a^i + \sqrt{-1}b^i$, where $a^i, b^i \in \mathbb{R}$. The functions x^i, y^i, $i = 1, \dots, m$, defined by

$$x^i(c) \equiv a^i, \qquad y^i(c) \equiv b^i,$$

are called the *standard coordinate functions on* $\mathbb{C}^m$.

Further, let V be an m-dimensional real vector space and let $\{e_i\}$ be a basis for V. It is clear that the pair (V, φ), with the mapping $\varphi : V \to \mathbb{R}^m$ described in example 2.2, defines an atlas on V consisting of one chart. Using different bases for V, we obtain C^∞-equivalent atlases on V. Hence, V has a natural differentiable structure. Such a differentiable structure is called the *standard differentiable structure* on V.

EXAMPLE 2.9 Recall that the standard sphere S^m is the subspace of the topological space $\mathbb{R}^{m+1}$ defined as

$$S^m \equiv \left\{ a \in \mathbb{R}^{m+1} \;\middle|\; \sum_{k=1}^{m+1} (a^k)^2 = 1 \right\}.$$

Note that S^m, being a subspace of a Hausdorff topological space, is Hausdorff. Define the open hemispheres $U^k_\pm$, $k = 1, \ldots, m+1$, by

$$U^k_+ \equiv \{a \in S^m \mid a^k > 0\}, \qquad U^k_- \equiv \{a \in S^m \mid a^k < 0\},$$

and introduce the mappings $\varphi^k_\pm : U^k_\pm \to \mathbb{R}^m$ by

$$\varphi^k_\pm(a) = (a^1, \ldots, \widehat{a^k}, \ldots, a^{m+1}).$$

Here the hat means that the corresponding term should be omitted. It can be shown that the mappings $\varphi^k_\pm$ are homeomorphisms from $U^k_\pm$ to the open ball $B^m_1(0)$ in $\mathbb{R}^m$. Consider the charts (U^1_+, φ^1_+) and (U^2_+, φ^2_+). The set $\varphi^1_+(U^1_+ \cap U^2_+)$ consists of all $(b^1, \ldots, b^m) \in \mathbb{R}^m$ such that $\sum_{i=1}^m (b^i)^2 < 1$ and $b^1 > 0$. Since

$$(\varphi^1_+)^{-1}(b^1, \ldots, b^m) = \left(\sqrt{1 - \sum_{i=1}^m (b^i)^2}, b^1, \ldots, b^m \right),$$

then for any $(b^1, \ldots, b^m) \in \varphi^1_+(U^1_+ \cap U^2_+)$ we have

$$\varphi^2_+ \circ (\varphi^1_+)^{-1}(b^1, \ldots, b^m) = \left(\sqrt{1 - \sum_{i=1}^m (b^i)^2}, b^2, \ldots, b^m \right).$$

It is clear now that the mapping $\varphi^2_+ \circ (\varphi^1_+)^{-1} : \varphi^1_+(U^1_+ \cap U^2_+) \to \varphi^2_+(U^1_+ \cap U^2_+)$ is of class C^∞. Considering all other pairs of charts, we come to the conclusion that $\{(U^k_\pm, \varphi^k_\pm)\}_{k=1,\ldots,m+1}$ is a C^∞-atlas which defines a C^∞-differentiable structure on S^m.

It is useful to introduce another atlas on S^m, consisting of two charts. To this end let us consider the stereographic projections of the standard sphere. Denote by H^0_{m+1} a subspace of $\mathbb{R}^{m+1}$ given by

$$H^0_{m+1} \equiv \{(a^1, \ldots, a^{m+1}) \in \mathbb{R}^{m+1} \mid a^{m+1} = 0\}.$$

This subspace can be naturally identified with the space $\mathbb{R}^m$. Denote also by n and s the 'north' and 'south' poles of S^m, defined as $n \equiv (0, \ldots, 0, 1)$, $s \equiv (0, \ldots, 0, -1)$.

Let $a = (a^1, \ldots, a^{m+1})$ be an arbitrary point of S^m which does not coincide with the north pole. Draw a straight line through the points a and n. This line intersects the subspace H^0_{m+1} at the point $(u^1(a), \dots, u^m(a))$, where

$$u^i(a) = \frac{a^i}{1 - a^{m+1}}.$$

The functions u^i can be considered as coordinate functions corresponding to some chart (U, φ) on S^m with $U = S^m - \{n\}$.

Suppose now that a is an arbitrary point of S^m which does not coincide with the south pole. Draw the line connecting the points a and s; this line intersects the subspace H^0_{m+1} at the point $(v^1(a), \dots, v^m(a))$, where

$$v^i(a) = \frac{a^i}{1 + a^{m+1}}.$$

The functions v^i define another chart (V, ψ) on S^m with $V = S^m - \{s\}$. It is not difficult to show that the coordinate functions, corresponding to the two charts, are connected by

$$u^i = \frac{v^i}{\sum_{j=1}^m (v^j)^2}.$$

Thus, we obtain an atlas on S^m of class C^∞. It can be shown that this atlas corresponds to the same differentiable structure on S^m as the atlas defined in the beginning of the example.

Let M be an m-dimensional manifold, $\{(U_\alpha, \varphi_\alpha)\}_{\alpha \in \mathcal{A}}$ be an atlas on M, and U be an open subset of M. Denote by $\mathcal{B}$ the set of all $\beta \in \mathcal{A}$ such that $U \cap U_\beta \neq \emptyset$. For any $\beta \in \mathcal{B}$, the pair (V_β, ψ_β), where $V_\beta \equiv U \cap U_\beta$, and $\psi_\beta = \varphi_\beta|_{V_\beta}$, is a chart on U. The set $\{(V_\beta, \psi_\beta)\}_{\beta \in \mathcal{B}}$ is an atlas which defines a smooth differentiable structure on U. This differentiable structure does not depend on the choice of an atlas on M, and the corresponding manifold U is called an *open submanifold* of M. Note that U has dimension m.

Let M and N be manifolds of dimensions m and n respectively. Consider an arbitrary atlas $\{(U_\alpha, \varphi_\alpha)\}_{\alpha \in \mathcal{A}}$ on M and an arbitrary atlas $\{(V_\iota, \psi_\iota)\}_{\iota \in \mathcal{I}}$ on N. For any $\alpha \in \mathcal{A}$ and $\iota \in \mathcal{I}$ denote $W_{\alpha\iota} \equiv U_\alpha \times V_\iota$ and define the mapping $\xi_{\alpha\iota} : W_{\alpha\iota} \to \mathbb{R}^{m+n}$ by

$$\xi_{\alpha\iota}(p, q) \equiv (\varphi_\alpha(p), \psi_\iota(q)).$$

The family $\{(W_{\alpha\iota}, \xi_{\alpha\iota})\}_{\alpha \in \mathcal{A}, \iota \in \mathcal{I}}$ is an atlas which defines on $M \times N$ a smooth differentiable structure. It can easily be shown that this differentiable structure does not depend on the choice of atlases

$\{(U_\alpha, \varphi_\alpha)\}$ and $\{(V_\iota, \psi_\iota)\}$. The corresponding manifold is called the *direct product* of the manifolds M and N and is denoted by $M \times N$. It is clear that $M \times N$ is an $(m+n)$-dimensional manifold.

2.2.2 Smooth functions and mappings

Let f be a real function on a manifold M and let (U, φ) be a chart on M. The function $f \circ \varphi^{-1} : \varphi(U) \to \mathbb{R}$ is called the *coordinate expression* for f with respect to the chart (U, φ). Let $\{(U_\alpha, \varphi_\alpha)\}_{\alpha \in \mathcal{A}}$ be an atlas on M. A function $f : M \to \mathbb{R}$ is called *smooth* if the coordinate expression $f_\alpha \equiv f \circ \varphi_\alpha^{-1}$ of the function f with respect to any chart $(U_\alpha, \varphi_\alpha)$ is a smooth function. It is evident that the coordinate expression of a smooth function f on a manifold M with respect to any admissible chart on M is a smooth function. From this fact it follows, in particular, that the notion of a smooth function does not depend on the choice of an atlas.

The set of smooth functions on a manifold M is denoted by $\mathfrak{F}(M)$. Define in $\mathfrak{F}(M)$ the algebraic operations by

$$(af)(p) \equiv a(f(p)),$$
$$(f+g)(p) \equiv f(p) + g(p), \quad (fg)(p) \equiv f(p)g(p),$$

for any $a \in \mathbb{R}$, and $f, g \in \mathfrak{F}(M)$. The set $\mathfrak{F}(M)$ with respect to the algebraic operations defined above is an associative commutative algebra over $\mathbb{R}$.

Let U be an open submanifold of a manifold M. It is clear that, for any function $f \in \mathfrak{F}(M)$, the restriction $f|_U$ of the function f to U is an element of $\mathfrak{F}(U)$. Note that, in general, each element of $\mathfrak{F}(U)$ cannot be obtained by this procedure. In other words, each smooth function on U cannot be extended to a smooth function on M.

Let M, N be two manifolds and let $\{(U_\alpha, \varphi_\alpha)\}_{\alpha \in \mathcal{A}}$, $\{(V_\iota, \psi_\iota)\}_{\iota \in \mathcal{I}}$ be some atlases on M and N respectively. Further, let χ be a mapping from M to N. For any two charts $(U_\alpha, \varphi_\alpha)$ and (V_ι, ψ_ι) such that $\chi(U_\alpha) \cap V_\iota \neq \emptyset$, we define the mapping $\chi_{\iota\alpha} : \varphi_\alpha(U_\alpha \cap \chi^{-1}(V_\iota)) \to \psi_\iota(\chi(U_\alpha) \cap V_\iota)$ by

$$\chi_{\iota\alpha} \equiv \psi_\iota \circ \chi \circ \varphi_\alpha^{-1}.$$

The mapping $\chi_{\iota\alpha}$ is called the *coordinate expression* for the mapping χ with respect to the charts $(U_\alpha, \varphi_\alpha)$ and (V_ι, ψ_ι). The mapping χ is said to be *smooth* if for any two charts $(U_\alpha, \varphi_\alpha)$ and

(V_ι, ψ_ι) such that $\chi(U_\alpha) \cap V_\iota \neq \emptyset$ the coordinate expression $\chi_{\iota\alpha}$ is smooth. It can be shown that if $\chi : M \to N$ is a smooth mapping, then its expression with respect to any pair of admissible charts is a smooth mapping. Thus the notion of a smooth mapping from a manifold M to a manifold N does not depend on the choice of atlases on M and N. The set of smooth mappings from a manifold M to a manifold N is denoted by $\mathfrak{F}(M, N)$.

A bijective mapping $\varphi \in \mathfrak{F}(M, N)$ is called a *diffeomorphism* if φ^{-1} is a smooth mapping. Manifolds M and N are called *diffeomorphic* if there exists a diffeomorphism $\varphi : M \to N$.

Let M, N be manifolds and let $\varphi : M \to N$ be a smooth mapping. Define the mapping $\varphi^* : \mathfrak{F}(N) \to \mathfrak{F}(M)$ by

$$\varphi^* f \equiv f \circ \varphi.$$

The mapping $\varphi^* : \mathfrak{F}(N) \to \mathfrak{F}(M)$ is an algebra homomorphism. Let $\psi \in \mathfrak{F}(M, N)$ and $\varphi \in \mathfrak{F}(N, K)$, then $\varphi \circ \psi \in \mathfrak{F}(M, K)$ and

$$(\varphi \circ \psi)^* = \psi^* \circ \varphi^*.$$

Exercises

2.1 Consider the set $M \equiv \{(x_1, x_2) \in \mathbb{R}^2 \mid x_1 > 0\}$ as an open submanifold of $\mathbb{R}^2$. Prove that M and $\mathbb{R}^2$ are diffeomorphic.

2.2 Show that the manifold $M \equiv \mathbb{R} \times S^1$ and the open submanifold $N \equiv \mathbb{R}^2 - \{(0, 0)\}$ of $\mathbb{R}^2$ are diffeomorphic.

2.3 Vector fields

2.3.1 Tangent vectors

An $\mathbb{R}$-linear mapping $v : \mathfrak{F}(M) \to \mathbb{R}$ is called a *tangent vector* to the manifold M at a point $p \in M$ if it satisfies the condition

$$v(fg) = v(f)g(p) + f(p)v(f) \qquad (2.1)$$

for any $f, g \in \mathfrak{F}(M)$. The totality of tangent vectors to M at p is called the *tangent space* to the manifold M at the point p. This space is denoted by $T_p(M)$.

It is quite evident that for any $v, u \in T_p(M)$ and $a \in \mathbb{R}$, the mappings $v + u$ and av, defined as

$$(v + u)(f) \equiv v(f) + u(f), \qquad (av)(f) \equiv a(v(f)),$$

belong to $T_p(M)$. Hence, for any $p \in M$ the tangent space $T_p(M)$ has a natural structure of a vector space.

Let U be an open submanifold of a manifold M and let v be a tangent vector to M at a point $p \in U$. It can be shown that for any function $f \in \mathfrak{F}(U)$ there is a neighbourhood V of the point p such that $V \subset U$ and a function $\tilde{f} \in \mathfrak{F}(M)$ such that $f|_V = \tilde{f}|_V$. Define the action of the tangent vector v on a function $f \in \mathfrak{F}(U)$ by

$$v(f) \equiv v(\tilde{f}).$$

It appears that this definition does not depend on the choice of the function $\tilde{f}$ and we have

$$v(f|_U) = v(f). \tag{2.2}$$

The action of v on the elements of $\mathfrak{F}(U)$ that we have just described defines a tangent vector to the manifold U at the point p. On the other hand, any tangent vector v to U at p induces a tangent vector to M at p, acting on $f \in \mathfrak{F}(M)$ by the rule

$$v(f) \equiv v(f|_U).$$

Thus, we can identify the tangent spaces $T_p(M)$ and $T_p(U)$.

Let M be a manifold and let $(U, x^1, \ldots, x^m)$ be a chart on M. For any point $p \in U$, define the tangent vectors $(\partial/\partial x^i)_p$, $i = 1, \ldots, m$, by

$$\left(\frac{\partial}{\partial x^i}\right)_p (f) \equiv \frac{\partial(f \circ \varphi^{-1})}{\partial x^i}(\varphi(p)).$$

In particular, for the coordinate functions, we have

$$\left(\frac{\partial}{\partial x^i}\right)_p (x^j) = \delta_i^j.$$

It can be shown that the vectors $(\partial/\partial x^i)_p$ form a basis of the vector space $T_p(M)$. Hence, any tangent vector $v \in T_p(M)$ can be represented as

$$v = v^i \left(\frac{\partial}{\partial x^i}\right)_p.$$

The action of v on the coordinate functions gives

$$v(x^i) = v^i.$$

Thus, for any tangent vector $v \in T_p(M)$ we have

$$v = v(x^i) \left(\frac{\partial}{\partial x^i}\right)_p.$$

Let M and N be two manifolds and let $\varphi \in \mathfrak{F}(M, N)$. For each $p \in M$ define the linear mapping $\varphi_{*p} : T_p(M) \to T_{\varphi(p)}(N)$ by

$$(\varphi_{*p}(v))(f) \equiv v(\varphi^* f)$$

for any $v \in T_p(M)$ and $f \in \mathfrak{F}(N)$. The mapping φ_{*p} is called the *differential of the mapping* φ at the point p. Let $\psi \in \mathfrak{F}(M, N)$ and $\varphi \in \mathfrak{F}(N, K)$. For any point $p \in M$ one has

$$(\varphi \circ \psi)_{*p} = \varphi_{*\psi(p)} \circ \psi_{*p}.$$

From this relation it follows that if a smooth mapping $\varphi : M \to N$ is a diffeomorphism, then for any $p \in M$ the mapping φ_{*p} is an isomorphism. A smooth mapping $\varphi : M \to N$ is called *regular* at a point $p \in M$ if the mapping φ_{*p} is an isomorphism. If a mapping $\varphi : M \to N$ is regular at a point $p \in M$, then there exists a neighbourhood U of p such that $\varphi(U)$ is an open set and the restriction $\varphi|_U$ of the mapping φ to U, considered as the mapping from U to $\varphi(U)$, is a diffeomorphism. This statement is called the *inverse mapping theorem.*

A smooth mapping from an open interval of $\mathbb{R}$ to a manifold M is called a *curve* in M. Let $\lambda : I \to M$ be a curve in M and let $t \in I$. Introduce the notation

$$\dot{\lambda}(t) \equiv \lambda_{*t}\left(\frac{d}{dx}\right),$$

where x is the standard coordinate function on I. It is clear that $\dot{\lambda}(t)$ is a tangent vector to M at the point $\lambda(t)$. This tangent vector is called the *tangent vector to the curve* λ at the point $p \equiv \lambda(t)$.

A curve $\lambda : I \to M$ with $0 \in I$ and $\lambda(0) = p$ is called a *curve at* p. It appears that for any $v \in T_p(M)$ there is a curve λ at p such that $\dot{\lambda}(0) = v$. In other words, any tangent vector to a manifold can be considered as the tangent vector to a curve.

Example 2.10 According to example 2.8, any real vector space has a natural smooth differentiable structure. Let V be a vector space and let u be an arbitrary vector of V. For each $v \in V$, the mapping $t \in \mathbb{R} \mapsto u + tv \in V$ is a curve at u. Denote by $\widetilde{v}$ the tangent vector to this curve at u. It can be shown that the mapping $v \mapsto \widetilde{v}$ is an isomorphism of vector spaces V and $T_u(V)$ which allows us to identify these vector spaces.

2.3.2 Vector fields and commutator

An assignment X of a tangent vector $X_p \in T_p(M)$ to each point p of M is called a *vector field* on M. The action of a vector field X on M on a function $f \in \mathfrak{F}(M)$ is defined as a function $X(f)$ given by the relation

$$(X(f))(p) \equiv X_p(f)$$

for each $p \in M$. A vector field on M is uniquely defined by its action on the elements of $\mathfrak{F}(M)$. A vector field X on M is called *smooth* if for any $f \in \mathfrak{F}(M)$ the function $X(f)$ is smooth. The totality of smooth vector fields on M is denoted by $\mathfrak{X}(M)$.

Define the sum of two vector fields and the multiplication of a vector field by a function as

$$(X+Y)(f) \equiv X(f) + Y(f),$$
$$(fX)(g) \equiv f(X(g)).$$

If X and Y are smooth vector fields, then $X+Y$ is a smooth vector field. If X is a smooth vector field and f is a smooth function, then fX is a smooth vector field. With respect to these operations, the set $\mathfrak{X}(M)$ is a module over the algebra $\mathfrak{F}(M)$.

From (2.1) it follows that for any $X \in \mathfrak{X}(M)$ and any $f, g \in \mathfrak{F}(M)$ we have

$$X(fg) = X(f)g + fX(g).$$

Thus, any element of $\mathfrak{X}(M)$ is a derivation of the algebra $\mathfrak{F}(M)$, and it can be shown that any derivation of $\mathfrak{F}(M)$ is generated by an element of $\mathfrak{X}(M)$. Recall that the set of derivations of an algebra is a Lie algebra with respect to the commutator of derivations. In particular. it is a vector space. Therefore, the set $\mathfrak{X}(M)$ is a Lie algebra over $\mathbb{R}$ with respect to the *commutator of vector fields*, defined as

$$[X, Y](f) = X(Y(f)) - Y(X(f))$$

for any $X, Y \in \mathfrak{X}(M)$ and $f \in \mathfrak{F}(M)$. In other words, the commutator of vector fields has the properties

$$[X, Y] = -[Y, X], \tag{2.3}$$
$$[aX + bY, Z] = a[X, Z] + b[Y, Z], \tag{2.4}$$
$$[X, [Y, Z]] + [Y, [Z, X]] + [Z, [X, Y]] = 0 \tag{2.5}$$

for any $X, Y, Z \in \mathfrak{X}(M)$ and $a, b \in \mathbb{R}$. Recall that equality (2.5) is called the Jacobi identity. Note also that the set $\mathfrak{X}(M)$ is an infinite-dimensional Lie algebra. The commutator $[X, Y]$ of the vector fields X and Y is often called the *Lie bracket* of X and Y.

Let M be a manifold, U an open submanifold of M, and X a vector field on M. As stated in the preceding section, the tangent vector $X_p \in T_p(M)$, $p \in U$, can be considered as an element of $T_p(U)$. Thus, a vector field on M induces a vector field on U which is denoted by $X|_U$. Here, if a vector field X is smooth, then the vector field $X|_U$ is also smooth. From (2.2) it follows that

$$X|_U(f|_U) = X(f)|_U.$$

Let $(U, x^1, \ldots, x^m)$ be a chart on a manifold M. The vectors $(\partial/\partial x^i)_p$, $p \in U$, specify a set of m smooth vector fields on U denoted by $\partial/\partial x^i$. Using the fact that for any $p \in U$ the vectors $(\partial/\partial x^i)_p$ form a basis for $T_p(M)$, we conclude that for any vector field X on M one can write

$$X|_U = X^i \frac{\partial}{\partial x^i},$$

where $X^i = X|_U(x^i)$. If the vector field X is smooth, then the functions X^i, $i = 1, \ldots, m$, are also smooth. In other words, the vector fields $\partial/\partial x^i$ form a basis for the $\mathfrak{F}(U)$-module $\mathfrak{X}(U)$.

Let M and N be two manifolds and let $\varphi : M \to N$ be a smooth mapping. Recall that for any point $p \in M$ there is defined the mapping φ_{*p} which connects the tangent spaces $T_p(M)$ and $T_{\varphi(p)}(N)$. In general, it is not possible, proceeding from the mappings φ_{*p}, $p \in M$, to define a mapping connecting vector field on M and N. Indeed, if for a given vector fields $X \in \mathfrak{X}(M)$ we try to define the vector field $Y \in \mathfrak{X}(N)$ by

$$Y_{\varphi(p)} \equiv \varphi_*(X_p),$$

we do not obtain, in general, a well-defined vector field on N. There are several reasons for this. Assume, for example, that φ is not an injective mapping; then there are at least two points $p, p' \in M$ such that $\varphi(p) = \varphi(p')$. In this case, to obtain a well-defined vector field Y, we must have the equality

$$\varphi_*(X_p) = \varphi_*(X_{p'}),$$

which is not valid for an arbitrary vector field X.

Nevertheless, let X and Y be two vector fields on M and N respectively. If for any point $p \in M$,

$$Y_{\varphi(p)} = \varphi_*(X_p), \tag{2.6}$$

we say that the vector fields X and Y are *φ-related*, and write $Y = \varphi_* X$. A vector field $X \in \mathfrak{X}(M)$ is called *φ-projectible* if there exists a vector field $Y \in \mathfrak{X}(N)$ such that the vector fields X and Y are φ-related. If the mapping φ is not surjective, then a φ-projectible vector field X does not uniquely define a vector field Y satisfying (2.6). In this case the expression $\varphi_* X$ means some vector field Y on N satisfying (2.6).

Let φ be a smooth mapping from a manifold M to a manifold N. It can be shown that vector fields $X \in \mathfrak{X}(M)$ and $Y \in \mathfrak{X}(N)$ are φ-related if and only if

$$\varphi^*(Y(f)) = X(\varphi^* f)$$

for any function $f \in \mathfrak{F}(N)$. Hence, if a vector field $X \in \mathfrak{X}(M)$ is φ-projectible, then for any function $f \in \mathfrak{F}(M)$ and for any choice of the vector field $\varphi_* X$ we have

$$\varphi^*(\varphi_* X(f)) = X(\varphi^* f). \tag{2.7}$$

If vector fields $X, Y \in \mathfrak{X}(M)$ are φ-projectible, then the vector fields $X + Y$ and $[X, Y]$ are also φ-projectible, and for any choice of the vector fields $\varphi_* X$ and $\varphi_* Y$, the vector fields $\varphi_*(X + Y)$ and $\varphi_*[X, Y]$ can be chosen in such a way that

$$\varphi_*(X + Y) = \varphi_* X + \varphi_* Y, \tag{2.8}$$

$$\varphi_*[X, Y] = [\varphi_* X, \varphi_* Y]. \tag{2.9}$$

If $\varphi : M \to N$ is a diffeomorphism, then any vector field $X \in \mathfrak{X}(M)$ is φ-projectible. In this case the vector field $\varphi_* X$ is unique, and from (2.7) it follows that

$$\varphi_* X(f) = \varphi^{-1*}(X(\varphi^* f))$$

for any $f \in \mathfrak{F}(N)$.

2.3.3 Integral curves and flows of vector fields

A smooth curve $\lambda : I \to M$, where $I = (a, b)$ with $a, b \in \mathbb{R} \cup \{-\infty, +\infty\}$, is called an *integral curve* of a smooth vector field X on M if

$$\dot{\lambda}(t) = X_{\lambda(t)}$$

for any $t \in I$. For any point p of a manifold M, and any vector field $X \in \mathfrak{X}(M)$, there is a unique maximal integral curve $\lambda_p^X : (a(p), b(p)) \to M$ of X, such that $0 \in (a(p), b(p))$ and $\lambda_p^X(0) = p$.

For an arbitrary $t \in \mathbb{R}$ we define the subset D_t^X of M by

$$D_t^X \equiv \{p \in M \mid t \in (a(p), b(p))\},$$

and the mapping $\Phi_t^X : D_t^X \to M$ by

$$\Phi_t^X(p) \equiv \lambda_p^X(t).$$

It is clear that $D_0^X = M$ and $\Phi_0^X = \mathrm{id}_M$. The mapping Φ_t^X, considered as a mapping from D_t^X to $\Phi_t^X(D_t^X)$, is a diffeomorphism, and

$$(\Phi_t^X)^{-1} = \Phi_{-t}^X.$$

This equality means, in particular, that $\Phi_t^X(D_t^X) = D_{-t}^X$. Moreover, for any $t, s \in \mathbb{R}$ the set $\Phi_s^X(D_s^X) \cap D_t^X$ is a subset of D_{t+s}^X, and

$$\Phi_t^X \circ \Phi_s^X = \Phi_{t+s}^X$$

on $\Phi_s^X(D_s^X) \cap D_t^X$.

Let $\varphi \in \mathfrak{F}(M, N)$ be a diffeomorphism, then

$$\varphi \circ \Phi_t^X \circ \varphi^{-1} = \Phi_t^{\varphi_* X}$$

for any $t \in R$. There is also a useful relation

$$\lim_{t \to 0} \frac{1}{t}[Y - (\Phi_t^X)_* Y] = [X, Y]. \tag{2.10}$$

A vector field X on a manifold M is called *complete* if $D_t^X = M$ for any $t \in \mathbb{R}$. It can be shown that any vector field on a compact manifold is complete.

Also introduce the subset D^X of $M \times \mathbb{R}$ given by

$$D^X \equiv \bigcup_{t \in \mathbb{R}} D_t^X \times \{t\},$$

and the mapping $\Phi^X : D^X \to M$ defined as

$$\Phi^X(p, t) \equiv \Phi_t^X(p) = \lambda_p^X(t).$$

The set D^X is an open subset of $M \times \mathbb{R}$ and the mapping Φ^X is smooth. The mapping Φ^X is called the *flow* induced by the vector field X. If X is a complete vector field, then $D^X = M \times \mathbb{R}$.

EXAMPLE 2.11 Consider the vector field

$$X = e^{-x} \frac{\partial}{\partial x}$$

defined on the manifold $\mathbb{R}$. A curve λ is an integral curve of X if

$$\frac{d\lambda(t)}{dt} = e^{-\lambda(t)}.$$

The solution of this differential equation is

$$\lambda(t) = \ln(t + c),$$

where c is the integration constant. Hence, for the maximal integral curves λ_a^X, $a \in \mathbb{R}$, we can write the expression

$$\lambda_a^X(t) = \ln(t + e^a).$$

Therefore, $D_t^X = (-\infty, +\infty)$ for any $t \geq 0$, while for $t < 0$ we have $D_t^X = (\ln|t|, +\infty)$. Thus, the vector field X is not complete.

Exercises

2.3 Consider the set $\mathbb{R}_+ \equiv (0, +\infty)$ as an open submanifold of $\mathbb{R}$ and define the mapping $\varphi : \mathbb{R} \to \mathbb{R}_+$ by

$$\varphi(a) = e^a.$$

Prove that φ is a diffeomorphism. For the vector field $X = e^{-x}\partial/\partial x$ on $\mathbb{R}$ find the vector field $\varphi_* X$.

2.4 Find the maximal integral curves for the vector fields

$$X = x^2 \frac{\partial}{\partial x}, \quad X = (1 + x^2)\frac{\partial}{\partial x}$$

on $\mathbb{R}$, and for the vector fields

$$X = x_1 \frac{\partial}{\partial x_1} + x_2 \frac{\partial}{\partial x_2}, \quad X = x_1 \frac{\partial}{\partial x_2} - x_2 \frac{\partial}{\partial x_1}$$

on $\mathbb{R}^2$.

2.5 Let Φ^X and Φ^{aX}, $a \in \mathbb{R}$, be the flows induced by the vector fields X and aX. Show that

$$\Phi_t^{aX} = \Phi_{at}^X$$

for any $t \in \mathbb{R}$.

2.4 Tensors

2.4.1 Cotangent space and covector fields

The dual of the tangent space $T_p(M)$ is called the *cotangent space* to the manifold M at the point p; this space is denoted by $T_p^*(M)$. The elements of $T_p^*(M)$ are called *cotangent vectors* or *covectors.*

An assignment ω of a covector $\omega_p \in T_p(M)$ to each point p of M is called a *covector field* on M. For any covector field ω on M we define the action of ω on a vector field X on M as a function $\omega(X)$ given by

$$(\omega(X))(p) \equiv \omega_p(X_p). \tag{2.11}$$

A covector field ω is called *smooth* if for any $X \in \mathfrak{X}(M)$ the function $\omega(X)$ is smooth. The totality of smooth covector fields on M is denoted by $\mathfrak{X}^*(M)$. From the definition of the action of a covector field on a vector field it follows that a smooth covector field induces an $\mathfrak{F}(M)$-linear mapping from the $\mathfrak{F}(M)$-module $\mathfrak{X}(M)$ to $\mathfrak{F}(M)$. Furthermore, any $\mathfrak{F}(M)$-linear mapping from $\mathfrak{X}(M)$ to $\mathfrak{F}(M)$ is generated by a covector field.

EXAMPLE 2.12 Let $f \in \mathfrak{F}(M)$, define the covector field df by

$$df(X) \equiv X(f).$$

It is evident that df is a smooth covector field on M, which is called the *differential* of the function f.

The sum of two covector fields and the multiplication of a covector field by a function are defined as

$$(\omega + \eta)(X) \equiv \omega(X) + \eta(X),$$
$$(f\omega)(X) \equiv f\omega(X).$$

If ω and η are smooth covector fields, then $\omega + \eta$ is a smooth covector field. If ω is a smooth covector field and f is a smooth function, then $f\omega$ is a smooth covector field. It is clear that the set $\mathfrak{X}^*(M)$ is a module over the algebra $\mathfrak{F}(M)$.

Let U be an open submanifold of M. Recall that for any $p \in U$ the tangent space $T_p(M)$ is naturally isomorphic to the tangent space $T_p(U)$. This fact allows us to identify the cotangent spaces $T_p^*(M)$ and $T_p^*(U)$ and to define the restriction $\omega|_U$ of a covector field ω on M to the submanifold U. Here we have

$$\omega|_U(X|_U) = \omega(X)|_U.$$

On the other hand, let U be an open submanifold of a manifold M, and $\omega \in \mathfrak{X}^*(U)$. For any point $p \in U$ there is a covector field $\widetilde{\omega} \in \mathfrak{X}(M)$ and an open neighbourhood $V \subset U$ of the point p such that $\omega|_V = \widetilde{\omega}|_V$.

Now let $(U, x^1, \ldots, x^m)$ be a chart on a manifold M. It can be shown that for any $p \in U$ the covectors $(dx^i)_p$ form a basis for $T^*_p(U)$. Actually, we have

$$(dx^i)_p((\partial/\partial x^j)_p) = \delta^i_j;$$

and hence, the basis, formed by the covectors $(dx^i)_p$, is the dual basis for $\{(\partial/\partial x^i)_p\}$. Further, for any covector field ω on M, we can write

$$\omega|_U = \omega_i dx^i,$$

where $\omega_i = \omega|_U(\partial/\partial x^i)$. If the covector field ω is smooth, then the functions ω_i are also smooth. In particular, for the differential of a smooth function f we have

$$(df)|_U = \frac{\partial f}{\partial x^i} dx^i.$$

Moreover, the covector fields dx^i form a basis of the $\mathfrak{F}(U)$-module $\mathfrak{X}^*(U)$.

Let $\varphi : M \to N$ be a smooth mapping. For any $p \in M$ define the mapping $\varphi^*_p : T^*_{\varphi(p)}(N) \to T^*_p(M)$ by

$$(\varphi^*_p(\alpha))(v) \equiv \alpha(\varphi_{*p}(v)).$$

Here for any $\psi \in \mathfrak{F}(M, N)$, $\varphi \in \mathfrak{F}(N, K)$ and $p \in M$ one has

$$(\varphi \circ \psi)^*_p = \psi^*_p \circ \varphi^*_{\psi(p)}.$$

For any covector field ω on N and a smooth mapping $\varphi : M \to N$ we define the covector field $\varphi^*\omega$ on M by the relation

$$(\varphi^*\omega)_p \equiv \varphi^*_p(\omega_{\varphi(p)}).$$

If the covector field ω is smooth, then the covector field $\varphi^*\omega$ is also smooth.

In the case where $\varphi : M \to N$ is a diffeomorphism, one has

$$(\varphi^*\omega)(X) = \varphi^*(\omega(\varphi_* X)) \tag{2.12}$$

for any vector field X on M.

2.4.2 Tensor fields

A *tensor of type* $\binom{k}{l}$ on a vector space V over a field $\mathbb{K}$ can be defined as a $\mathbb{K}$-multilinear mapping from the set

$$\underbrace{V^* \times V^* \times \ldots \times V^*}_{k} \times \underbrace{V \times V \times \ldots \times V}_{l}$$

to $\mathbb{K}$. From this point of view, the tensor product of tensors $t \in T^k_l(V)$ and $s \in T^m_n(V)$ is the tensor $t \otimes s \in T^{k+m}_{l+n}(V)$ given by

$$t \otimes s(\mu_1, \dots, \mu_k, \nu_1, \dots, \nu_m, v_1, \dots, v_l, u_1, \dots, u_n)$$
$$\equiv t(\mu_1, \dots, \mu_k, v_1, \dots, v_l) s(\nu_1, \dots, \nu_m, u_1, \dots, u_n).$$

Recall that if $\{e_i\}$ is a basis of an m-dimensional vector space V, and $\{\alpha^i\}$ is the corresponding dual basis of V^*, then the tensors

$$e_{i_1} \otimes \dots \otimes e_{i_k} \otimes \alpha^{j_1} \otimes \dots \otimes \alpha^{j_l},$$

where $1 \leq i_1, \dots, i_k, j_1, \dots j_l \leq m$, form a basis for $T^k_l(V)$.

Now let M be an m-dimensional manifold. The space of tensors of type $\binom{k}{l}$ on the tangent space $T_p(M)$ is denoted by $T^k_{lp}(M)$. Let $(U, x^1, \dots, x^m)$ be a chart on M; the tensors

$$\left(\frac{\partial}{\partial x^{i_1}}\right)_p \otimes \dots \otimes \left(\frac{\partial}{\partial x^{i_k}}\right)_p \otimes (dx^{j_1})_p \otimes \dots \otimes (dx^{j_l})_p,$$

where $1 \leq i_1, \dots, i_k, j_1, \dots j_l \leq m$, form a basis for $T^k_{lp}(M)$.

An assignment T of an element of $T_p \in T^k_{lp}(M)$ to each point $p \in M$ is called a *tensor field* of type $\binom{k}{l}$ on M. In particular, a vector field is a tensor field of type $\binom{1}{0}$, and a covector field is a tensor field of type $\binom{0}{1}$.

A tensor field T of type $\binom{k}{l}$ on M generates an $\mathfrak{F}(M)$-multilinear mapping from the set

$$\underbrace{\mathfrak{X}^*(M) \times \dots \times \mathfrak{X}^*(M)}_{k} \times \underbrace{\mathfrak{X}(M) \times \dots \times \mathfrak{X}(M)}_{l}$$

to $\mathfrak{F}(M)$, given by

$$(T(\omega_1, \dots, \omega_k, X_1, \dots, X_l))(p) \equiv T_p(\omega_{1p}, \dots, \omega_{kp}, X_{1p}, \dots, X_{lp}),$$

and any such a mapping is generated by a tensor field of type $\binom{k}{l}$ on M. A tensor field is called *smooth* if its action on the corresponding number of smooth covector and vector fields is a smooth function. The set of smooth tensor fields of type $\binom{k}{l}$ on M is denoted $\mathfrak{X}^k_l(M)$.

The pointwise definition of the corresponding operations turns the space $\mathfrak{X}^k_l(M)$ into a module over the algebra $\mathfrak{F}(M)$. Similarly, defining the tensor product of tensor fields by

$$(R \otimes S)_p \equiv R_p \otimes S_p,$$

we endow the space

$$\mathfrak{T}(M) \equiv \bigoplus_{k,l=0}^{\infty} \mathfrak{X}^k_l(M),$$

where $\mathfrak{X}^0_0(M) \equiv \mathfrak{F}(M)$, with the structure of an associative algebra over $\mathfrak{F}(M)$, called the *tensor algebra* of M.

Any tensor on a manifold can be naturally restricted to any of its open submanifolds. Let $(U, x^1, \dots, x^m)$ be a chart on a manifold M. It is clear that the tensors

$$\frac{\partial}{\partial x^{i_1}} \otimes \dots \otimes \frac{\partial}{\partial x^{i_k}} \otimes dx^{j_1} \otimes \dots \otimes dx^{j_l},$$

where $1 \le i_1, \dots, i_k, j_1, \dots j_l \le m$, form a basis for the $\mathfrak{F}(M)$-module $\mathfrak{X}^k_l(M)$. For any tensor T of type $\binom{k}{l}$ on M we have

$$T|_U = T^{i_1 \cdots i_k}_{j_1 \cdots j_l} \frac{\partial}{\partial x^{i_1}} \otimes \dots \otimes \frac{\partial}{\partial x^{i_k}} \otimes dx^{j_1} \otimes \dots \otimes dx^{j_l},$$

where

$$T^{i_1 \cdots i_k}_{j_1 \cdots j_l} = T|_U(dx^{i_1}, \dots, dx^{i_k}, \partial/\partial x^{j_1}, \dots, \partial/\partial x^{j_l}).$$

Let $\varphi : M \to N$ be a diffeomorphism and let $T \in \mathfrak{X}^k_l(M)$. The tensor field $\varphi_* T \in \mathfrak{X}^k_l(M)$, defined by

$$\begin{aligned}(\varphi_* T)&(\omega_1, \dots, \omega_k, X_1, \dots, X_l) \\ &\equiv \varphi^{-1*}(T(\varphi^* \omega_1, \dots, \varphi^* \omega_k, \varphi_*^{-1} X_1, \dots, \varphi_*^{-1} X_l)),\end{aligned}$$

is called the *push-forward* of T by φ. Similarly, for any $T \in \mathfrak{X}^k_l(N)$ the *pull-back* $\varphi^* T \in \mathfrak{X}^k_l(M)$ of T by φ is defined as

$$\begin{aligned}(\varphi^* T)&(\omega_1, \dots, \omega_k, X_1, \dots, X_l) \\ &\equiv \varphi^*(T(\varphi^{-1*} \omega_1, \dots, \varphi^{-1*} \omega_k, \varphi_* X_1, \dots, \varphi_* X_l)).\end{aligned}$$

Here, for any diffeomorphisms $\psi : M \to N$ and $\varphi : N \to K$, one has

$$(\varphi \circ \psi)_* = \varphi_* \circ \psi_*, \qquad \varphi_*^{-1} = \varphi^*.$$

Furthermore, for any tensor fields R and S on M,

$$\varphi_*(R \otimes S) = \varphi_* R \otimes \varphi_* S.$$

2.4.3 Differential forms

Define an action of the symmetric group S_k on the space of the tensors of type $\binom{0}{k}$ by

$$t_\sigma(v_1, \dots, v_k) \equiv t(v_{\sigma(1)}, \dots, v_{\sigma(k)}).$$

Denote by ϵ_σ the sign of the permutation σ. A tensor t of type $\binom{0}{k}$ is called skew-symmetric if

$$t_\sigma = \epsilon_\sigma t$$

for any $\sigma \in S_k$. It is customary to call skew-symmetric tensors of type $\binom{0}{k}$ on a vector space V *exterior k-forms*. The totality of exterior k-forms on V is denoted $\Lambda_k(V)$. It is convenient to consider $\Lambda_0(V) \equiv \mathbb{K}$ and $\Lambda_1(V) \equiv V^*$. In other words, the elements of the field $\mathbb{K}$ are 0-forms, and the covectors are 1-forms. The set $\Lambda_k(V)$ is a subspace of $T_k^0(V)$. Clearly, $\Lambda_k(V) = \{0\}$ if k is greater that the dimension of V.

For a given tensor t of type $\binom{0}{k}$, we can construct an exterior k-form, applying to t the *alternation mapping* Alt defined as

$$\operatorname{Alt} t \equiv \frac{1}{k!} \sum_{\sigma \in S_k} \epsilon_\sigma t_\sigma.$$

The tensor product of two forms is not a skew-symmetric tensor. Define the *exterior product* of $\alpha \in \Lambda_k(V)$ and $\beta \in \Lambda_l(V)$ by

$$\alpha \wedge \beta \equiv \frac{(k+l)!}{k!\, l!} \operatorname{Alt}(\alpha \otimes \beta).$$

The exterior product is a bilinear associative operation satisfying the relation

$$\alpha \wedge \beta = (-1)^{kl} \beta \wedge \alpha$$

for any $\alpha \in \Lambda_k(V)$ and $\beta \in \Lambda_l(V)$.

The direct sum $\Lambda(V)$ of the spaces $\Lambda_k(V)$ is an associative algebra with respect to the multiplication induced by the exterior product operation. This algebra is called the *exterior algebra* of V, or the *Grassmann algebra* of V.

Let V be an m-dimensional vector space. For $0 < k \leq m$ the dimension of the vector space $\Lambda_k(V)$ is equal to $\binom{m}{k}$. The exterior algebra $\Lambda(V)$ has dimension 2^m. Let $\{e_i\}$ be a basis for V and let $\{\alpha^i\}$ be the corresponding dual basis for V^*. The k-forms

$$\alpha^{i_1} \wedge \cdots \wedge \alpha^{i_k}, \qquad 1 \leq i_1 < i_2 \cdots < i_k \leq m,$$

form a basis for $\Lambda_k(V)$.

Let M be an m-dimensional manifold. We denote the space of exterior k-forms on $T_p(M)$ by $\Lambda_{kp}(M)$. Let $(U, x^1, \ldots, x^m)$ be a chart on M, such that $p \in U$. The tensors

$$(dx^{i_1})_p \wedge \cdots \wedge (dx^{i_k})_p, \qquad 1 \leq i_1 < \cdots < i_k \leq m,$$

form a basis for $\Lambda_{kp}(M)$.

An assignment ω of an exterior k-form $\omega_p \in \Lambda_{kp}(M)$ to each point p of M is called a differential k-form on M, or simply a

k-form on M. Since differential forms on M are a special case of tensor fields on M, they possess the properties of tensor fields. In particular, we can define the notion of a smooth differential form. The totality of smooth differential k-forms on M is denoted by $\Omega_k(M)$. The pointwise definition of the algebraic operations turns $\Omega_k(M)$ into the module over the algebra $\mathfrak{F}(M)$. Defining the *exterior product* of differential forms by

$$(\omega \wedge \eta)_p \equiv \omega_p \wedge \eta_p,$$

we provide the direct sum

$$\Omega(M) \equiv \bigoplus_{k=0}^{m} \Omega_k(M),$$

where $\Omega_0(M) \equiv \mathfrak{F}(M)$, with the structure of an associative algebra over $\mathfrak{F}(M)$, called the *algebra of differential forms* on M.

Let $(U, x^1, \ldots, x^m)$ be a chart on a manifold M. The k-forms

$$dx^{i_1} \wedge \cdots \wedge dx^{i_k}, \qquad 1 \le i_1 < \cdots < i_k \le m,$$

form a basis of the $\mathfrak{F}(M)$-module $\Omega_k(M)$. For any k-form ω on M, we have

$$\omega|_U = \frac{1}{k!} \omega_{i_1 \ldots i_k} dx^{i_1} \wedge \cdots \wedge dx^{i_k}. \tag{2.13}$$

Here the functions $\omega_{i_1 \ldots i_k}$ are unique if we suppose that they are skew-symmetric with respect to any transposition of the indices $i_1, \ldots, i_k$. In this case we have

$$\omega_{i_1 \ldots i_k} = \omega|_U(\partial/\partial x^{i_1}, \ldots, \partial/\partial x^{i_k}).$$

The differential of a function can be considered as a particular case of the *exterior derivative* operation. In the general case, the exterior derivative is defined as an $\mathbb{R}$-linear mapping $d : \Omega(M) \to \Omega(M)$ satisfying the properties

(ED1) $d\Omega_k(M) \subset \Omega_{k+1}(M)$;

(ED2) if $f \in \Omega_0(M)$, then $df(X) = X(f)$ for any $X \in \mathfrak{X}(M)$;

(ED3) $d \circ d = 0$;

(ED4) $d(\omega \wedge \eta) = d\omega \wedge \eta + (-1)^k \omega \wedge d\eta$ for any $\omega \in \Omega_k(M)$ and $\eta \in \Omega(M)$.

It can be shown that these properties define the mapping d uniquely.

If U is an open submanifold of a manifold M, then for any $\omega \in \Omega(M)$ one has

$$d(\omega|_U) = (d\omega)|_U.$$

Let $(U, x^1, \ldots, x^m)$ be a chart on a manifold M. Using the properties of the external derivative and relation (2.13), for any k-form ω on M we obtain

$$(d\omega)|_U = \frac{1}{k!}\frac{\partial \omega_{i_1 \cdots i_k}}{\partial x^i} dx^i \wedge dx^{i_1} \wedge \cdots \wedge dx^{i_k}.$$

It can also be shown that for any k-form ω there is a useful formula:

$$d\omega(X_1, \ldots, X_{k+1}) = \sum_{i=1}^{k+1} (-1)^{i+1} X_i(\omega(X_1, \ldots, \widehat{X}_i, \ldots, X_{k+1}) \\ + \sum_{1 \le i < j \le k+1} (-1)^{i+j} \omega([X_i, X_j], X_1, \ldots, \widehat{X}_i, \ldots, \widehat{X}_j, \ldots, X_{k+1}). \tag{2.14}$$

In particular, for the case where $k = 1$ we have

$$d\omega(X, Y) = X(\omega(Y)) - Y(\omega(X)) - \omega([X, Y]). \tag{2.15}$$

It is important that the pull-back of a differential form is defined not only for diffeomorphisms. Let $\varphi : M \to N$ be a smooth mapping. Define the *pull-back* $\varphi^*\omega$ of the differential form ω on N as a differential form on M given by

$$(\varphi^*\omega)_p \equiv \varphi_p^*(\omega_p).$$

It can be shown that if the differential form ω is smooth, then the differential form $\varphi^*\omega$ is also smooth. Here

$$d(\varphi^*\omega) = \varphi^*(d\omega),$$

and for any differential forms $\omega, \eta \in \Omega(N)$ one has

$$\varphi^*(\omega \wedge \eta) = \varphi^*\omega \wedge \varphi^*\eta.$$

Exercises

2.6 Let ω and η be 1-forms. Show that

$$\omega \wedge \eta(X_1, X_2) = \omega(X_1)\eta(X_2) - \omega(X_2)\eta(X_1).$$

In other words, demonstrate that

$$\omega \wedge \eta = \omega \otimes \eta - \eta \otimes \omega.$$

2.7 Let ω be a 2-form and let η be a 1-form. Prove that

$$\omega \wedge \eta(X_1, X_2, X_3) = \omega(X_1, X_2)\eta(X_3) \\ + \omega(X_2, X_3)\eta(X_1) + \omega(X_3, X_1)\eta(X_2).$$

2.8 Show that for the case where $k = 2$, equality (2.14) takes the form

$$\begin{aligned} d\omega(X_1, X_2, X_3) = {} & X_1(\omega(X_2, X_3)) + X_2(\omega(X_3, X_1)) \\ & + X_3(\omega(X_1, X_2)) - \omega([X_1, X_2], X_3) \\ & - \omega([X_2, X_3], X_1) - \omega([X_3, X_1], X_2). \end{aligned}$$

2.5 Complex manifolds

2.5.1 Definition of a complex manifold

Let f be a continuous complex function defined in an open subset U of the space $\mathbb{C}^m$. Denote by z^i, $i = 1, \ldots, m$, the standard complex coordinate functions in $\mathbb{C}^m$, and by x^i and y^i the corresponding real coordinate functions such that $z^i = x^i + \sqrt{-1}y^i$. Represent f as

$$f = g + \sqrt{-1}h,$$

where g and h are real functions on U. Suppose that for any i there exist the partial derivatives of g and h over x^i and y^i, and define the partial derivatives of f over x^i and y^i by

$$\frac{\partial f}{\partial x^i} \equiv \frac{\partial g}{\partial x^i} + \sqrt{-1}\frac{\partial h}{\partial x^i}, \quad \frac{\partial f}{\partial y^i} \equiv \frac{\partial g}{\partial y^i} + \sqrt{-1}\frac{\partial h}{\partial y^i}.$$

Also introduce the notation

$$\frac{\partial f}{\partial z^i} \equiv \frac{1}{2}\left(\frac{\partial f}{\partial x^i} - \sqrt{-1}\frac{\partial f}{\partial y^i}\right), \quad \frac{\partial f}{\partial \bar{z}^i} \equiv \frac{1}{2}\left(\frac{\partial f}{\partial x^i} + \sqrt{-1}\frac{\partial f}{\partial y^i}\right).$$

The function f is called *(anti)holomorphic* in U if

$$\frac{\partial f}{\partial \bar{z}^i} = 0 \quad \left(\frac{\partial f}{\partial z^i} = 0\right), \quad i = 1, \ldots, i.$$

Let φ be a continuous mapping from an open subset U of $\mathbb{C}^m$ to $\mathbb{C}^n$. Denote by z^i, $i = 1, \ldots, m$, and w^a, $a = 1, \ldots, n$, the standard coordinate functions on $\mathbb{C}^m$ and $\mathbb{C}^n$ respectively. The mapping φ is called *(anti)holomorphic* if the functions $\varphi^a \equiv \varphi^* w^a$ are (anti)holomorphic.

Let M be a topological space. A pair (U, φ), where U is an open subset of M and φ is a homeomorphism from U to an open subset $\varphi(U)$ of the space $\mathbb{C}^m$, is called a *complex chart* on M. Let M be provided with the set of complex charts $\{(U_\alpha, \varphi_\alpha)\}_{\alpha \in \mathcal{A}}$ such that the mappings

$$\varphi_\alpha \circ \varphi_\beta^{-1} : \varphi_\beta(U_\alpha \cap U_\beta) \to \varphi_\alpha(U_\alpha \cap U_\beta), \quad \alpha, \beta \in \mathcal{A},$$

are holomorphic and

$$\bigcup_{\alpha \in \mathcal{A}} U_\alpha = M.$$

In this case M is said to be an m-dimensional *complex manifold*, and the set $\{(U_\alpha, \varphi_\alpha)\}_{\alpha \in \mathcal{A}}$ is called an *atlas* of M. It is quite obvious how to define the notions of the maximal atlas and of an admissible chart in this case.

EXAMPLE 2.13 The standard structure of a complex manifold on the space $\mathbb{C}^m$ is specified by an atlas consisting of just one chart $(\mathbb{C}^m, \mathrm{id}_{\mathbb{C}^m})$. The standard coordinate functions z^i, $i = 1, \ldots, m$, in this case are defined as

$$z^i(c) \equiv c^i$$

for any $c = (c^1, \ldots, c^m) \in \mathbb{C}^m$.

EXAMPLE 2.14 Consider the standard two-dimensional sphere S^2. As a real manifold it has an atlas consisting of two charts defined in example 2.9. Introducing the mappings

$$z \equiv u^1 + \sqrt{-1}u^2, \qquad w \equiv v^1 - \sqrt{-1}v^2,$$

we obtain two complex charts on S^2. In the corresponding domain we have $w = 1/z$. Thus, we see that the manifold S^2 can be endowed with the structure of a complex manifold.

Since the space $\mathbb{C}^m$ can be identified with the space $\mathbb{R}^{2m}$, and the real and imaginary parts of a holomorphic function are real analytic functions of the corresponding real arguments, any complex manifold M can be considered as a real manifold of class C^ω. This manifold is called the *realification* of M and is denoted by $M_\mathbb{R}$. It is clear that

$$\dim M_\mathbb{R} = 2 \dim M.$$

A complex function on a real manifold M is called smooth if its real and imaginary parts are smooth functions. A function f on a complex manifold M is said to be *smooth* if it is smooth as a function on $M_\mathbb{R}$. The set of complex smooth functions on a real or complex manifold M is denoted by $\mathfrak{F}^\mathbb{C}(M)$. This set has the natural structure of a complex associative commutative algebra.

A complex function f on a complex manifold M is called *holomorphic* (*antiholomorphic*) if the local expression of f with respect to any admissible chart on M is a holomorphic (antiholomorphic) function. Note that not all complex manifolds admit the existence of holomorphic or antiholomorphic functions different from constants.

Let M and N be complex manifolds. A mapping $\varphi : M \to N$ is called *smooth* if it is smooth as the mapping from the manifold $M_{\mathbb{R}}$ to $N_{\mathbb{R}}$. A mapping $\varphi : M \to N$ is said to be *holomorphic* (*antiholomorphic*) if its local expressions with respect to all admissible charts on M and N are described by holomorphic (antiholomorphic) functions.

2.5.2 Vector fields on complex manifolds

Let M be a real manifold. A tangent vector $v \in T_p(M)$ defines a mapping from $\mathfrak{F}^{\mathbb{C}}(M)$ to $\mathbb{C}$ in the following way. Let g and h be the real and imaginary parts of the function $f \in \mathfrak{F}^{\mathbb{C}}(M)$; define the action of v on f by

$$v(f) \equiv v(g) + \sqrt{-1}v(h).$$

It is useful to consider the complexification of the space $T_p(M)$, which is usually denoted by $T_p^{\mathbb{C}}(M)$. The elements of $T_p^{\mathbb{C}}(M)$ are called *complex tangent vectors.* If it is necessary to distinguish complex tangent vectors and ordinary ones, we use the term *real tangent vectors* for the elements of $T_p(M)$. Representing a complex tangent vector $v \in T^{\mathbb{C}}(M)$ as $v = u + \sqrt{-1}w$, where $u, w \in T_p(M)$, we define the action of v on a real or complex smooth function f by

$$v(f) = u(f) + \sqrt{-1}w(f).$$

A notion of a tangent vector to a complex manifold is introduced in the same way as for a real manifold. The real vector space of all real tangent vectors to a complex manifold M at a point $p \in M$ will be denoted $T_{\mathbb{R}p}(M)$. Since we identify the set of smooth functions on a complex manifold M with the set of smooth functions on the manifold $M_{\mathbb{R}}$, we can identify the space $T_{\mathbb{R}p}(M)$ with the tangent space $T_p(M_{\mathbb{R}})$. Let $(U, z^1, \ldots, z^m)$ be a chart on a complex manifold M. For each $i = 1, \ldots, m$, we can write a unique representation

$$z^i = x^i + \sqrt{-1}y^i, \tag{2.16}$$

where x^i and y^i are real functions which can be considered as coordinate functions on $M_{\mathbb{R}}$. For any $p \in U$, the tangent vectors $(\partial/\partial x^i)_p$ and $(\partial/\partial y^i)$ form a basis of $T_p(M_{\mathbb{R}})$, and, therefore, they form a basis of $T_{\mathbb{R}p}(M)$. Define the linear operator $J_p^M : T_{\mathbb{R}p}(M) \to T_{\mathbb{R}p}(M)$ by

$$J_p^M((\partial/\partial x^i)_p) \equiv (\partial/\partial y^i)_p, \quad J_p^M((\partial/\partial y^i)_p) \equiv -(\partial/\partial x^i)_p.$$

It is evident that J_p^M is a complex structure on $T_{\mathbb{R}p}(M)$. This complex structure does not depend on the choice of the chart containing the point p. We define the *tangent space* $T_p(M)$ to the complex manifold M at the point p as

$$T_p(M) \equiv \widetilde{T_{\mathbb{R}p}(M)}.$$

As it follows from (1.28), we have for the space

$$T_p^{\mathbb{C}}(M) \equiv (T_{\mathbb{R}p}(M))^{\mathbb{C}}$$

the following direct sum decomposition:

$$T_p^{\mathbb{C}}(M) = T_p^{(1,0)}(M) \oplus T_p^{(0,1)}(M). \tag{2.17}$$

The operator J_p^M can be uniquely extended to the linear operator acting in $T_p^{\mathbb{C}}(M)$, which will also be denoted by J_p^M. The spaces $T_p^{(1,0)}(M)$ and $T_p^{(0,1)}(M)$ can be characterised as the eigenspaces of the operator J_p^M, corresponding to the eigenvalues $+\sqrt{-1}$ and $-\sqrt{-1}$ respectively. Moreover, there are natural isomorphisms of the spaces $T_p^{(1,0)}(M)$ and $T_p^{(0,1)}(M)$ with the spaces $T_p(M)$ and $\overline{T_p(M)}$; in other words, we have

$$T_p^{\mathbb{C}}(M) \simeq T_p(M) \oplus \overline{T_p(M)}.$$

Since the space $T_p^{\mathbb{C}}(M)$ is, by definition, the complexification of $T_{\mathbb{R}p}(M)$, the complex conjugation is defined for the elements of $T_p^{\mathbb{C}}(M)$. This conjugation transforms elements of $T_p^{(1,0)}(M)$ to elements of $T_p^{(0,1)}(M)$ and vice versa. The linear operators P_p^M and $\overline{P}_p^M$ on $T_p^{\mathbb{C}}(M)$, defined by

$$P_p^M x \equiv \frac{1}{2}(x - \sqrt{-1}J_p^M x), \quad \overline{P}_p^M x \equiv \frac{1}{2}(x + \sqrt{-1}J_p^M x), \tag{2.18}$$

are projection operators on $T_p^{(1,0)}(M)$ and $T_p^{(0,1)}(M)$ respectively.

Let (U, φ) be a complex chart on a complex manifold M with the coordinate functions z^i. Introduce the real coordinate functions x^i, y^i by (2.16), and for any $p \in U$ define the following complex tangent vectors:

$$(\partial/\partial z^i)_p \equiv \frac{1}{2}\left((\partial/\partial x^i)_p - \sqrt{-1}(\partial/\partial y^i)_p\right),$$

$$(\partial/\partial \bar{z}^i)_p \equiv \frac{1}{2}\left((\partial/\partial x^i)_p + \sqrt{-1}(\partial/\partial y^i)_p\right).$$

The vectors $(\partial/\partial z^i)_p$ and $(\partial/\partial \bar{z}^i)_p$ form bases for the spaces $T_p^{(1,0)}(M)$ and $T_p^{(0,1)}(M)$.

Let φ be a smooth mapping from a complex manifold M to a complex manifold N. For any $p \in M$ we have the linear mapping $\varphi_{*p} : T_{\mathbb{R}p}(M) \to T_{\mathbb{R}\varphi(p)}(M)$. The mapping φ is (anti)holomorphic if and only if

$$\varphi_{*p} \circ J_p^M = J_{\varphi(p)}^N \circ \varphi_{*p} \quad \left(\varphi_{*p} \circ J_p^M = -J_{\varphi(p)}^N \circ \varphi_{*p}\right)$$

for all $p \in M$. Note also that each mapping $\varphi_{*p} : T_{\mathbb{R}p}(M) \to T_{\mathbb{R}\varphi(p)}(M)$ can be uniquely extended to the linear mapping from $T_p^{\mathbb{C}}(M)$ to $T_{\varphi(p)}^{\mathbb{C}}$, which will also be denoted by φ_{*p}.

Return now to the case of real manifolds. Using the definition of the action of a tangent vector on a smooth complex function, we can define the action of a vector field on such a function. Moreover, the notion of a complex tangent vector leads to a natural definition of a *complex vector field* on a manifold. A complex vector field is called smooth if its action on any smooth function is a smooth function. The set of smooth complex vector fields on a manifold M is denoted by $\mathfrak{X}^{\mathbb{C}}(M)$. To distinguish complex vector fields and real ones we use the term *real vector fields*. The notion of the commutator of vector fields can easily be extended to the case of complex vector fields. The same can be done for the notion of φ-related vector fields and for the notion of a φ-projectible vector field.

Proceed now to the case of complex manifolds. It is clear that we can identify vector fields on a complex manifold M with vector fields on the manifold $M_{\mathbb{R}}$, and we say that a vector field on M is smooth if the corresponding vector field on $M_{\mathbb{R}}$ is smooth. Further, a complex vector field X on M is said to be *of type* $(1,0)$ if $X_p \in T^{(1,0)}(M)$ for any $p \in M$, and it is said to be *of type* $(0,1)$ if $X_p \in T^{(0,1)}(M)$ for any $p \in M$. The set of smooth vector fields of type $(1,0)$ (of type $(0,1)$) on M is denoted by $\mathfrak{X}^{(1,0)}(M)$ ($\mathfrak{X}^{(0,1)}(M)$). For any vector field X we can write a unique decomposition

$$X = X^{(1,0)} + X^{(0,1)}, \tag{2.19}$$

where $X^{(1,0)}$ and $X^{(0,1)}$ are vector fields of types $(1,0)$ and $(0,1)$ respectively.

A complex vector field X of type $(1,0)$ (of type $(0,1)$) is called *holomorphic* (*antiholomorphic*) if $X(f)$ is a holomorphic (antiholomorphic) function for any locally defined holomorphic (antiholomorphic) function f. Note that the complex conjugate of a holomorphic vector field is an antiholomorphic vector field. The commutator of holomorphic (antiholomorphic) vector fields is a holomorphic (antiholomorphic) vector field. Thus, the holomorphic (antiholomorphic) vector fields form a complex Lie algebra which is a subalgebra of the Lie algebra of complex vector fields.

Let $(U, z^1, \ldots, z^m)$ be a chart on a complex manifold M. The vectors $(\partial/\partial z^i)_p$ and $(\partial/\partial \bar{z}^i)_p$, $p \in U$, specify a set of complex vector fields on M, denoted by $\partial/\partial z^i$ and $\partial/\partial \bar{z}^i$. Let X be a vector field X on M. It can easily be shown that

$$X|_U = X^i \frac{\partial}{\partial z^i} + X^{\bar{\imath}} \frac{\partial}{\partial \bar{z}^i},$$

where

$$X^i = X|_U(z^i), \qquad X^{\bar{\imath}} = X|_U(\bar{z}^i).$$

Here and below we mark the indices corresponding to $\bar{z}^i$ by a bar. For a vector field of type $(1,0)$ (of type $(0,1)$) we obtain $X^i = 0$ ($X^{\bar{\imath}} = 0$). A vector field X of type $(1,0)$ (of type $(0,1)$) on U is holomorphic (antiholomorphic) if and only if the functions X^i ($X^{\bar{\imath}}$) are holomorphic (antiholomorphic).

2.5.3 *Almost complex structures and their automorphisms*

Let M be a smooth manifold of even dimension $m = 2n$. Suppose that each tangent space $T_p(M)$, $p \in M$, is provided with a complex structure J^M_p. For any vector field X on M define the vector field $J^M(X)$ by

$$(J^M(X))_p \equiv J^M_p(X_p).$$

For any smooth real vector field X let the vector field $J^M(X)$ also be smooth. In this case we call the $\mathfrak{F}(M)$-linear mapping $J^M : \mathfrak{X}(M) \to \mathfrak{X}(M)$ an *almost complex structure* on M. A manifold equipped with an almost complex structure is called an *almost complex manifold.* By definition, the mapping J^M satisfies the condition

$$(J^M)^2 = -\operatorname{id}_{\mathfrak{X}(M)}.$$

Now let M be a complex manifold. The linear mappings J_p^M, $p \in M$, specify an almost complex structure on the real manifold $M_{\mathbb{R}}$. Thus, for any complex manifold M there is the natural almost complex structure on $M_{\mathbb{R}}$ which is called the *canonical* almost complex structure on $M_{\mathbb{R}}$.

Let M be an almost complex manifold with an almost complex structure J^M. Define a mapping $S : \mathfrak{X}(M) \times \mathfrak{X}(M) \to \mathfrak{X}(M)$ by

$$\begin{aligned} S(X,Y) \equiv 2\{&[J^M(X), J^M(Y)] - [X,Y] \\ &- J^M([X, J^M(Y)]) - J^M([J^M(X), X])\}. \end{aligned}$$

The almost complex structure J^M is called *integrable* if $S(X,Y) = 0$ for all $X, Y \in \mathfrak{X}(M)$. For any complex manifold M, the canonical almost complex structure on the manifold $M_{\mathbb{R}}$ is integrable. From the other hand, if the almost complex structure J^M on an almost complex manifold M is integrable, then there exists a unique complex manifold N, such that $M = N_{\mathbb{R}}$ and J^M coincides with the canonical almost complex structure on M. In other words, if we have an almost complex manifold M and the almost complex structure J^M is integrable, then we can always introduce on M the structure of a complex manifold, such that J^M will coincide with the corresponding canonical almost complex structure.

The tangent vector to a smooth curve is always a real vector, so we can find flows for real vector fields only. Considering the case of complex manifolds, it is interesting to specify those real vector fields whose flows are holomorphic (antiholomorphic) mappings. Suppose that for some vector field $X \in \mathfrak{X}(M)$ and for any $t \in \mathbb{R}$ the mappings Φ_t^X are holomorphic. Then for any $p \in D_t^X$ and any $Y \in \mathfrak{X}(M)$ we have

$$\Phi_{t*p}^X(J_p^M(Y_p)) = J_{p'}^M(\Phi_{t*p}^X(Y_p)), \quad p' = \Phi_t^X(p).$$

This relation, taking account of (2.1), implies that

$$[X, J^M(Y)] = J^M([X,Y]). \tag{2.20}$$

Thus, if a real vector field X has the holomorphic flow, then for any real vector field Y relation (2.20) is valid. It can be shown that this condition is also sufficient, i.e., if relation (2.20) is valid for any real vector field Y, then the vector field X has a holomorphic flow.

A smooth vector field X on an almost complex manifold M, satisfying relation (2.20), is called an infinitesimal automorphism of the complex structure J^M. Suppose that the almost complex structure J^M is integrable, then for any infinitesimal automorphism X of J^M and any $Y \in \mathfrak{X}(M)$ we have

$$[J^M(X), J^M(Y)] = J^M([J^M(X), Y]).$$

Hence, the vector field $J^M(X)$ is also an infinitesimal automorphism of J^M. Moreover, the commutator of any two infinitesimal automorphisms of J^M is an infinitesimal automorphism of J^M. Thus, infinitesimal automorphisms of the complex structure form a Lie algebra. According to (2.20), the operator J^M is a Lie complex structure for this Lie algebra. Therefore, for any almost complex manifold M whose complex structure J^M is integrable, the Lie algebra of infinitesimal automorphisms of J^M can be considered as a complex Lie algebra. For the case of a complex manifold M, this Lie algebra is isomorphic to the Lie algebra of holomorphic vector fields on M. The corresponding isomorphism is realised in the following way. Let X be an infinitesimal automorphism of J^M; the corresponding holomorphic vector field is $X^{(1,0)}$. On the other hand, if X is a holomorphic vector field, then the real vector field $X + \overline{X}$ is the corresponding infinitesimal automorphism of J^M.

2.5.4 Complex covectors and covector fields

Let us begin with the case of a real manifold. In this case we can naturally define the action of a cotangent vector field on complex vector fields. Moreover, we can consider the complexification $T_p^{*\mathbb{C}}(M)$ of the cotangent space $T_p^*(M)$ and define complex covectors, complex covector fields, and the differential of a complex function. A complex covector field is called smooth if its action on any smooth complex vector field is a smooth function. The set of smooth complex covector fields on a manifold M is denoted $\mathfrak{X}^{*\mathbb{C}}(M)$. This set is a module over the algebra $\mathfrak{F}^{\mathbb{C}}(M)$.

In the case of a complex manifold M, we denote the set of real cotangent vectors to M at a point p by $T_{\mathbb{R}p}^*(M)$. This space is the dual of $T_{\mathbb{R}p}(M)$, and the operator J_p^M on $T_{\mathbb{R}p}(M)$ induces the dual operator $(J^M)_p^*$ on $T_{\mathbb{R}p}^*(M)$. The operator $(J^M)_p^*$ satisfies the relation $(J^M)_p^{*2} = -1$; hence, it is a complex structure on $T_{\mathbb{R}p}^*(M)$.

The cotangent space $T_p^*(M)$ to the complex manifold M at the point p is defined as

$$T_p^*(M) \equiv \widetilde{T_{\mathbb{R}p}^*(M)}.$$

The linear space $T_p^{*\mathbb{C}}(M)$, defined by

$$T_p^{*\mathbb{C}}(M) \equiv (T_{\mathbb{R}p}^*(M))^{\mathbb{C}},$$

has the direct sum decomposition

$$T_p^{*\mathbb{C}}(M) = T_{(1,0)p}(M) \oplus T_{(0,1)p}(M),$$

where the spaces $T_{(1,0)p}(M)$ and $T_{(0,1)p}(M)$ can be characterised as the eigenspaces of the operator $(J^M)_p^*$ corresponding to the eigenvalues $+\sqrt{-1}$ and $-\sqrt{-1}$ respectively. Let $(U, z^1, \ldots, z^m)$ be a chart on M and let $p \in U$. The set $\{(dz^i)_p, (d\bar{z}^i)_p\}$ is a basis of $T_p^{*\mathbb{C}}(M)$.

A complex covector field ω on M is said to be *of type* $(1,0)$ if $\omega_p \in T_{(1,0)p}(M)$ for any $p \in M$, and it is said to be *of type* $(0,1)$ if $\omega_p \in T_{(0,1)p}(M)$ for any $p \in M$. The set of smooth covector fields of type $(1,0)$ (of type $(0,1)$) on M is denoted by $\mathfrak{X}_{(1,0)}(M)$ ($\mathfrak{X}_{(0,1)}(M)$). For any covector field ω we have a unique decomposition

$$\omega = \omega^{(1,0)} + \omega^{(0,1)},$$

where $\omega^{(1,0)}$ and $\omega^{(0,1)}$ are covector fields of types $(1,0)$ and $(0,1)$, respectively. Here

$$\omega^{(1,0)}(X) = \omega\left(X^{(1,0)}\right), \qquad \omega^{(0,1)}(X) = \omega\left(X^{(0,1)}\right)$$

for any vector field X.

A complex covector field ω of type $(1,0)$ (of type $(0,1)$) is called *holomorphic* (*antiholomorphic*) if $\omega(X)$ is a holomorphic (antiholomorphic) function for any holomorphic (antiholomorphic) vector field X.

Let $(U, z^1, \ldots, z^m)$ be a complex chart on a complex manifold M. The differentials dz^i ($d\bar{z}^i$) are holomorphic (antiholomorphic) covector fields on U. For any complex covector field ω on M we have

$$\omega|_U = \omega_i dz^i + \omega_{\bar{\imath}} d\bar{z}^i.$$

Moreover, the covector fields dz^i and $d\bar{z}^i$, $i = 1, \ldots, m$, form the basis of $\mathfrak{F}^{\mathbb{C}}(U)$-module $\mathfrak{X}^{*\mathbb{C}}(U)$.

Using the relations

$$dz^i(\partial/\partial z^j) = \delta^i_j, \qquad dz^i(\partial/\partial \bar{z}^j) = 0;$$
$$d\bar{z}^i(\partial/\partial z^j) = 0, \qquad d\bar{z}^i(\partial/\partial \bar{z}^j) = \delta^{\bar{\imath}}_{\bar{\jmath}},$$

we see that

$$\omega_i = \omega|_U(\partial/\partial z^i) \qquad \omega_{\bar{\imath}} = \omega|_U(\partial/\partial \bar{z}^i).$$

The covector fields dz^i form a basis for $\mathfrak{X}_{(1,0)}(M)$, while the covector fields $d\bar{z}^i$ form a basis for $\mathfrak{X}_{(0,1)}(M)$. A complex covector field ω on M is holomorphic (antiholomorphic) if and only if $\omega^{\bar{\imath}} = 0$ ($\omega^i = 0$) and ω^i ($\omega^{\bar{\imath}}$) are holomorphic (antiholomorphic) functions for any complex chart on M.

2.5.5 Complex differential forms

Starting with the complexified tangent spaces $T^{\mathbb{C}}_p(M)$, we obtain complex tensor fields on M. Actually, a complex tensor field of type $\binom{k}{l}$ can be considered as an $\mathfrak{F}^{\mathbb{C}}(M)$-multilinear mapping from

$$\underbrace{\mathfrak{X}^{*\mathbb{C}}(M) \times \cdots \times \mathfrak{X}^{*\mathbb{C}}(M)}_{k} \times \underbrace{\mathfrak{X}^{\mathbb{C}}(M) \times \cdots \times \mathfrak{X}^{\mathbb{C}}(M)}_{l}$$

to $\mathfrak{F}^{\mathbb{C}}(M)$. The set of smooth complex tensor fields of type $\binom{k}{l}$ on M is denoted by $\mathfrak{X}^{k\mathbb{C}}_l(M)$.

We are especially interested in complex differential forms. Denote by $\Omega^{\mathbb{C}}_k(M)$ the subspace of $\mathfrak{X}^{0\mathbb{C}}_k(M)$ formed by all smooth totally skew-symmetric complex tensors on the complex manifold M.

Let ω be a complex 2-form. Using decomposition (2.19), we obtain the relation

$$\omega = \omega^{(2,0)} + \omega^{(1,1)} + \omega^{(0,2)}$$

where

$$\omega^{(2,0)}(X,Y) \equiv \omega\left(X^{(1,0)}, Y^{(1,0)}\right), \tag{2.21}$$

$$\omega^{(1,1)}(X,Y) \equiv \omega\left(X^{(1,0)}, Y^{(0,1)}\right) + \omega\left(X^{(0,1)}, Y^{(1,0)}\right), \tag{2.22}$$

$$\omega^{(0,2)}(X,Y) \equiv \omega\left(X^{(0,1)}, Y^{(0,1)}\right). \tag{2.23}$$

It can easily be shown that $\omega^{(2,0)}$, $\omega^{(1,1)}$ and $\omega^{(0,2)}$ are 2-forms.

For an arbitrary complex k-form ω we have

$$\omega = \sum_{p+q=k} \omega^{(p,q)}, \qquad (2.24)$$

where the k-forms $\omega^{(p,q)}$ are defined by the relations similar to (2.21)-(2.23). Here p and q are the numbers of the arguments of ω having types $(1,0)$ and $(0,1)$ respectively. The k-form ω is said to be *of type* (p,q) if in decomposition (2.24) only the component $\omega^{(p,q)}$ is different from zero. The set of all k-forms of type (p,q) forms a linear subspace of $\Omega_k^{\mathbb{C}}(M)$ which is denoted $\Omega_{(p,q)}(M)$ and there is the following direct sum decomposition

$$\Omega_k^{\mathbb{C}}(M) = \bigoplus_{p+q=k} \Omega_{(p,q)}(M).$$

Applying the operator of the external derivative d to a form of type (p,q), we obtain a sum of two differential forms, one form of type $(p+1,q)$ and the other one of type $(p,q+1)$. Define the operators $d^{(1,0)} : \Omega_{(p,q)}(M) \to \Omega_{(p+1,q)}(M)$ and $d^{(0,1)} : \Omega_{(p,q)}(M) \to \Omega_{(p,q+1)}(M)$ by

$$d^{(1,0)}\omega \equiv (d\omega)^{(p+1,q)}, \qquad d^{(0,1)}\omega \equiv (d\omega)^{(p,q+1)}.$$

From this definition it follows immediately that

$$d = d^{(1,0)} + d^{(0,1)}.$$

Furthermore, it can be shown that

$$d^{(1,0)} \circ d^{(1,0)} = 0, \qquad d^{(0,1)} \circ d^{(0,1)} = 0,$$
$$d^{(1,0)} \circ d^{(0,1)} + d^{(0,1)} \circ d^{(1,0)} = 0.$$

Let $(U, z^1, \dots, z^m)$ be a chart on M. For any k-form ω we can write

$$\omega|_U = \sum_{p+q=k} \frac{1}{p!\,q!} \omega^{(p,q)}_{i_1 \dots i_p \bar{\jmath}_1 \dots \bar{\jmath}_q} dz^{i_1} \wedge \dots \wedge dz^{i_p} \wedge d\bar{z}^{\bar{\jmath}_1} \wedge \dots \wedge d\bar{z}^{\bar{\jmath}_q}, \qquad (2.25)$$

and this expansion is unique if we suppose that the functions $\omega^{(p,q)}_{i_1 \dots i_p \bar{\jmath}_1 \dots \bar{\jmath}_q}$ are skew-symmetric with respect to any transposition either of the indices $i_1, \dots, i_p$, or of the indices $\bar{\jmath}_1, \dots, \bar{\jmath}_q$. In such a case we have

$$\omega^{(p,q)}_{i_1 \dots i_p \bar{\jmath}_1 \dots \bar{\jmath}_q} = \omega|_U(\partial/\partial z^{i_1}, \dots, \partial/\partial z^{i_p}, \partial/\partial \bar{z}^{\bar{\jmath}_1}, \dots, \partial/\partial \bar{z}^{\bar{\jmath}_q}).$$

For the action of the operators $d^{(1,0)}$ and $d^{(0,1)}$ on ω we obtain the following expressions:

$$(d^{(1,0)}\omega)|_U = \sum_{p+q=k} \frac{1}{p!\,q!} \frac{\partial \omega^{(p,q)}_{i_1 \dots \bar{\jmath}_q}}{\partial z^i} dz^i \wedge dz^{i_1} \wedge \dots \wedge dz^{\bar{\jmath}_q}, \quad (2.26)$$

$$(d^{(0,1)}\omega)|_U = \sum_{p+q=k} \frac{1}{p!\,q!} \frac{\partial \omega^{(p,q)}_{i_1 \dots \bar{\jmath}_q}}{\partial \bar{z}^{\bar{\jmath}}} d\bar{z}^{\bar{\jmath}} \wedge dz^{i_1} \wedge \dots \wedge dz^{\bar{\jmath}_q}. \quad (2.27)$$

A complex differential form ω of type $(p,0)$ $((0,q))$ is called *holomorphic* (*antiholomorphic*) if $d^{(0,1)}\omega = 0$ $(d^{(1,0)}\omega = 0)$. Let ω be a complex differential form of type $(p,0)$. In this case relation (2.25) takes the form

$$\omega|_U = \frac{1}{p!} \omega_{i_1 \dots i_p} dz^{i_1} \wedge \dots \wedge dz^{i_p}.$$

Now using (2.27), we conclude that the form ω is holomorphic if and only if the functions $\omega_{i_1 \dots i_p}$ are holomorphic for any choice of the chart $(U, z^1, \dots, z^m)$. For a differential form ω of type $(0,q)$ we obtain

$$\omega|_U = \frac{1}{q!} \omega_{\bar{\jmath}_1 \dots \bar{\jmath}_q} d\bar{z}^{\bar{\jmath}_1} \wedge \dots \wedge d\bar{z}^{\bar{\jmath}_q},$$

and relation (2.26) implies that the form ω is antiholomorphic if and only if the functions $\omega_{\bar{\jmath}_1 \dots \bar{\jmath}_q}$ are antiholomorphic for any choice of the chart $(U, z^1, \dots, z^m)$.

Exercises

2.9 Prove that the operators $\partial/\partial z^i$ and $\partial/\partial \bar{z}^i$ are $\mathbb{C}$-linear and satisfy the Leibnitz rule, i.e.,

$$\frac{\partial(f_1 f_2)}{\partial z^i} = \frac{\partial f_1}{\partial z^i} f_2 + f_1 \frac{\partial f_2}{\partial z^i},$$
$$\frac{\partial(f_1 f_2)}{\partial \bar{z}^i} = \frac{\partial f_1}{\partial \bar{z}^i} f_2 + f_1 \frac{\partial f_1}{\partial \bar{z}^i}$$

for any complex functions f_1 and f_2.

2.10 Prove that

$$\frac{\partial z^i}{\partial z^j} = \delta^i_j, \qquad \frac{\partial \bar{z}^i}{\partial \bar{z}^j} = \delta^i_j,$$
$$\frac{\partial \bar{z}^i}{\partial z^j} = 0, \qquad \frac{\partial z^i}{\partial \bar{z}^j} = 0.$$

2.11 Show that a complex function $f = u + \sqrt{-1}v$ is holomorphic if and only if

$$\frac{\partial u}{\partial x^i} - \frac{\partial v}{\partial y^i} = 0, \quad \frac{\partial u}{\partial y^i} + \frac{\partial v}{\partial x^i} = 0.$$

2.12 For any complex function $f = g + \sqrt{-1}h$ define the function $\bar{f}$ by $\bar{f} \equiv g - \sqrt{-1}h$. Show that

$$\overline{\frac{\partial f}{\partial z^i}} = \frac{\partial \bar{f}}{\partial \bar{z}^i}.$$

2.6 Submanifolds

2.6.1 Definition of a submanifold

A mapping $\varphi \in \mathfrak{F}(M, N)$ is called *immersive* at a point $p \in M$ if the mapping $\varphi_{*p} : T_p(M) \to T_{\varphi(p)}(M)$ is injective. A smooth mapping $\varphi : M \to N$ is called an *immersion* if it is immersive at any point of M. Let M and N be two manifolds such that $M \subset N$. The manifold M is called a *submanifold* of the manifold N if the inclusion mapping $\iota : M \to N$ is an immersion. It is clear that an open submanifold of a manifold is a submanifold. Since for any point p of a submanifold M of a manifold N, the mapping ι_{*p} is injective, we will identify the tangent space $T_p(M)$ with the subspace $\iota_{*p}(T_p(M))$ of the tangent space $T_{\iota(p)}(N) = T_p(N)$.

Now let M and N be complex manifolds. The manifold M is said to be a (*complex*) *submanifold* of the manifold N if the inclusion mapping $\iota : M \to N$ is a holomorphic immersion. From this definition we see, in particular, that for any point p of a submanifold M of a complex manifold N,

$$J_p^N(T_p(M)) = T_p(M).$$

There is an inverse statement. Let N be a complex manifold and let $\widetilde{M}$ be a submanifold of $N_{\mathbb{R}}$ such that for any $p \in \widetilde{M}$ the relation

$$J_p^N(T_p(\widetilde{M})) = T_p(\widetilde{M})$$

is valid. In this case there exists a complex submanifold M of N coinciding with $\widetilde{M}$ as a set. In other words, the set of all complex submanifolds of a complex manifold N coincides with the set of all submanifolds $\widetilde{M}$ of $N_{\mathbb{R}}$ such that for any $p \in \widetilde{M}$ the tangent space $T_p(\widetilde{M})$ is stable under the action of the operator J_p^N.

Let M be a manifold and let φ be an injective immersion from M to a manifold N. Denote by $\widetilde{\varphi}$ the mapping φ considered as a mapping from M to $\varphi(M)$. The mapping $\widetilde{\varphi}$ is bijective. Introduce a topology on $\varphi(M) = \widetilde{\varphi}(M)$ supposing that the subset $U \subset \widetilde{\varphi}(M)$ is open if the set $\widetilde{\varphi}^{-1}(U)$ is open. Further, endow $\varphi(M)$ with a differentiable structure with respect to which the mapping $\widetilde{\varphi}$ is a diffeomorphism. It can be shown that such a differentiable structure exists and it is unique. Denote the resulting manifold by $\widetilde{M}$. It is clear that $\varphi = \iota \circ \widetilde{\varphi}$, where $\iota : \varphi(M) \to N$ is the inclusion mapping. Since $\widetilde{\varphi}$ is a diffeomorphism, the mapping ι is an immersion, and $\widetilde{M}$ is a submanifold of N. Thus, any injective immersion from a manifold M to a manifold N defines a submanifold of N.

EXAMPLE 2.15 Define a mapping φ from the manifold $\mathbb{R}$ to the standard two-dimensional torus T^2 by

$$\varphi(a) = ((\cos \alpha a, \sin \alpha a), (\cos \beta a, \sin \beta a)),$$

where α and β are some nonzero real numbers. The mapping φ definitely is an immersion. Suppose that for some $a_1, a_2 \in \mathbb{R}$ such that $a_1 \neq a_2$ we have $\varphi(a_1) = \varphi(a_2)$. It is clear that it is possible only if $\alpha(a_2 - a_1) = 2k\pi$ and $\beta(a_2 - a_1) = 2l\pi$ for some integers k and l. In this case we have the equality $\alpha/\beta = k/l$ which implies that α/β is a rational number. Therefore, the mapping φ is injective if and only if α/β is an irrational number. In such a case $\varphi(\mathbb{R})$ is a submanifold of T^2 called an *irrational winding of the two-dimensional torus.*

Consider a submanifold M of a manifold N. The inclusion mapping $\iota : M \to N$ is a smooth mapping; hence, it is continuous. Therefore, for any subset U open in N, the subset $M \cap U = \iota^{-1}(U)$ is open in M, but, in general, not any open subset $V \subset M$ can be represented as $M \cap U$ for some U open in N. Thus, the topology of M is, in general, stronger than the induced topology. For example, an irrational winding of the two-dimensional torus has topology stronger than the induced topology. An abstract subset M of a manifold N may have different differentiable structures allowing one to consider it as a submanifold of N. Actually, a topology on M fixes its differentiable structure. Namely, if two submanifolds coincide as topological spaces, then they coincide as manifolds. In particular, if the topology of a submanifold M of a manifold N

coincides with the induced topology, then there is no other differentiable structures on the set M, endowed with the induced topology, which turn it into a submanifold of N.

A submanifold is called an *embedded submanifold* if its topology coincides with the induced topology. An injective immersion $\varphi : M \to N$ is called an *embedding* if it is a homeomorphism from M to $\varphi(M)$, where $\varphi(M)$ is considered as a topological space with respect to the induced topology. If $\varphi : M \to N$ is an embedding, then the corresponding submanifold $\widetilde{M}$ of the manifold N is an embedded submanifold.

A mapping $\varphi \in \mathfrak{F}(M, N)$ is called *submersive* at a point $p \in M$ if the mapping $\varphi_{*p} : T_p(M) \to T_{\varphi(p)}(N)$ is surjective. If a mapping $\varphi \in \mathfrak{F}(M, N)$ is submersive at any point of M, it is called a *submersion*. If a smooth mapping $\varphi : M \to N$ is submersive at a point $p \in M$, then the point p is called a *regular point* of the mapping φ. A point $q \in M$ is called a *regular value* of a smooth mapping $\varphi : M \to N$ if any point $p \in M$ such that $f(p) = q$ is a regular point of φ. If a point $q \in N$ is a regular value of a smooth mapping $\varphi : M \to N$, then $\varphi^{-1}(p)$ can be endowed with the structure of an embedded submanifold of M.

EXAMPLE 2.16 Let M and N be two manifolds. Consider the manifold $M \times N$ and define the projections $\mathrm{pr}_M : M \times N \to M$ and $\mathrm{pr}_N : M \times N \to N$ by

$$\mathrm{pr}_M(p, q) \equiv p, \qquad \mathrm{pr}_N(p, q) \equiv q.$$

Any point $p \in M$ is a regular value of the mapping pr_M; hence, the set $\mathrm{pr}_M^{-1}(p) = \{p\} \times N$ has the natural structure of an embedded submanifold of $M \times N$, which is certainly diffeomorphic to the manifold M. Similarly, for any $q \in N$ the set $\mathrm{pr}_N^{-1}(q) = M \times \{q\}$ is an embedded submanifold of $M \times N$ diffeomorphic to M.

Let us also show that for any point $(p, q) \in M \times N$ there is the following isomorphism

$$T_{(p,q)}(M \times N) \simeq T_p(M) \oplus T_q(N). \tag{2.28}$$

To this end, define the inclusions $\iota_q^M : M \to M \times N$ and $\iota_p^N : N \to M \times N$ as

$$\iota_q^M(p) \equiv (p, q), \qquad \iota_p^N(q) \equiv (p, q).$$

These mapping are immersions; so one has

$$\iota_{q*p}^M(T_p(M)) \simeq T_p(M), \qquad \iota_{p*q}^N(T_q(N)) \simeq T_q(N). \tag{2.29}$$

Further, the sum $\iota^M_{q*p}(T_p(M)) + \iota^N_{p*q}(T_p(N))$ is direct. Indeed, suppose that for some nonzero $u \in T_p(M)$ and $v \in T_q(n)$ we have

$$\iota^M_{q*p}(u) = \iota^N_{p*q}(v). \tag{2.30}$$

For any $f \in \mathfrak{F}(M)$ the function $\mathrm{pr}^*_M f$ belongs to $\mathfrak{F}(M \times N)$, and (2.30) implies

$$u(\iota^{M*}_q(\mathrm{pr}^*_M f)) = v(\iota^{N*}_p(\mathrm{pr}^*_M f)).$$

On the other hand, $\mathrm{pr}_M \circ \iota^M_q = \mathrm{id}_M$ and $\mathrm{pr}_M \circ \iota^N_p$ is a constant mapping. Therefore,

$$u(\iota^{M*}_q(\mathrm{pr}^*_M f)) = u(f), \quad v(\iota^{N*}_p(\mathrm{pr}^*_M f)) = 0,$$

that contradicts to our assumption. Since the dimension of $M \times N$ coincides with the sum of the dimensions of M and N, we obtain

$$T_{(p,q)}(M \times N) = \iota^M_{q*p}(T_p(M)) \oplus \iota^N_{p*q}(T_q(N)).$$

Now, taking (2.29) into account, we obtain (2.28).

Let N be an n-dimensional manifold and let $\{f^1, \ldots, f^k\}$, $k < n$, be a set of smooth functions on N. Consider a subset M of N defined as

$$M \equiv \{q \in N \mid f^1(q) = 0, \ldots, f^k(q) = 0\}, \tag{2.31}$$

and suppose that the differentials $df^1_p, \ldots, df^k_p$ are linearly independent at any point $p \in M$. Define the mapping $\varphi : N \to \mathbb{R}^k$ by

$$\varphi(q) \equiv (f^1(q), \ldots, f^k(q)),$$

then $M = \varphi^{-1}(0, \ldots, 0)$. According to our assumption, for any $p \in M$ the mapping φ^*_p is injective, therefore the mapping φ_{*p} is surjective. Hence $(0, \ldots, 0)$ is a regular value of φ, and M can be provided with the structure of an embedded submanifold of N. In such a case we say that M is an embedded submanifold of N, defined by the equations

$$f^1 = 0, \quad \ldots \quad , \quad f^k = 0.$$

The dimension of the manifold M is equal to $n - k$.

EXAMPLE 2.17 The standard sphere S^m can be considered as a submanifold of $\mathbb{R}^{m+1}$, defined by the equation

$$\sum_{i=1}^{m+1} (x^i)^2 - 1 = 0,$$

where x^i, $i = 1, \ldots, m+1$, are the standard coordinate functions on $\mathbb{R}^{m+1}$.

Now let N be an n-dimensional complex manifold and let $\{f^1, \ldots, f^k\}$, $k < n$, be a set of holomorphic functions on N. Define the set M by relation (2.31) and suppose that the complex differentials $df_p^1, \ldots, df_p^k$ are linearly independent at any point $p \in M$. In this case M can be endowed with the structure of a complex embedded submanifold of N of dimension $n - k$.

2.6.2 Distributions and the Frobenius theorem

Let M be an m-dimensional manifold and let n be an integer such that $1 \leq n < m$. An assignment $\mathcal{D}$ of an n-dimensional subspace $\mathcal{D}_p \subset T_p(M)$ to each point $p \in M$ is called an n-dimensional *distribution* on M. An n-dimensional distribution $\mathcal{D}$ on M is said to be *smooth* if for any point $p \in M$ there is an open subset U containing p, and the set of smooth vector fields $X_1, \ldots, X_n$ on U such that for any $q \in U$ the set $\{X_{iq}\}_{i=1,\ldots,n}$ is a basis for $\mathcal{D}_q$. We say that a vector field X *belongs* to a distribution $\mathcal{D}$ if $X_p \in \mathcal{D}_p$ for any $p \in M$. A smooth mapping φ from a manifold M to a manifold N is said to be tangent to a distribution $\mathcal{D}$ on N if $\varphi_{*p}(v) \in \mathcal{D}_{\varphi(p)}$ for any $p \in M$ and $v \in T_p(M)$.

A smooth distribution $\mathcal{D}$ is called *involutive* if for any smooth vector fields X and Y belonging to $\mathcal{D}$ the commutator $[X, Y]$ belongs to $\mathcal{D}$. A submanifold N of a manifold M is called an *integral manifold* of a distribution $\mathcal{D}$ on M if

$$T_p(N) = \mathcal{D}_p$$

for any $p \in N$. Suppose that $\mathcal{D}$ is a smooth distribution on a manifold M such that for any point $p \in M$ there is an integral manifold of $\mathcal{D}$ containing the point p. In this case $\mathcal{D}$ is an involutive distribution.

An integral manifold of a distribution $\mathcal{D}$ is said to be *maximal* if it is not a proper subset of any other connected integral manifold of $\mathcal{D}$. The *Frobenius theorem* states that for any n-dimensional smooth involutive distribution $\mathcal{D}$ on a manifold M and for any point $p \in M$ there exists a unique maximal n-dimensional integral manifold of $\mathcal{D}$ containing the point p.

In the case of a complex manifold M, a smooth distribution $\mathcal{D}$ on M is said to be *complex* if $J^M(\mathcal{D}_p) = \mathcal{D}_p$ for any $p \in M$.

Let a real vector space V be endowed with a complex structure J and let W be a linear subspace of V such that $J(W) = W$. In this case the restriction of J to W is a complex structure on W, and the complex vector space $\widetilde{W}$ can naturally be considered as a linear subspace of the complex vector space $\widetilde{V}$. Recall that in the case under consideration the vector space $V^{\mathbb{C}}$ has a direct sum decomposition $V^{\mathbb{C}} = V^{(0,1)} \oplus V^{(1,0)}$, where the subspaces $V^{(0,1)}$ and $V^{(1,0)}$ are naturally isomorphic to the vector spaces $\widetilde{V}$ and $\overline{\widetilde{V}}$ respectively. The corresponding isomorphism allows one to identify the subspace $\widetilde{W}$ of $\widetilde{V}$ with a subspace of $V^{(1,0)}$.

The above discussion shows that if $\mathcal{D}$ is a complex distribution on a complex manifold M, then for any $p \in M$ we can identify the subspace $\mathcal{D}_p$ with a subspace of $T_p^{(1,0)}(M)$ which will also be denoted by $\mathcal{D}_p$.

A complex distribution $\mathcal{D}$ on a complex manifold M is called *holomorphic* if for any point $p \in M$ there is an open neighbourhood U of p and a set of smooth holomorphic vector fields $X_1, \ldots, X_n$ on U such that for any $q \in U$ the set $\{X_{iq}\}_{i=1,\ldots,n}$ is a basis for $\mathcal{D}_q$. The maximal integral manifolds of an involutive holomorphic distribution are complex submanifolds.

2.7 Lie groups

2.7.1 Definition of a Lie group

A group G is called a *Lie group* if it is a manifold and the mapping

$$(a, b) \in G \times G \mapsto ab^{-1}$$

is smooth. From this definition it follows that for any Lie group G the mappings

$$(a, b) \in G \times G \mapsto ab \in G,$$
$$a \in G \mapsto a^{-1} \in G$$

are smooth.

EXAMPLE 2.18

(i) The set $\mathbb{R}^{\times} \equiv \mathbb{R} - \{0\}$ is an open submanifold of $\mathbb{R}$; it is a Lie group with respect to the multiplication of real numbers.

(ii) The vector space $\mathrm{Mat}(m, \mathbb{R})$ of all real $m \times m$ matrices can naturally be considered as a smooth manifold diffeomorphic to $\mathbb{R}^{m^2}$. Here the standard coordinate functions $g^i{}_j$, $i, j = 1, \ldots, n$, are defined by

$$g^i{}_j(a) \equiv a^i{}_j, \tag{2.32}$$

where $a = (a^i{}_j)$. Denote by $\mathrm{GL}(m, \mathbb{R})$ the set of all real nondegenerate $m \times m$ matrices. Considering $\mathrm{GL}(m, \mathbb{R})$ as an open submanifold of the manifold $\mathrm{Mat}(m, \mathbb{R})$, and as a group with respect to the matrix multiplication, we see that it is a Lie group. This Lie group is called the *real general linear group*.

(iii) Let V be an m-dimensional real vector space. Consider the set $\mathrm{GL}(V)$ of all nondegenerate linear operators on V. It is a group with respect to the product of linear operators. Fixing a basis for V, we associate with any nonsingular linear operator on V a real nondegenerate $m \times m$ matrix. Hence, one can introduce on $\mathrm{GL}(V)$ a smooth differentiable structure which obviously does not depend on the choice of a basis for V. With respect to this differentiable structure, the group $\mathrm{GL}(V)$ is a Lie group.

Let G and H be Lie groups. The direct product $G \times H$ considered as the direct product of the groups G and H and the direct product of the manifolds G and H is a Lie group which is called the *direct product* of the Lie groups G and H.

A group G is called a *complex Lie group* if it is a complex manifold, and the mapping

$$(a, b) \in G \times G \mapsto ab^{-1} \in G$$

is holomorphic. Since any complex manifold can be considered as a real smooth manifold, it is clear that any complex Lie group G is simultaneously a real Lie group which will be denoted by $G_{\mathbb{R}}$.

EXAMPLE 2.19

(i) The set $\mathbb{C}^\times \equiv \mathbb{C} - \{0\}$ is an open submanifold of $\mathbb{C}$; it is a Lie group with respect to the multiplication operation.

(ii) The set $\mathrm{Mat}(m, \mathbb{C})$ of all complex $m \times m$ matrices can be considered as a complex manifold diffeomorphic to $\mathbb{C}^{m^2}$ with the standard coordinate functions

$$g^i{}_j(c) \equiv c^i{}_j,$$

where $c = (c^i{}_j)$. The *complex general linear group* $\mathrm{GL}(m, \mathbb{C})$ is, by definition, the set of all complex nondegenerate $m \times m$ matrices considered as an open submanifold of the manifold of all complex $m \times m$ matrices, and as a group with respect to the matrix multiplication. It is clear that $\mathrm{GL}(m, \mathbb{C})$ is a complex Lie group.

(iii) The set of all nondegenerate linear operators on a complex vector space V is a complex Lie group isomorphic to the complex general linear group $\mathrm{GL}(m, \mathbb{C})$, where $m = \dim V$.

2.7.2 Lie algebra of a Lie group

With any element a of a Lie group G we can associate the following differentiable mappings of G onto itself:

left translation $L_a : b \in G \mapsto ab \in G$;

right translation $R_a : b \in G \mapsto ba \in G$.

For any $a, b \in G$ we have

$$L_a \circ L_b = L_{ab}, \qquad R_a \circ R_b = R_{ba}, \tag{2.33}$$

$$L_a \circ R_b = R_b \circ L_a. \tag{2.34}$$

From these equalities if follows, in particular, that

$$(L_a)^{-1} = L_{a^{-1}}, \qquad (R_a)^{-1} = R_{a^{-1}}.$$

Thus, the mappings L_a and R_a are diffeomorphisms.

A vector field X on a Lie group G is said to be *left invariant* if

$$L_{a*}X = X$$

for any $a \in G$. The left invariant vector fields form a subspace of the vector space of vector fields on G.

Denote, as usual, the identity element of the Lie group G by e. Let v be an arbitrary element of $T_e(G)$ and define the vector field X_v by

$$X_{va} \equiv L_{a*}(v). \tag{2.35}$$

The vector field X_v defined by (2.35) is left invariant. It is not difficult to show that the mapping $v \mapsto X_v$ from $T_e(G)$ to the vector space of left invariant vector fields on G is linear and injective. Let us show that any left invariant vector space on G can be obtained by the above procedure. Let X be a left invariant vector field on G, then $X_a = L_{a*}(X_e)$ and, hence, $X = X_v$ for $v = X_e \in T_e(G)$. In other words, if the mapping $v \mapsto X_v$ is surjective, then it is an isomorphism.

It can be shown that for any $v \in T_e(G)$, the vector field X_v is smooth; hence, any left invariant vector field on G is smooth. From (2.9) it follows that the commutator of two left invariant vector fields is left invariant. Therefore, the vector space of left invariant vector fields on G is a Lie algebra which is a subalgebra of the Lie algebra of smooth vector fields on G. This Lie algebra is called the *Lie algebra of the Lie group* G and is denoted by $\mathfrak{g}$.

The isomorphism of $T_e(G)$ and the space of left invariant fields on G becomes an isomorphism of Lie algebras if we introduce in $T_e(G)$ the structure of a Lie algebra by

$$[v, u] \equiv [X_v, X_u]_e. \tag{2.36}$$

Thus, we can identify the Lie algebra $\mathfrak{g}$ with $T_e(G)$. Actually, it is the interpretation of the Lie algebra of a Lie group which is used in the present book.

EXAMPLE 2.20 Consider the case of the Lie group $\mathrm{GL}(m, \mathbb{R})$. Using the standard coordinate functions on $\mathrm{GL}(m, \mathbb{R})$, defined by (2.32), we can write an arbitrary tangent vector v at a point $b \in \mathrm{GL}(m, \mathbb{R})$ in the form

$$v = v^i{}_j (\partial/\partial g^i{}_j)_b,$$

where

$$v^i{}_j = v(g^i{}_j). \tag{2.37}$$

For any $a \in \mathrm{GL}(m, \mathbb{R})$ we obtain

$$L_a^* g^i{}_j = a^i{}_k g^k{}_j. \tag{2.38}$$

By definition, for any $f \in \mathfrak{F}(G)$ we have

$$(L_{a*}(v))(f) = v(L_a^* f);$$

therefore, using (2.37) and (2.38), we obtain the relation

$$(L_{a*}(v))^i{}_j = a^i{}_k v^k{}_j.$$

From this relation we obtain the equality

$$(X_v)^i{}_j(a) = a^i{}_k v^k{}_j,$$

which implies

$$(X_v)^i{}_j = g^i{}_k v^k{}_j. \tag{2.39}$$

Recall that $\mathrm{GL}(m, \mathbb{R})$ is an open submanifold of the vector space $\mathrm{Mat}(m, \mathbb{R})$, provided with the standard differentiable structure. According to example 2.10, we can identify the tangent space $T_{I_m}(\mathrm{GL}(m,\mathbb{R}))$ with $\mathrm{Mat}(m,\mathbb{R})$. Here, a tangent vector $v = v^i{}_j(\partial/\partial g^i{}_j)_e$ is identified with the matrix $(v^i{}_j)$, which will also be denoted by v. Hence, for any matrix $v \in \mathrm{Mat}(m,\mathbb{R})$ there is defined a left invariant vector field on $\mathrm{GL}(m,\mathbb{R})$ which coincides with v at the identity element of $\mathrm{GL}(m,\mathbb{R})$. Using (2.37) and (2.39), we see that

$$X_v(X_u(g^i{}_j)) = X_v(g^i{}_k u^k{}_j) = g^i{}_l v^l{}_k u^k{}_j,$$

and from this relation we obtain

$$[X_v, X_u]^i{}_j(e) = v^i{}_k u^k{}_j - u^i{}_k v^k{}_j.$$

Therefore, the Lie algebra of the general linear group $\mathrm{GL}(m,\mathbb{R})$ can be identified with the Lie algebra $\mathfrak{gl}(m,\mathbb{R})$.

A *representation* of a Lie group G in a real or complex vector space V is defined as a smooth homomorphism from G to the Lie group $\mathrm{GL}(V)$ or $\mathrm{GL}(V)_\mathbb{R}$ respectively. In the former case we call the corresponding representation *real*, while in the latter case we say that the representation is *complex*. The adjoint representation $\mathrm{Ad} : a \in G \mapsto \mathrm{Ad}(a) \in \mathrm{GL}(\mathfrak{g})$ of the Lie group G is defined by the relation

$$\mathrm{Ad}(a)v \equiv (L_a \circ R_{a^{-1}})_{*e}(v).$$

EXAMPLE 2.21 Continue the consideration of the Lie group $\mathrm{GL}(m,\mathbb{R})$, that we started in example 2.20. Using the argument similar to those given there for left translations, for any $a \in \mathrm{GL}(m,\mathbb{R})$ and $v \in T_b(\mathrm{GL}(m,\mathbb{R}))$ we have the relation

$$(R_{a*}(v))^i{}_j = v^i{}_k a^k{}_j,$$

where $v^i{}_j$ are the coordinates of v with respect to the basis $(\partial/\partial g^i{}_j)_b$. From this relation we immediately obtain

$$\mathrm{Ad}(a)v = ava^{-1}. \tag{2.40}$$

This is the well-known relation for the adjoint representation of matrix Lie groups.

Any left invariant vector field on a Lie group G is complete. Therefore, for any $v \in \mathfrak{g}$ and any $t \in \mathbb{R}$ there is the diffeomorphism $\Phi_t^{X_v} : G \to G$. Define the *exponential mapping* from $\mathfrak{g}$ to G by

$$\exp_G : v \in \mathfrak{g} \mapsto \exp(v) \equiv \Phi_1^{X_v}(e) \in G.$$

When it is clear which Lie group is being considered, we write simply exp for $\exp_G$. The exponential mapping has the properties

$$\exp(tv)\exp(sv) = \exp((t+s)v),$$
$$(\exp(v))^{-1} = \exp(-v)$$

for all $t, s \in \mathbb{R}$ and $v \in \mathfrak{g}$. Note also that

$$a\exp(v)a^{-1} = \exp(\mathrm{Ad}(a)v) \tag{2.41}$$

for any $a \in G$ and $v \in \mathfrak{g}$. Note that the exponential mapping allows one to establish the relation between representations of the group G and representations of the Lie algebra $\mathfrak{g}$. Indeed, let ρ_G be a representation of the Lie group G in a vector space V. It can be shown that the mapping $\rho_\mathfrak{g}$ defined by

$$\rho_\mathfrak{g}(x)v \equiv \left.\frac{d}{dt}\rho_G(\exp(tx))v\right|_{t=0},$$

is a representation of the Lie algebra $\mathfrak{g}$ in V. In particular, we have the following relation between the adjoint representations of a Lie group and the corresponding Lie algebra:

$$\mathrm{ad}(v)u = \left.\frac{d}{dt}\mathrm{Ad}(\exp(tv))u\right|_{t=0}. \tag{2.42}$$

EXAMPLE 2.22 Consider the Lie group $\mathrm{GL}(m,\mathbb{R})$. Recall the definition of the exponential function of the matrix argument. The exponential function $a \in \mathrm{Mat}(m,\mathbb{R}) \mapsto e^a \in \mathrm{Mat}(m,\mathbb{R})$ is defined by

$$e^a \equiv \sum_{k=0}^{\infty}\frac{a^k}{k!}.$$

The series in this definition is absolutely convergent with respect to any norm in $\mathrm{Mat}(m,\mathbb{R})$. It can be shown that for any $a \in \mathrm{Mat}(m,\mathbb{R})$ the function $t \in \mathbb{R} \mapsto e^{ta} \in \mathrm{Mat}(m,\mathbb{R})$ is infinitely differentiable and satisfies the differential equation

$$\frac{de^{ta}}{dt} = e^{ta}a. \tag{2.43}$$

Let λ be an integral curve of the left invariant vector field X_v, where v is an arbitrary element of the Lie algebra $\mathfrak{gl}(m,\mathbb{R})$. As follows from the definition of an integral curve and from (2.39), the matrix valued function $t \in \mathbb{R} \mapsto \lambda(t) \in \mathrm{GL}(m, \mathbb{R})$ satisfies the equation

$$\frac{d\lambda(t)}{dt} = \lambda(t)v.$$

In accordance with (2.43), the general solution of this equation is

$$\lambda(t) = ae^{tv},$$

where a is an arbitrary element of $\mathrm{GL}(m, \mathbb{R})$. Therefore, we have

$$\Phi_t^{X_v}(a) = ae^{tv},$$

hence,

$$\exp(v) = e^v. \tag{2.44}$$

In other words, in the case under consideration the exponential mapping coincides with the exponential function. Note that relation (2.41) takes the form

$$ae^v a^{-1} = e^{ava^{-1}}.$$

Proceed now to the case of complex Lie groups. It is natural to require that the Lie algebra of a complex Lie group G be a complex Lie algebra $\mathfrak{g}$. Moreover, it is desirable that the Lie algebra of the real Lie group $G_{\mathbb{R}}$ be the realification of $\mathfrak{g}$. So, let us denote the Lie algebra of $G_{\mathbb{R}}$ by $\mathfrak{g}_{\mathbb{R}}$ and try to find the corresponding complex Lie algebra $\mathfrak{g}$. Since the group operation in a complex Lie group G is holomorphic, we have for all $a \in G$ the following relations:

$$L_{a*} \circ J^G = J^G \circ L_{a*}, \tag{2.45}$$

$$R_{a*} \circ J^G = J^G \circ R_{a*}. \tag{2.46}$$

The restriction $J \equiv J_e^G$ of the complex structure J^G to $T_{\mathbb{R}e}(G) = T_e(G_{\mathbb{R}})$ generates a complex structure on the Lie algebra $\mathfrak{g}_{\mathbb{R}}$. It follows from (2.45) and (2.46) that

$$\mathrm{Ad}(a) \circ J = J \circ \mathrm{Ad}(a)$$

for any $a \in G$. From this relation and from (2.42) one sees that

$$\mathrm{ad}(v) \circ J = J \circ \mathrm{ad}(v)$$

for any $v \in \mathfrak{g}_{\mathbb{R}}$. This equality can be written in the form

$$[v, Ju] = J([v, u]).$$

Hence, J is a Lie complex structure on $\mathfrak{g}_{\mathbb{R}}$, and $\mathfrak{g}_{\mathbb{R}}$ has the structure of a complex Lie algebra $\mathfrak{g} \equiv \widetilde{\mathfrak{g}_{\mathbb{R}}}$. It is clear that the Lie algebra $\mathfrak{g}_{\mathbb{R}}$ is the realification of the Lie algebra $\mathfrak{g}$. Therefore, $\mathfrak{g}$ is the required complex Lie algebra. It is the Lie algebra which we call the (complex) Lie algebra of a (complex) Lie group G. There exists another, more convenient, realisation of the Lie algebra of a complex Lie group.

Consider the complex Lie algebra $(\mathfrak{g}_{\mathbb{R}})^{\mathbb{C}} = (T_{\mathbb{R}e}(G))^{\mathbb{C}} = T_e^{\mathbb{C}}(G)$. Write the direct sum decomposition (2.17) in the form

$$(\mathfrak{g}_{\mathbb{R}})^{\mathbb{C}} = \mathfrak{g}^{(1,0)} \oplus \mathfrak{g}^{(1,0)},$$

where $\mathfrak{g}^{(1,0)} \equiv T_e^{(1,0)}(G)$ and $\mathfrak{g}^{(0,1)} = T_e^{(0,1)}(G)$. In fact, $\mathfrak{g}_{(1,0)}$ and $\mathfrak{g}^{(0,1)}$ are ideals of $(\mathfrak{g}_{\mathbb{R}})^{\mathbb{C}}$, such that the former is isomorphic to $\mathfrak{g}$, while the latter is isomorphic to $\bar{\mathfrak{g}}$. Here $\bar{\mathfrak{g}}$ is the Lie algebra which is obtained from $\mathfrak{g}_{\mathbb{R}}$ with the help of the Lie complex structure $-J$.

Thus, we can identify the Lie algebra of a complex Lie group G with the space $T_e^{(1,0)}(G)$. Here any element of $T_e^{(1,0)}(G)$ generates a left invariant complex vector field of type (1,0) on G, given by the relation of form (2.35); and the Lie algebra operation in $\mathfrak{g}$ is related to the commutator of the corresponding left invariant vector fields by (2.36). Note that any left invariant vector field on G of type (1,0) is a holomorphic vector field.

EXAMPLE 2.23 Repeating the arguments used in example 2.20, we see that the Lie algebra of the complex Lie group $\mathrm{GL}(m, \mathbb{C})$ can be identified with the Lie algebra $\mathfrak{gl}(m, \mathbb{C})$. Here the element of $T_{I_m}^{(1,0)}(\mathrm{GL}(m, \mathbb{C}))$ corresponding to the element $(v^i{}_j)$ of $\mathfrak{gl}(m, \mathbb{C})$ is $v^i{}_j(\partial/\partial g^i{}_j)_{I_m}$. The adjoint representation and the exponential mapping are described by relations (2.40) and (2.44).

Let G be a real Lie group. Suppose that the Lie algebra $\mathfrak{g}$ has a Lie complex structure J, so that we can define the complex Lie algebra $\tilde{\mathfrak{g}}$. Define the almost complex structure on G by

$$J_a^G(v) \equiv J(L_{a^{-1}*}(v)), \qquad v \in T_a(G).$$

This almost complex structure is integrable. Hence, it is possible to provide G with the structure of a complex manifold $\widetilde{G}$. It can be shown that the group operations in $\widetilde{G}$ are holomorphic and, therefore, that $\widetilde{G}$ has the structure of a complex Lie group.

2.7.3 Lie subgroups

First consider the case of real Lie groups. A subgroup H of a Lie group G is called a *Lie subgroup* of G if it is a Lie group and a submanifold of G. In other words, a subgroup H of a Lie group G is called a Lie subgroup of G if it is endowed with a smooth differentiable structure such that the group operation in H is smooth and the inclusion mapping of H into G is an immersion.

Let H be a Lie subgroup of a Lie group G and let ι be the inclusion mapping of H into G. It can be shown that $\iota_*(\mathfrak{h})$ is a subalgebra of the Lie algebra $\mathfrak{g}$ which is isomorphic to $\mathfrak{h}$, and it can be proved that for any subalgebra $\tilde{\mathfrak{h}}$ of the Lie algebra $\mathfrak{g}$ there exists a unique connected Lie subgroup H of G such that $\iota_*(\mathfrak{h}) = \tilde{\mathfrak{h}}$, with ι being the inclusion mapping of H into G. In other words, there is a bijective correspondence between the Lie subgroups of a Lie group and the subalgebras of its Lie algebra. It is customary to identify the Lie algebra $\mathfrak{h}$ of the Lie subgroup H of a Lie group G with the corresponding subalgebra $\iota_*(\mathfrak{h})$ of the Lie algebra $\mathfrak{g}$. In particular, we say that $\mathfrak{h}$ is a subalgebra of $\mathfrak{g}$.

Not every subgroup of a Lie group can be considered as a Lie subgroup. There are a few useful criteria for a subgroup of a Lie group to be a Lie subgroup. Let us discuss them briefly.

Recall that a subset of a manifold may have different differentiable structures turning it into a submanifold. This is not the case for subgroups of a Lie group. If a subgroup H of a Lie group G has the structure of a submanifold of G, then this structure is unique, and H, with respect to this structure, is a Lie subgroup of G. From this fact it follows that a subgroup of a Lie group is a Lie subgroup if and only if it can be provided with the structure of a submanifold; and the statement that a subgroup of a Lie group is a Lie subgroup is not ambiguous. In other words, we do not need to explain which differentiable structure we have in mind. The corresponding differentiable structure, if it exists, is unique.

Any open subgroup H of a Lie group G can be considered as an open submanifold of G; therefore, it is a Lie subgroup of G. Moreover, any closed subgroup H of a Lie group G is a Lie subgroup of G. Note that any open subgroup of a Lie group is also its closed subgroup.

It can also be shown that a closed Lie subgroup H of a Lie group

G is an embedded submanifold of G. Therefore, the topology of a Lie subgroup H of a Lie group G coincides with the induced topology if and only if H is a closed subgroup.

Proceed now to the case of complex Lie groups. It is natural to call a subgroup H of a complex Lie group G a Lie subgroup of G if H is a complex Lie group and a submanifold of G. Let G be a complex Lie group and let H be a Lie subgroup of a real Lie group $G_{\mathbb{R}}$ such that $J(\mathfrak{h}) = \mathfrak{h}$. The restriction of the natural Lie complex structure J on $\mathfrak{g}_{\mathbb{R}}$ to $\mathfrak{h}$ gives a Lie complex structure on $\mathfrak{h}$. Hence, the Lie group H has the natural structure of a complex Lie group $\widetilde{H}$. It can be shown that the inclusion mapping of $\widetilde{H}$ into G is holomorphic. Therefore, the Lie group $\widetilde{H}$ is a Lie subgroup of G. The complex Lie algebra $\widetilde{\mathfrak{h}}$ can be considered as a subalgebra of the complex Lie algebra $\mathfrak{g}$. On the other hand, any subalgebra of $\mathfrak{g}$, considered as a subalgebra of $\mathfrak{g}_{\mathbb{R}}$, is invariant with respect to the action of the operator J. Thus, we have a bijective correspondence between the Lie subgroups of a complex Lie group and the subalgebras of its Lie algebra.

Let H be a Lie subgroup of a real or complex Lie group G. An element $v \in \mathfrak{g}$ belongs to $\mathfrak{h}$ if and only if $\exp_G(tv) \in H$ for all $t \in \mathbb{R}$.

A Lie subgroup H of a Lie group G is an invariant subgroup of G if and only if the Lie algebra $\mathfrak{h}$ is an ideal of the Lie algebra $\mathfrak{g}$.

EXAMPLE 2.24 The set of elements a of the Lie group $\mathrm{GL}(m, \mathbb{R})$, satisfying the requirement $\det a = 1$, is a closed subgroup of $\mathrm{GL}(m, \mathbb{R})$. Hence, this subgroup has the natural structure of a Lie group denoted by $\mathrm{SL}(m, \mathbb{R})$ and is called the *real special linear group*. The *complex special linear group* $\mathrm{SL}(m, \mathbb{C})$ is defined similarly. Using the well-known relation

$$\det e^a = e^{\operatorname{tr} a}, \qquad a \in \mathrm{Mat}(m, \mathbb{K}),$$

we conclude that the Lie algebra of $\mathrm{SL}(m, \mathbb{K})$ is the special linear algebra $\mathfrak{sl}(m, \mathbb{K})$.

EXAMPLE 2.25 Let B be a bilinear form on an m-dimensional vector space V over a field $\mathbb{K}$. Denote by $\mathrm{GL}_B(V)$ the subset of $\mathrm{GL}(V)$ formed by the nondegenerate linear operators A satisfying the relation

$$B(Av, Au) = B(v, u)$$

for all $v, u \in V$. It can easily be shown that $\mathrm{GL}_B(V)$ is a closed subgroup of $\mathrm{GL}(V)$. Hence, it has the natural structure of a Lie algebra, and the Lie algebra of $\mathrm{GL}_B(V)$ coincides with the Lie algebra $\mathfrak{gl}_B(V)$ defined in example 1.8.

Let $\{e_i\}$ be a basis of V. It is clear that $A \in \mathrm{GL}_B(V)$ if and only if the matrix a of A with respect to $\{e_i\}$ satisfies the relation

$$a^t b a = b, \tag{2.47}$$

where b is the matrix of the bilinear form B with respect to $\{e_i\}$. This relation gives the isomorphism of $\mathrm{GL}_B(V)$ with some subgroup of the group $\mathrm{GL}(m, \mathbb{K})$.

Consideration of nondegenerate bilinear forms having definite symmetry leads to the following Lie subgroups of $\mathrm{GL}(m, \mathbb{K})$. Note that for a nondegenerate bilinear form B it follows from (2.47) that $(\det a)^2 = 1$. Thus, either $\det a = +1$, or $\det a = -1$.

The *pseudo-orthogonal* group $\mathrm{O}(k, l)$ is defined as the subgroup formed by the elements $a \in \mathrm{GL}(m, \mathbb{R})$, $m = k + l$, satisfying relation (2.47), with $b = I_{k,l}$, where $I_{k,l}$ is defined by (1.5). The Lie groups $\mathrm{O}(k, l)$ and $\mathrm{O}(l, k)$ are obviously isomorphic. For the Lie group $\mathrm{O}(m, 0)$ we use the notation $\mathrm{O}(m)$. The Lie group $\mathrm{O}(m)$ is called the *real orthogonal* group. The condition $\det a = +1$ singles out the Lie subgroup of $\mathrm{O}(k, l)$ called the *special pseudo-orthogonal* group and denoted $\mathrm{SO}(k, l)$. The Lie group $\mathrm{SO}(m, 0)$ is called the *real special orthogonal* group and is denoted by $\mathrm{SO}(m)$. The Lie algebra of $\mathrm{O}(k, l)$ and $\mathrm{SO}(k, l)$ is the pseudo-orthogonal Lie algebra $\mathfrak{o}(k, l)$.

The *complex orthogonal* group $\mathrm{O}(m, \mathbb{C})$ is, by definition, a Lie subgroup of $\mathrm{GL}(m, \mathbb{C})$ specified by condition (2.47) with $b = I_m$. The *complex special orthogonal* group is a Lie subgroup of $\mathrm{O}(m, \mathbb{C})$ singled out by the condition $\det a = +1$. This group is denoted by $\mathrm{SO}(m, \mathbb{C})$. The Lie algebra of the Lie groups $\mathrm{O}(m, \mathbb{C})$ and $\mathrm{SO}(m, \mathbb{C})$ is the complex orthogonal algebra $\mathfrak{o}(m, \mathbb{C})$.

The *real symplectic* group $\mathrm{Sp}(n, \mathbb{R})$ is defined as the Lie subgroup of the Lie group $\mathrm{GL}(2n, \mathbb{R})$ consisting of the elements which satisfy relation (2.47) with $b = J_n$. Here the matrix J_n is the matrix defined by (1.6). The Lie algebra of $\mathrm{Sp}(n, \mathbb{R})$ is the real symplectic algebra $\mathfrak{sp}(n, \mathbb{R})$. The *complex symplectic* group $\mathrm{Sp}(n, \mathbb{C})$ having $\mathfrak{sp}(n, \mathbb{C})$ as its Lie algebra is defined similarly. Note that for any $a \in \mathrm{Sp}(n, \mathbb{K})$ we have $\det a = +1$.

EXAMPLE 2.26 Let V be an m-dimensional complex vector space endowed with a sesquilinear form B. Consider the subset $\mathrm{GL}_B(V)$ of the Lie group $\mathrm{GL}(V)$ formed by the elements A satisfying the relation

$$B(Av, Au) = (v, u) \tag{2.48}$$

for all $v, u \in V$. This subset, considered as a subset of the real Lie group $\mathrm{GL}(V)_\mathbb{R}$, is a Lie subgroup. Let $\{e_i\}$ be a basis of V and let b be a matrix of B with respect to $\{e_i\}$. It is obvious that relation (2.48) is equivalent to the matrix relation

$$a^\dagger b a = b, \tag{2.49}$$

where a is the matrix of A with respect to $\{e_i\}$. This relation defines a subalgebra of $\mathrm{GL}(m, \mathbb{C})_\mathbb{R}$ which is isomorphic to $\mathrm{GL}_B(V)$.

Consider the case of a nondegenerate hermitian form. In this case relation (2.48) gives a Lie group isomorphic to the *unitary* group $\mathrm{U}(m)$ defined by (2.49) with $b = I_m$. From (2.49) we also obtain $|\det a|^2 = 1$. The Lie subgroup of $\mathrm{U}(m)$ specified by the condition $\det a = 1$ is called the *special unitary* group and is denoted by $\mathrm{SU}(m)$.

2.7.4 Maurer–Cartan form of a Lie group

Let M be a manifold and let V be a real vector space. A mapping f from M to V is called a function on M taking values in V, or a vector valued function on M. A function on M taking values in V is said to be smooth if it is smooth as a mapping from the manifold M to the vector space V, that is provided with the standard differentiable structure. Let $\{e_r\}$ be a basis for the vector space V. For any point $p \in M$ we can write $f(p) = e_r f^r(p)$, thus defining a set $\{f^r\}$ of functions on M. In this case we write $f = e_r f^r$. The vector valued function f is smooth if and only if all the functions f^r are smooth. For any vector field X on M we define the action of X on a vector valued function f as

$$X(f) \equiv e_r X(f^r).$$

Let M be a smooth manifold; a totally skew-symmetric $\mathfrak{F}(M)$-linear mapping from $\underbrace{\mathfrak{X}(M) \times \cdots \times \mathfrak{X}(M)}_{k}$ to the space of functions on M taking values in a real vector space V is called a k-form on M taking values in V, or a vector valued k-form on M.

Let ω be a k-form on a manifold M taking values in a real vector space V. Writing the relation

$$\omega(X_1, \ldots, X_k) \equiv e_r \omega^r(X_1, \ldots, X_k),$$

where $\{e_r\}$ is a basis for V, we define a set $\{\omega^r\}$ of ordinary k-forms on M. In such a situation we write

$$\omega = e_r \omega^r.$$

The exterior derivative of a k-form ω taking values in V is defined as

$$d\omega \equiv e_r d\omega^r,$$

and its behaviour under a smooth mapping φ from M to a manifold N is governed by the relation

$$\varphi^* \omega \equiv e_r \varphi^* \omega^r.$$

Here we have

$$d\varphi^* \omega = \varphi^* d\omega. \tag{2.50}$$

It is easy to show that the above definitions do not depend on the choice of a basis for V.

It is interesting to consider the case of the forms taking values not just in a vector space but also in a Lie algebra. In this case we can define the commutator of forms in the following way. Let ω and η be, respectively, a k-form and an l-form on a manifold M, taking values in a real Lie algebra $\mathfrak{g}$. The commutator of ω and η is defined by

$$[\omega, \eta] = [e_r \omega^r, e_s \eta^s] \equiv [e_r, e_s] \omega^r \wedge \eta^s,$$

where $\{e_r\}$ is a basis for $\mathfrak{g}$. It is clear that this definition does not depend on the choice of a basis for $\mathfrak{g}$, and

$$[\omega, \eta] = -(-1)^{kl} [\eta, \omega],$$
$$d[\omega, \eta] = [d\omega, \eta] + (-1)^k [\omega, d\eta].$$

Furthermore, for any smooth mapping φ from M to a manifold N we have

$$\varphi^* [\omega, \eta] = [\varphi^* \omega, \varphi^* \eta]. \tag{2.51}$$

When ω and η are 1-forms, we obtain

$$[\omega, \eta](X, Y) = [\omega(X), \eta(Y)] - [\omega(Y), \eta(X)]. \tag{2.52}$$

The *Maurer–Cartan form* θ of a Lie group G is a 1-form on G, taking values in $\mathfrak{g}$, and defined by the relation

$$\theta(v) \equiv L_{a^{-1}*a}(v), \qquad v \in T_a(G). \tag{2.53}$$

Let us show that the form θ is left invariant. Indeed, for any $v \in T_b(G)$ and $a \in G$ we have

$$(L_a^*\theta)(v) = \theta(L_{a*b}(v)) = L_{(ab)^{-1}*ab}(L_{a*b}(v)) = \theta(v).$$

Therefore,

$$L_a^*\theta = \theta \tag{2.54}$$

for any $a \in \mathfrak{g}$. The behaviour of θ under the right translations is described by the formula

$$R_a^*\theta = \mathrm{Ad}(a^{-1}) \circ \theta. \tag{2.55}$$

Further, for any $v \in \mathfrak{g}$ we have

$$\theta(X_v) = v, \tag{2.56}$$

where v on the right-hand side is understood as a $\mathfrak{g}$-valued function on G taking the value v at any point of G. Note that equality (2.56) can be taken as an alternative definition of the Maurer–Cartan form.

It can be shown that relations (2.14) and (2.15) are valid for the forms taking values in a vector space. In particular, from (2.15) for the Maurer–Cartan form θ of a Lie group G we obtain

$$d\theta(X_v, X_u) = -\theta(X_{[v,u]}) = -[v, u].$$

On the other hand, using (2.52), we arrive at the equality

$$[\theta, \theta](X_v, X_u) = 2[\theta(X_v), \theta(X_u)] = 2[v, u].$$

It is not difficult to show that these equalities imply

$$d\theta + \frac{1}{2}[\theta, \theta] = 0. \tag{2.57}$$

Another interesting case is the case of matrix valued differential forms. Let ω be a k-form taking values in the vector space $\mathrm{Mat}(n, \mathbb{R})$. Such a differential form is called a *matrix valued differential form*. We can associate with ω the set $\{\omega^r{}_s\}_{r,s=1,\ldots,n}$ consisting of n^2 ordinary k-forms defined by

$$\omega^r{}_s(X_1, \ldots, X_k) \equiv (\omega(X_1, \ldots, X_k))^r{}_s. \tag{2.58}$$

On the other hand, reversing equality (2.58), we associate with a set $\{\omega^r{}_s\}_{r,s=1,\ldots,n}$ of ordinary k-forms a matrix valued k-form.

The exterior derivative of a matrix valued k-form ω is defined by the relation

$$(d\omega)^r{}_s \equiv d(\omega^r{}_s),$$

while the exterior product of two matrix valued differential forms ω and η is given by

$$(\omega \wedge \eta)^r{}_s \equiv \omega^r{}_t \wedge \eta^t{}_s.$$

If we consider $\mathrm{Mat}(n, \mathbb{R})$ as a Lie algebra, that is, $\mathfrak{gl}(n, \mathbb{R})$, we can write

$$[\omega, \eta] = 2\omega \wedge \eta. \tag{2.59}$$

EXAMPLE 2.27 Consider the general linear group $\mathrm{GL}(m, \mathbb{R})$. The standard coordinate functions $g^i{}_j$ define a matrix valued function on $\mathrm{GL}(n, \mathbb{R})$ which is denoted by g. Following example 2.20, we identify the Lie algebra of $\mathrm{GL}(m, \mathbb{R})$ with the Lie algebra $\mathfrak{gl}(m, \mathbb{R})$. Therefore, the Maurer–Cartan form in the case in question can be considered as a matrix valued 1-form. Using relation (2.39), for any $v \in \mathfrak{gl}(m, \mathbb{R})$ we obtain

$$dg(X_v) = gv.$$

Define the matrix valued function g^{-1} on $\mathrm{GL}(m, \mathbb{R})$ by

$$g^{-1}(a) \equiv a^{-1}.$$

Then, we can write

$$g^{-1}dg(X_v) = v.$$

Comparing this equality with the alternative definition of the Maurer–Cartan form (2.56), we conclude that in our case

$$\theta = g^{-1}dg.$$

Taking the exterior derivative of the obvious equality

$$g^{-1}g = I_m,$$

we arrive at the relation

$$dg^{-1} = -g^{-1}(dg)g^{-1}.$$

Using this relation, we obtain

$$d\theta = dg^{-1} \wedge dg = -g^{-1}dg \wedge g^{-1}dg.$$

Hence, we have the equality

$$d\theta + \theta \wedge \theta = 0,$$

which, by virtue of (2.59), coincides with (2.57).

Now let G be a complex Lie group. The real Lie group $G_{\mathbb{R}}$ has the Maurer–Cartan form $\theta_{\mathbb{R}}$ taking values in $\mathfrak{g}_{\mathbb{R}}$. Note that

$$\theta_{\mathbb{R}}(J^G(X)) = J^G_e \circ \theta_{\mathbb{R}}(X) \tag{2.60}$$

for any vector field X on $G_{\mathbb{R}}$. The Maurer–Cartan form θ of the complex Lie group G is defined by

$$\theta(X) \equiv (\theta_{\mathbb{R}}(X))^{(1,0)}.$$

It follows from this definition that θ takes values in the Lie algebra $\mathfrak{g}$ of G. Using (2.60), we obtain

$$\theta(X) = \theta_{\mathbb{R}}(X^{(1,0)});$$

hence, θ is a 1-form of type $(1,0)$. Moreover, it can be shown that it is a holomorphic 1-form.

Actually, we can define θ as a unique 1-form of type $(1,0)$ satisfying (2.56), where X_v is the left invariant vector field on G corresponding to the element $v \in \mathfrak{g} = T^{(1,0)}_e(G)$. It can be shown that in the case under consideration we still have relations (2.54), (2.55) and (2.57) while, instead of (2.53), we obtain

$$\theta(v) = L_{a^{-1}*a}(P^G_a v), \tag{2.61}$$

where $v \in T^{\mathbb{C}}_a(G)$ and the linear operator P^G_a projects v to its (1,0)-component; see (2.18).

EXAMPLE 2.28 Repeating the arguments of example 2.27, we conclude that the Maurer–Cartan form of the Lie group $\mathrm{GL}(m, \mathbb{C})$ is given by

$$\theta = g^{-1}dg, \tag{2.62}$$

where g is the matrix valued function formed by the standard coordinate functions $g^i{}_j$ on $\mathrm{GL}(m, \mathbb{C})$.

Suppose now that H is a Lie subgroup of a real or complex Lie group G. Let θ_G be a Maurer–Cartan form of G and let ι be the inclusion mapping of H into G. It is not difficult to show that the 1-form $\iota^*\theta_G$ takes values in the Lie algebra $\mathfrak{h}$ of H and, moreover, that this form coincides with the Maurer–Cartan form of H.

EXAMPLE 2.29 Let G be a Lie subgroup of the general linear group $\mathrm{GL}(m, \mathbb{K})$. It is clear that the Maurer–Cartan form of G is given by relation (2.62), where g is now the matrix valued function on G formed by the restrictions of the standard coordinate functions on $\mathrm{GL}(m, \mathbb{K})$ to G.

2.7.5 Lie transformation groups

Let G be a Lie group and M a manifold. A smooth mapping $\Phi : M \times G \to M$, satisfying the conditions

(RA1) $\Phi(p, e) = p$ for all $p \in M$,

(RA2) $\Phi(\Phi(p, a), b) = \Phi(p, ab)$ for all $a, b \in G$ and $p \in M$,

is called a *right action* of the group G on the manifold M. Here the manifold M is often called a *right G-manifold.* It is customary to write $p \cdot a$ for $\Phi(p, a)$. For the case where G is a complex Lie group and M is a complex manifold, we require Φ to be a holomorphic mapping. Sometimes in such a situation we also say that Φ is a holomorphic action.

EXAMPLE 2.30 Let X be a complete vector field on a manifold M. The flow Φ^X, induced by X is a right action of $\mathbb{R}$ considered as an abelian group with respect to the addition operation.

A right action of a Lie group G on the manifold M is said to be *effective* if, for any $a \in G$ such that $a \neq e$, there is an element $p \in M$ such that $p \cdot a \neq p$.

Let M and N be two right G-manifolds. A mapping $\varphi : M \to N$ is called *equivariant* if $\varphi(a \cdot p) = a \cdot \varphi(p)$ for all $a \in G$ and $p \in M$.

Let Φ be a right action of a Lie group G on the manifold M. For each $a \in G$ define the smooth mapping $\Phi_a : M \to M$ by

$$\Phi_a(p) \equiv \Phi(p, a).$$

From the definition of a right action we have

$$R_e = \mathrm{id}_M, \qquad R_b \circ R_a = R_{ab}.$$

These equalities imply that

$$R_{a^{-1}} = (R_a)^{-1}.$$

Hence, for any $a \in G$ the mapping R_a is a diffeomorphism.

Let G be a group, M a right G-manifold, and S a subset of M. Introduce the notation

$$S \cdot G \equiv \{p \cdot a \mid p \in S, \ a \in G\}.$$

The set $p \cdot G$, $p \in M$, is called the *orbit* of the point p. Two orbits are either coincident or disjoint, and M is the union of the orbits. The set of orbits is denoted by M/G. There is a natural surjective mapping π from M to M/G. This mapping sends a point $p \in M$ to the corresponding orbit $p \cdot G$. The mapping π is called the canonical projection. We shall always endow M/G with the quotient topology with respect to the canonical projection π. Hence, π is a continuous mapping. It can be shown that π is also an open mapping. Moreover, the topological space M/G is second countable, but it need not be Hausdorff.

EXAMPLE 2.31 Let V be a vector space over a field $\mathbb{K}$. The set $\mathbb{P}(V)$ of all one-dimensional subspaces of V is called the *projective space* of V. The projective space $\mathbb{P}(\mathbb{K}^{m+1})$ is denoted by $\mathbb{KP}^m$. Choose a basis $\{e_\alpha\}_{\alpha=1}^n$ of V. Any point p of $\mathbb{P}(V)$ is uniquely characterised by any nonzero vector v belonging to the corresponding one-dimensional subspace. The coordinates $v^1, \ldots, v^n$ of the vector v are called the *homogeneous coordinates* of p. These coordinates are defined up to multiplication by a nonzero element of $\mathbb{K}$. Due to this fact it is customary to denote the point p by $(v^1 : v^2 : \cdots : v^n)$.

Thus, we can define the projective space as the orbit space of the right action of the Lie group $\mathbb{K}^\times$ on V defined as $v \cdot a \mapsto av$ The orbit space here is a Hausdorff topological space. It is possible to define a natural differentiable structure on $\mathbb{P}(V)$. Consider, for example, the projective space $\mathbb{KP}^m$. For each $\alpha = 1, \ldots, m+1$, denote by U_α the set of all one-dimensional subspaces intersecting the hyperplane

$$H_\alpha \equiv \{(a^1, \ldots, a^{m+1}) \in \mathbb{K}^{m+1} \mid a^\alpha = 1\}.$$

Let $p \in U_\alpha$ and let a point $a = (a^1, \ldots, a^{m+1})$ belong to the subspace p. Define the mapping $\varphi_\alpha : U_\alpha \to \mathbb{K}^m$ by

$$\varphi_\alpha(p) \equiv (a^1/a^\alpha, \ldots, a^{\alpha-1}/a^\alpha, a^{\alpha+1}/a^\alpha, \ldots, a^{m+1}/a^\alpha).$$

It is clear that this definition does not depend on the choice of the point a in p. It can be shown that $\{(U_\alpha, \varphi_\alpha)\}$ is an atlas of $\mathbb{KP}^m$.

Let G be a Lie group and let H be a Lie subgroup of G. The mapping $(a, b) \in G \times H \mapsto ab \in G$ is a right action of H on G. It appears that the space of orbits G/H is Hausdorff if and only if H is a closed subgroup. Moreover, in this case there is a unique smooth differentiable structure on G/H such that the natural projection $\pi : G \to G/H$ is a smooth mapping. We shall always consider G/H as a smooth manifold with respect to this differentiable structure. In the case where G is a complex manifold, the manifold G/H has the unique structure of a complex manifold with respect to which the natural projection π is holomorphic.

A *left action* of a Lie group G on a manifold M is specified by a mapping $\Psi : G \times M \to M$ satisfying the conditions

(LA1) $\Psi(e,p) = p$ for all $p \in M$;

(LA2) $\Psi(a, \Psi(b,p)) = \Psi(ab,p)$ for all $a, b \in G$ and $p \in M$.

Here M is called a *left G-manifold.*. The left action Ψ defines the set of diffeomorphisms $\Psi_a^X : p \in M \mapsto a \cdot p \in M$. The corresponding space of orbits is denoted here by $G\backslash M$.

Any right action Φ of a Lie group G on a manifold M generates a left action Ψ of G on M defined by

$$\Psi(a,p) \equiv \Phi(p, a^{-1}).$$

Actually, any definition or a statement which uses a right action of a Lie group on a manifold has its analogue for a left action.

Let M be a right (left) G-manifold. The corresponding action of G on M is called *transitive* if for any $p, q \in M$ there is $a \in G$ such that $p = q \cdot a$ ($p = a \cdot q$). If an action is transitive, then there is just one orbit. A manifold endowed with a transitive action of a Lie group G is called a *homogeneous space* of G.

Let H be a closed subgroup of a Lie group G. It can be shown that the mapping $\Psi : G \times G/H \to G/H$, defined by

$$\Psi(a, bH) \equiv abH, \tag{2.63}$$

is a left action of G on G/H. This action is obviously transitive, therefore G/H is a homogeneous space of G. In the case where G is a complex manifold, the action Ψ defined by (2.63) is holomorphic.

Let Ψ be a left action of a Lie group G on a manifold M, and let $p \in M$. The set

$$G_p \equiv \{a \in G \mid a \cdot p = p\}$$

is a subgroup of G called the *isotropy subgroup* of Ψ at p. Denote by Ψ_p the mapping from G to M, defined by

$$\Psi_p(a) \equiv a \cdot p.$$

The mapping Ψ_p is smooth and certainly continuous. Hence, the subgroup $G_p = \Psi_p^{-1}(p)$ is a closed subgroup and G/G_p has a natural smooth differentiable structure. Define a surjective mapping $\psi_p : G/G_p \to G \cdot p$ by

$$\psi_p(aG_p) \equiv a \cdot p.$$

Show that this mapping is injective. Indeed, suppose that for some $a, b \in G$ we have $\psi_p(aG_p) = \psi_p(bG_p)$. Here $a \cdot p = b \cdot p$, and, therefore, $a^{-1}b \cdot p = p$. From this fact it follows that $a^{-1}b \in G_p$, and $aG_p = bG_p$. Thus, the mapping ψ_p is a bijection. It can be shown that ψ_p, considered as a mapping from G/G_p to M, is an immersion. This means that the orbit $G \cdot p$ has the structure of a submanifold of M which is diffeomorphic to G/G_p. Note that the mapping ψ_p is equivariant.

If Ψ is a transitive action, then it can be shown that the mapping ψ_p is a diffeomorphism. It is important here that we consider only manifolds that are second countable topological spaces. In other words, any homogeneous space of a Lie group G is diffeomorphic to a manifold of the type G/H, where H is a closed subgroup of G. In the case where G is a complex Lie group, the corresponding diffeomorphism is a holomorphic mapping.

Actually, we can consider actions of Lie groups on arbitrary sets. For example, a right action of a Lie group G on a set M is defined as a mapping $\Phi : M \times G \to M$ satisfying conditions (RA1) and (RA2). If Ψ is a transitive action, then for any $p \in M$ the mapping ψ_p is bijective and we can identify the set M with G/G_p, thus equipping M with the structure of a manifold.

Let M be a right G-manifold. Construct a mapping from the Lie algebra $\mathfrak{g}$ to the Lie algebra of smooth vector fields on M, assigning to an element $v \in \mathfrak{g}$ the vector field X_v^M acting on a function $f \in \mathfrak{F}(M)$ as

$$X_{vp}^M(f) \equiv \left. \frac{d}{dt} f(p \cdot \exp(tv)) \right|_{t=0} . \tag{2.64}$$

It can be shown that for any $v, u \in \mathfrak{g}$ we have

$$[X_v^M, X_u^M] = X_{[v,u]}^M.$$

Thus the mapping $v \mapsto X_v^M$ is a homomorphism of Lie algebras.

In the case of a left action of a Lie group G on a manifold M, we define a mapping from the Lie algebra $\mathfrak{g}$ to the Lie algebra of smooth vector fields on M, assigning to an element $v \in \mathfrak{g}$ the vector field $\overline{X}_v^M$ given by

$$\overline{X}_{vp}^M(f) \equiv \left. \frac{d}{dt} f(\exp(tv) \cdot p) \right|_{t=0} .$$

It can be shown that in this case for any $v, u \in \mathfrak{g}$ we have

$$[\overline{X}_v^M, \overline{X}_u^M] = -\overline{X}_{[v,u]}^M.$$

Thus the mapping $v \mapsto \overline{X}^M_v$ is an antihomomorphism of Lie algebras.

A right action of a Lie group G on a manifold M is said to be *free* if for any fixed $p \in M$ the equality $p \cdot a = p$ is valid only for $a = e$. A necessary condition for a right action of a Lie group G to be free is the injectivity of the mapping $v \in \mathfrak{g} \mapsto X^M_{vp} \in T_p(M)$ for any $p \in M$.

EXAMPLE 2.32 Let V be a vector space and let $A \in \mathrm{GL}(V)$. For any basis $\{e_i\}$ of V, the set $\{Ae_i\}$ is also a basis of V. Thus, we have a left action of the Lie group $\mathrm{GL}(V)$ on the set of all bases of V. It is obvious that this action is free and transitive. Using this fact we identify the set of all bases of V with the Lie group $\mathrm{GL}(V)$, equipping it with the structure of a manifold.

EXAMPLE 2.33 Let V be an n-dimensional vector space. The set of all k-dimensional subspaces of V is denoted by $\mathbb{G}^k(V)$. For any $A \in \mathrm{GL}(V)$ and $W \in \mathbb{G}^k(V)$, the set AW is a k-dimensional subspace of V. Therefore, we have a left action of $\mathrm{GL}(V)$ on $\mathbb{G}^k(V)$, which is obviously transitive. Let $\{e_i\}$ be a basis of V. Consider the linear span of the vectors e_a, $a = 1, \dots, k$. It is a k-dimensional subspace of V. This subspace is invariant with respect to the subgroup H of $\mathrm{GL}(V)$, formed by the linear operators $A \in \mathrm{GL}(V)$ whose matrix representation a with respect to the basis $\{e_i\}$ has the block form

$$a = \begin{pmatrix} X_1 & Y \\ 0 & X_2 \end{pmatrix}.$$

Here X_1 and X_2 are nondegenerate $k \times k$ and $(n-k)\times(n-k)$ matrices, and Y is an arbitrary $k\times(n-k)$ matrix. Using this fact, we identify $\mathbb{G}^k(V)$ with the homogeneous manifold $H\backslash\mathrm{GL}(V)$, and consider it as a manifold. Any such a manifold is called a *Grassmann manifold* or a *Grassmannian*. The Grassmann manifold $\mathbb{G}^k(V)$ is $k(n-k)$-dimensional and compact. The Grassmann manifold $\mathbb{G}^1(V)$ is the projective space $\mathbb{P}(V)$.

Consider another representation of a Grassmann manifold. Let V be an n-dimensional vector space over a field $\mathbb{K}$. Fix a basis $\{e_i\}$ of V. Any basis $\{f_a\}_{a=1}^k$ of a k-dimensional subspace of V is uniquely characterised by the coordinates $f^i{}_a$ of the vectors f_a with respect to the basis $\{e_i\}$. The $n\times k$ matrix $f \equiv (f^i{}_a)$ is of rank k. It is clear that we have a bijective correspondence between the bases of k-dimensional subspaces of V and $n \times k$ matrices of rank k. The set of such matrices is an open submanifold of the manifold of $n \times k$ matrices. Two bases $\{f_a\}$ and $\{f'_a\}$ of the same k-dimensional subspace V lead to the matrices f and f' connected by the relation $f' = fa$, where $a \in \mathrm{GL}(k, \mathbb{K})$. Thus, different matrices corresponding to the same subspace are connected by a right action of the Lie group $\mathrm{GL}(k, \mathbb{K})$. The orbits of this action are in a bijective correspondence with the elements of $\mathbb{G}^k(V)$. It can be shown that in actual fact the Grassmann manifold $\mathbb{G}^k(V)$ is diffeomorphic to the manifold obtained by factorising the manifold of all $n \times k$ matrices of rank k by the right action of the Lie group $\mathrm{GL}(k, \mathbb{K})$.

The Grassmann manifold $\mathbb{G}^k(\mathbb{K}^n)$ is denoted by $\mathbb{K}\mathbb{G}^{k,n-k}$. The Grassmann manifold $\mathbb{K}\mathbb{G}^{1,m}$ is the projective space $\mathbb{K}\mathbb{P}^m$.

Exercises

2.13 Consider the mapping ψ from $\mathbb{C} \times \mathbb{C} \times \mathbb{C}^\times \subset \mathbb{C}^3$ to the Lie group $\mathrm{SL}(2, \mathbb{C})$ defined as

$$\psi(a, b, c) = \begin{pmatrix} 1 & a \\ 0 & 1 \end{pmatrix} \begin{pmatrix} 1 & 0 \\ b & 1 \end{pmatrix} \begin{pmatrix} c & 0 \\ 0 & 1/c \end{pmatrix}.$$

Prove that $(\psi(\mathbb{C} \times \mathbb{C} \times \mathbb{C}^\times), \psi^{-1})$ is a chart on $\mathrm{SL}(2, \mathbb{C})$. Denote the corresponding coordinate functions by α, β and γ. Show that the functions $g^i{}_j$ are connected with the functions α, β and γ by

$$\begin{pmatrix} g^1{}_1 & g^1{}_2 \\ g^2{}_1 & g^2{}_2 \end{pmatrix} = \begin{pmatrix} 1 & \alpha \\ 0 & 1 \end{pmatrix} \begin{pmatrix} 1 & 0 \\ \beta & 1 \end{pmatrix} \begin{pmatrix} \gamma & 0 \\ 0 & 1/\gamma \end{pmatrix}.$$

Find the expression for the Maurer–Cartan form in terms of the coordinate functions α, β and γ.

2.14 Show that the projective space $\mathbb{R}\mathbb{P}^1$ is diffeomorphic to the sphere S^1.

2.15 Prove that the projective space $\mathbb{C}\mathbb{P}^1$ and the sphere S^2 are diffeomorphic as complex manifolds.

2.8 Smooth fibre bundles

2.8.1 Definition of a fibre bundle

Let E and M be manifolds and let π be a smooth mapping from E to M. The triple $\xi \equiv (E, \pi, M)$ is called a *bundle* with the *bundle manifold* E, the *base manifold* M and the *bundle projection* π. The counter image $F_p \equiv \pi^{-1}(p)$, $p \in M$, is called the *fibre* over the point p. The fibre bundle (E, π, M) is also denoted by $E \xrightarrow{\pi} M$, or simply by $E \to M$.

Let $\xi = (E, \pi, M)$ and $\xi' = (E', \pi', M')$ be two bundles. A smooth mapping $\varphi : E \to E'$ is called *fibre preserving* if it maps each fibre of $F_p = \pi^{-1}(p)$, $p \in M$, of the bundle ξ into some fibre $F_{p'}$ of the bundle ξ'. Any such a mapping, with the help of the relation $p' = \psi(p)$, defines the mapping $\psi : M \to M'$ satisfying the relation

$$\psi \circ \pi = \pi' \circ \varphi.$$

If the mapping ψ is smooth, the mapping φ is called a *bundle morphism*. In this case we also write $\varphi : \xi \to \xi'$.

A morphism $\varphi : \xi \to \xi'$ is called an isomorphism if the mapping $\varphi : E \to E'$ and the corresponding mapping $\psi : M \to M'$ are diffeomorphisms. In such a case the inverse mapping $\varphi^{-1} : E \to E'$ can be defined, and this mapping is also a bundle morphism. The bundles ξ and ξ' are called isomorphic if there exists an isomorphism $\varphi : \xi \to \xi'$.

In the case of $M = M'$ and $\psi = \mathrm{id}_M$, any fibre preserving mapping $\varphi : E \to E'$ is a morphism called a morphism over M. In this case we have

$$\pi = \varphi \circ \pi'.$$

A morphism over M, being an isomorphism, is called an isomorphism over M. Two smooth bundles ξ and ξ' with the same base manifold are called isomorphic if there exists an isomorphism $\varphi : \xi \to \xi'$ over M.

Let U be an open subset of the base manifold of a bundle $\xi = (E, \pi, M)$. It is clear that $\xi|_U \equiv (\pi^{-1}(U), \pi|_{\pi^{-1}(U)}, U)$ is a bundle. This bundle is called the *restriction* of the bundle ξ to U.

In general, a fibre of a bundle cannot be provided with the structure of a submanifold of the bundle manifold. A bundle (E, π, M)

isomorphic to the bundle $(M \times F, \mathrm{pr}_M, M)$, where F is a manifold, is called a *trivial* bundle. It is clear that the fibres of a trivial bundle are embedded submanifolds of the bundle manifold. Furthermore, they are diffeomorphic one to another. Such a situation also occurs in the case of a fibre bundle.

Let $\xi = (E, \pi, M)$ be a bundle and let F be a manifold. The bundle ξ is said to be a *fibre bundle* with the *typical fibre* F if for any point $p \in M$ there exists a neighbourhood U of p such that $\xi|_U$ is isomorphic to the trivial bundle $(U \times F, \mathrm{pr}_U, U)$. In other words, there exists a diffeomorphism $\psi : \pi^{-1}(U) \to U \times F$ such that

$$\mathrm{pr}_U \circ \psi = \pi|_{\pi^{-1}(U)}.$$

Here the pair (U, ψ) is called a *bundle chart* on E. A set $\{U_\alpha, \psi_\alpha\}_{\alpha \in \mathcal{A}}$ of bundle charts on E is called a *bundle atlas* on E if $\bigcup_{\alpha \in \mathcal{A}} U_\alpha = M$. A fibre bundle is also called a *locally trivial bundle*.

A trivial bundle is a fibre bundle which has an atlas consisting of just one chart.

Let $\xi = (E, \pi, M)$ be a fibre bundle and let $\{U_\alpha, \psi_\alpha\}_{\alpha \in \mathcal{A}}$ be a bundle atlas on E. For each $p \in U_\alpha$ the mapping

$$\psi_{\alpha p} \equiv \mathrm{pr}_F \circ \psi_\alpha|_{\pi^{-1}(p)}$$

is a diffeomorphism from the fibre $\pi^{-1}(p)$ to the typical fibre F.

Let (U_α, ψ_α) and (U_β, ψ_β) be two elements of the bundle atlas $\{(U_\alpha, \psi_\alpha)\}$ such that $U_\alpha \cap U_\beta \neq \emptyset$. For any point $p \in U_\alpha \cap U_\beta$ we define a diffeomorphism of the typical fibre $\psi_{\alpha\beta}(p) : F \to F$ by

$$\psi_{\alpha\beta}(p) \equiv \psi_{\alpha p} \circ \psi_{\beta p}^{-1}. \tag{2.65}$$

From this definition it follows that $\psi_{\alpha\alpha} = \mathrm{id}_F$. Furthermore, it is clear that

$$\psi_{\alpha\beta}(p) = \psi_{\alpha\gamma}(p) \circ \psi_{\gamma\beta}(p) \tag{2.66}$$

for any $\alpha, \beta, \gamma \in \mathcal{A}$ such that $U_\alpha \cap U_\beta \cap U_\gamma \neq \emptyset$, and $p \in U_\alpha \cap U_\beta \cap U_\gamma$. Suppose that F is a complex manifold. If all diffeomorphisms $\psi_{\alpha\beta}(p)$ are holomorphic mappings, then the fibre bundle (E, π, M) is called a *complex fibre bundle*. The fibre bundle (E, π, M) is called *holomorphic* if E and M are complex manifolds and the mappings π, ψ_α and ψ_α^{-1}, $\alpha \in \mathcal{A}$ are holomorphic.

The group of diffeomorphisms of a differential manifold cannot be endowed with the structure of a finite-dimensional smooth manifold; therefore, we can say nothing about the differentiability of the mappings $\psi_{\alpha\beta}$. To remain in the framework of the differential geometry of finite-dimensional manifolds, let us suppose that a left action of a Lie group G on the typical fibre F is given, and that there exist smooth mappings $g_{\alpha\beta} : U_\alpha \cap U_\beta \to G$ such that

$$\psi_{\alpha\beta}(p) = L^F_{g_{\alpha\beta}(p)} \tag{2.67}$$

for any $p \in U_\alpha \cap U_\beta$. If the action L^F of the Lie group G on F is effective, then from (2.66) we obtain

$$g_{\alpha\beta}(p) = g_{\alpha\gamma}(p) g_{\gamma\beta}(p), \qquad p \in U_\alpha \cap U_\gamma \cap U_\beta. \tag{2.68}$$

If the left action L^F is not effective, we will suppose that the mappings $g_{\alpha\beta}$, $\alpha, \beta \in \mathcal{A}$ can be chosen in such a way that (2.68) is valid. The Lie group G is said to be the *structure group* of the fibre bundle ξ, and we say that ξ is a fibre G-bundle. The mappings $g_{\alpha\beta}$ are called the *transition functions* of the bundle atlas $\{(U_\alpha, \psi_\alpha)\}$.

It can be shown that for any fibre G-bundle $\xi = (E, \pi, M)$ and any open subset U of M the bundle $\xi|_U$ has a natural structure of a fibre G-bundle. Indeed, let $\{(U_\alpha, \psi_\alpha)\}_{\alpha \in \mathcal{A}}$ be a bundle atlas on E. Denote by $\mathcal{B}$ the set of all $\beta \in \mathcal{A}$ such that $U \cap U_\beta \neq \emptyset$. For any $\beta \in \mathcal{B}$ the pair (V_β, χ_β), where

$$V_\beta \equiv U \cap U_\beta, \qquad \chi_\beta \equiv \psi_\beta|_{\pi^{-1}(V_\beta)},$$

is a bundle chart on $\pi^{-1}(U)$, and the set $\{(V_\beta, \chi_\beta)\}_{\beta \in \mathcal{B}}$ is a bundle atlas on $\pi^{-1}(U)$.

Example 2.34 Let M be an m-dimensional manifold. Denote by $T(M)$ the union of all tangent spaces to M, i.e.,

$$T(M) \equiv \bigcup_{p \in M} T_p(M).$$

Define the projection π by the relation

$$\pi(v) = p, \qquad v \in T_p(M).$$

Let $(U_\alpha, \varphi_\alpha)_{\alpha \in \mathcal{A}}$ be an atlas of the manifold M. Denote the coordinate functions corresponding to the chart $(U_\alpha, \varphi_\alpha)$ by x^i_α. Consider the set of mappings $\psi_\alpha : \pi^{-1}(U_\alpha) \to U_\alpha \times \mathbb{R}^m$,

$$\psi_\alpha(v) \equiv (\pi(v), (dx^1_\alpha(v), \ldots, dx^m_\alpha(v))), \qquad v \in \pi^{-1}(U_\alpha).$$

It can be shown that there exists a unique differentiable structure on $T(M)$, such that the mappings ψ_α are diffeomorphisms. This differentiable structure does not depend on the choice of an atlas on M, and the projection π is smooth with respect to it. Thus, we see that $(T(M), \pi, M)$ is a fibre bundle called the *tangent bundle* of M. The typical fibre of the tangent bundle $T(M)$ is the space $\mathbb{R}^m$. It is clear that the diffeomorphisms $\psi_{\alpha\beta}(p)$, $p \in U_\alpha \cap U_\beta$, are linear transformations of $\mathbb{R}^m$, defined by the matrices

$$(a_{\alpha\beta})^i{}_j(p) = \frac{\partial x^i_\alpha}{\partial x^j_\beta}(p).$$

Therefore, the structure group of the fibre bundle $T(M)$ is $\mathrm{GL}(m, \mathbb{R})$.

Similar arguments show that for any complex manifold the manifold

$$T^{(1,0)}(M) \equiv \bigcup_{p \in M} T^{(1,0)}_p(M)$$

has the natural structure of a holomorphic fibre bundle with the typical fibre $\mathbb{C}^m$ and the structure group $\mathrm{GL}(m, \mathbb{C})$.

Let $\xi = (E, \pi, M)$ be a bundle. A mapping $s : M \to E$ is called a *section* of ξ if

$$\pi \circ s = \mathrm{id}_M .$$

In particular, the sections of the tangent bundle $T(M) \xrightarrow{\pi} M$ are vector fields on M. Note that not each smooth fibre bundle has smooth sections.

A section of a fibre bundle $\xi|_U$ is called a *local section* of ξ. Let $\{U_\alpha\}_{\alpha \in \mathcal{A}}$ be an open cover of M and let s_α, $\alpha \in \mathcal{A}$, be sections of the bundles $\xi|_{U_\alpha}$; in such a situation we say that the set of the sections $\{s_\alpha\}_{\alpha \in \mathcal{A}}$ is a family of local sections of ξ *covering* M. It is clear that for any fibre bundle there is a family of smooth local sections covering its base.

2.8.2 Principal fibre bundles and connections

Let (P, π, M) be a fibre G-bundle. Suppose that the typical fibre of ξ coincides with the Lie group G which acts on itself by left translations. In such a situation (P, π, M) is called a *principal fibre G-bundle*.

EXAMPLE 2.35 Let G be a Lie group and let H be a closed Lie subgroup of G. The triple $(G, \pi, G/H)$, where π is the canonical projection, can be naturally considered as a fibre H-bundle. If G is a complex Lie group, then (G, π, H) is a holomorphic fibre H-bundle.

Let (P, π, M) be a principal fibre G-bundle. Define a right action of the Lie group G on P in the following way. Let $p \in P$; suppose that $q \equiv \pi(p) \in U_\alpha$, where (U_α, ψ_α) is a chart on the fibre bundle P. For any $a \in G$ we put

$$R_a^P(p) \equiv \psi_{\alpha q}^{-1} \circ R_a \circ \psi_{\alpha q}(p). \tag{2.69}$$

It can easily be shown that this definition does not depend on the choice of a chart. Indeed, let $p \in U_\beta$, where (U_β, ψ_β) is another chart on $P \xrightarrow{\pi} M$. From (2.65) and (2.67) we have

$$\psi_{\alpha q} = L_{g_{\alpha\beta}(q)} \circ \psi_{\beta q}.$$

Substituting this relation into (2.69), we obtain

$$R_a^P(p) = \psi_{\beta q}^{-1} \circ \left(L_{g_{\alpha\beta}(q)}\right)^{-1} \circ R_a \circ L_{g_{\alpha\beta}(q)} \circ \psi_{\beta q}(p).$$

Using (2.34) now, we come to the equality

$$\psi_{\alpha q}^{-1} \circ R_a \circ \psi_{\alpha q}(p) = \psi_{\beta q}^{-1} \circ R_a \circ \psi_{\beta q}(p).$$

Each fibre of the fibre bundle $P \xrightarrow{\pi} M$ is invariant with respect to the transformations defined by (2.69), which can be written as

$$\pi \circ R_g^P = \pi, \qquad g \in G.$$

It is clear that R^P is a free action, and the orbits of this action coincide with the fibres of the fibre bundle $P \xrightarrow{\pi} M$.

Let $P \xrightarrow{\pi} M$ be a principal G-bundle and let $P' \xrightarrow{\pi'} M'$ be a principal G'-bundle. A smooth mapping $\varphi : P \to P'$ is called a *principal bundle morphism* if there exists such a group homomorphism $\varphi' : G \to G'$ that $\varphi(p \cdot a) = \varphi(p) \cdot \varphi'(a)$ for any $p \in P$, $a \in G$. Any principal bundle morphism is a fibre preserving mapping; in other words, it is a bundle morphism. In the case where $P = P'$, $M = M'$, $\varphi' = \mathrm{id}_G$, and the principal bundle morphism $\varphi : P \to P$ is an isomorphism over M, we call φ a *principal bundle isomorphism*.

Let $P \xrightarrow{\pi} M$ be a principal G-bundle. For an arbitrary $p \in P$, denote by $\mathcal{V}_p$ a tangent space to the fibre through p. Since the fibres of P are diffeomorphic to the Lie group G, then $\dim \mathcal{V}_p = \dim G$. A smooth distribution $\mathcal{H}$ on P is called a *connection* on P, if

(C1) $T_p(P) = \mathcal{V}_p \oplus \mathcal{H}_p$, for any $p \in P$;

(C2) $\mathcal{H}_{p\cdot a} = R^P_{a*}(\mathcal{H}_p)$ for any $a \in G$, and $p \in P$.

A tangent vector $v \in T_p(P)$ is called *vertical* (*horizontal*) if $v \in \mathcal{V}_p$ ($v \in \mathcal{H}_p$). A vector field X on P is called *vertical* (*horizontal*) if for any $p \in P$ the vector X_p is vertical (horizontal). For any $v \in T_p(P)$ there is a unique expansion

$$v = v^{\mathcal{V}} + v^{\mathcal{H}},$$

where $v^{\mathcal{V}} \in \mathcal{V}_p$, $v^{\mathcal{H}} \in \mathcal{H}_p$.

With a connection $\mathcal{H}$ on a principal fibre G-bundle $P \xrightarrow{\pi} M$ we can associate a 1-form ω taking values in $\mathfrak{g}$ in the following way. Let $v \in T_p(P)$; there is a unique element $\tilde{v} \in \mathfrak{g}$, such that

$$v^{\mathcal{V}} = X^P_{\tilde{v}p},$$

where the vector field $X^P_{\tilde{v}}$ is defined by relation (2.64). Define the form ω by

$$\omega(v) \equiv \tilde{v}.$$

A vector $v \in T_p(P)$ is horizontal if and only if

$$\omega(v) = 0.$$

The form ω is called the *connection* form of the connection $\mathcal{H}$. The connection form ω has the properties

$$\omega(X^P_v) = v \tag{2.70}$$

for any $v \in \mathfrak{g}$, and

$$R^{P*}_a \omega = \mathrm{Ad}(a^{-1}) \circ \omega \tag{2.71}$$

for any $a \in G$.

Conversely, any $\mathfrak{g}$-valued 1-form on a principal fibre G-bundle $P \xrightarrow{\pi} M$ satisfying conditions (2.70) and (2.71) is a connection form of a unique connection on $P \xrightarrow{\pi} M$. Here the corresponding subspaces $\mathcal{H}_p$, $p \in P$, are defined by

$$\mathcal{H}_p \equiv \{v \in T_p(P) \mid \omega(v) = 0\}.$$

Let (P, π, M) be a principal fibre G-bundle; consider some set $\{s_\alpha\}_{\alpha \in \mathcal{A}}$ of local sections $s_\alpha : U_\alpha \to \pi^{-1}(U_\alpha)$ covering M. The mappings $\psi_\alpha : \pi^{-1}(U_\alpha) \to U_\alpha \times G$, defined with the help of the relation

$$\psi_\alpha^{-1}(q, a) \equiv s_\alpha(q) \cdot a,$$

are diffeomorphisms satisfying the condition

$$\mathrm{pr}_{U_\alpha} \circ \psi_\alpha = \pi.$$

It is clear that the set $\{(U_\alpha, \psi_\alpha)\}_{\alpha\in\mathcal{A}}$ is a bundle atlas on P. The transition functions in this case have the form

$$g_{\alpha\beta}(p) = g_\alpha(s_\beta(p)), \qquad p \in U_\alpha \cap U_\beta,$$

where the mappings $g_\alpha : \pi^{-1}(U_\alpha) \to G$ are defined by

$$g_\alpha \equiv \mathrm{pr}_G \circ \psi_\alpha.$$

Thus, any set $\{s_\alpha\}_{\alpha\in\mathcal{A}}$ of local sections $s_\alpha : U_\alpha \to \pi^{-1}(U_\alpha)$ of a principal fibre G-bundle $P \xrightarrow{\pi} M$ covering M generates a bundle atlas on P. On the other hand, let $\{U_\alpha, \psi_\alpha\}_{\alpha\in\mathcal{A}}$ be a bundle atlas on P. The set $\{s_\alpha\}_{\alpha\in\mathcal{A}}$ of local sections defined by

$$s_\alpha(q) \equiv \psi_\alpha^{-1}(q, e), \quad q \in U_\alpha$$

generates the bundle atlas $\{U_\alpha, \psi_\alpha\}$.

Now let ω be a connection form of some connection on P. On $\pi^{-1}(U_\alpha)$ we obtain

$$\omega = \mathrm{Ad}(g_\alpha^{-1}) \circ \pi^*\omega_\alpha + g_\alpha^*\theta, \tag{2.72}$$

where θ is the Maurer–Cartan form of G and

$$\omega_\alpha \equiv s_\alpha^*\omega.$$

It can be also shown that

$$\omega_\alpha = \mathrm{Ad}(g_{\alpha\beta}) \circ \omega_\beta + g_{\alpha\beta}^*\theta \tag{2.73}$$

on $\pi^{-1}(U_\alpha \cap U_\beta)$.

On the other hand, if there is given a set of $\mathfrak{g}$-valued 1-forms $\{\omega_\alpha\}_{\alpha\in\mathcal{A}}$ satisfying (2.73), we can construct a unique $\mathfrak{g}$-valued 1-form ω satisfying (2.72). In other words, a set of $\mathfrak{g}$-valued 1-forms $\{\omega_\alpha\}_{\alpha\in\mathcal{A}}$ satisfying (2.73) define a unique connection on P.

Let η be a k-form on a principal fibre bundle $P \xrightarrow{\pi} M$ provided with a connection $\mathcal{H}$. The horizontal component $\eta^\mathcal{H}$ of η is defined by

$$\eta^\mathcal{H}(X_1, \dots, X_k) \equiv \eta(X_1^\mathcal{H}, \dots, X_k^\mathcal{H})$$

for all $X_1, \dots, X_k \in \mathfrak{X}(P)$. It is clear that $\eta^\mathcal{H}$ is a k-form on P. The horizontal component of a form taking values in a vector space can be defined similarly. Here, for the connection form ω corresponding to the connection $\mathcal{H}$, we have $\omega^\mathcal{H} = 0$ but, in general, $(d\omega)^\mathcal{H} \neq 0$. The $\mathfrak{g}$-valued 2-form

$$\Omega \equiv (d\omega)^\mathcal{H}$$

is called the *curvature form* of the connection $\mathcal{H}$. It can be shown that

$$\Omega = d\omega + \frac{1}{2}[\omega, \omega]. \tag{2.74}$$

Using a set $\{s_\alpha\}$ of local sections covering M, we obtain the relations

$$\Omega = \mathrm{Ad}(g_\alpha^{-1}) \circ \pi^* \Omega_\alpha, \tag{2.75}$$

where

$$\Omega_\alpha \equiv s_\alpha^* \Omega.$$

The 2-forms Ω_α are connected with the 1-forms ω_α by

$$\Omega_\alpha = d\omega_\alpha + \frac{1}{2}[\omega_\alpha, \omega_\alpha],$$

and

$$\Omega_\alpha = \mathrm{Ad}(g_{\alpha\beta}) \circ \Omega_\beta$$

on $\pi^{-1}(U_\alpha \cap U_\beta)$.

Note also that the following relation is valid:

$$(d\Omega)^{\mathcal{H}} = 0;$$

this is called the *Bianchi identity*.

3
Differential geometry of Toda-type systems

3.1 More about semisimple Lie algebras

In the next four sections we present some additional information on semisimple Lie algebras which will be needed in our consideration of the Toda-type systems. Here we mainly follow the remarkable books by Helgason (1978); Bourbaki (1975); Gorbatsevich, Onishchik & Vinberg (1994) and the original papers of Dynkin (1975a,b) and Kostant (1959).

3.1.1 Groups of automorphisms

The group $\mathrm{Aut}(A)$ of automorphisms of an algebra A is a closed subgroup of the Lie group $\mathrm{GL}(A)$. Therefore, we can consider this group as a Lie group. It can be shown that the Lie algebra of the group $\mathrm{Aut}(A)$ coincides with the Lie algebra $\mathrm{Der}(A)$ of derivations of A.

Let $\mathfrak{g}$ be a Lie algebra. The mapping $x \in \mathfrak{g} \mapsto \mathrm{ad}(x) \in \mathrm{Der}(\mathfrak{g})$ is a homeomorphism of $\mathfrak{g}$ onto some subalgebra of $\mathrm{Der}(\mathfrak{g})$ denoted by $\mathrm{ad}(\mathfrak{g})$. The corresponding connected Lie subgroup of $\mathrm{Aut}(\mathfrak{g})$ is called the *group of inner automorphisms* of $\mathfrak{g}$ and is denoted by $\mathrm{Int}(\mathfrak{g})$. The elements of $\mathrm{Int}(\mathfrak{g})$ are called *inner automorphisms* of $\mathfrak{g}$. Sometimes $\mathrm{Int}(\mathfrak{g})$ is called the *adjoint group* of $\mathfrak{g}$. An inner automorphism can be represented as a product of a finite number of automorphisms of the form $\exp(\mathrm{ad}(x))$, $x \in \mathfrak{g}$. Since $\mathrm{ad}(\mathfrak{g})$ is an ideal of $\mathrm{Der}(\mathfrak{g})$, the group $\mathrm{Int}(\mathfrak{g})$ is an invariant subgroup of $\mathrm{Aut}(\mathfrak{g})$. Elements or subsets of $\mathfrak{g}$ connected by an inner automorphism are called *conjugated* in $\mathfrak{g}$.

Let $\mathfrak{g}$ be a Lie algebra and let G be a connected Lie group having $\mathfrak{g}$ as its Lie algebra. For any $a \in G$ the mapping $\varphi_a : \mathfrak{g} \to \mathfrak{g}$ defined by

$$\varphi_a(v) \equiv \mathrm{Ad}(a)v$$

is an automorphism of $\mathfrak{g}$. Any element of G can be represented as a product of a finite number of elements of the form $\exp(x)$, $x \in \mathfrak{g}$. Using the relation

$$\mathrm{Ad}(\exp(x))y = \exp(\mathrm{ad}(x))y,$$

we see that

$$\varphi_{\exp(x)} = \exp(\mathrm{ad}(x)).$$

Therefore, for any $a \in G$ the automorphism φ_a is an inner automorphism.

On the other hand, since any inner automorphism of $\mathfrak{g}$ can be represented as a product of a finite number of inner automorphisms of the form $\exp(\mathrm{ad}(x))$, $x \in \mathfrak{g}$, we conclude that any inner automorphism of $\mathfrak{g}$ has the form φ_a for some $a \in G$.

Thus, we have $\mathrm{Ad}(G) = \mathrm{Int}(\mathfrak{g})$. From relation (2.41) it follows that the centre $Z(G)$ of the group G coincides with the kernel of the adjoint representation of G. Therefore, the adjoint group of $\mathfrak{g}$ is isomorphic to the group $G/Z(G)$.

Any derivation of a semisimple Lie algebra is an inner derivation. Hence, for any semisimple Lie algebra $\mathfrak{g}$, the component of $\mathrm{Aut}(\mathfrak{g})$, containing the identity element, coincides with $\mathrm{Int}(\mathfrak{g})$.

Let $\mathfrak{g}$ be a complex semisimple Lie algebra, $\mathfrak{h}$ a Cartan subalgebra of $\mathfrak{g}$, and Δ a root system of $\mathfrak{g}$ with respect to $\mathfrak{h}$. Consider a base Π of Δ. Denote by $\mathrm{Aut}(\Pi)$ a subgroup of a symmetric group of Π, consisting of the permutations which do not change the Cartan matrix. The elements of $\mathrm{Aut}(\Pi)$ are naturally identified with the symmetry transformations of the corresponding Dynkin diagram. It can be shown that the group $\mathrm{Aut}(\Pi)$ is isomorphic to the subgroup of $\mathrm{Aut}(\Delta)$ consisting of the automorphisms which transform Π onto itself; this subgroup is also denoted by $\mathrm{Aut}(\Pi)$. Actually, the group $\mathrm{Aut}(\Delta)$ is the semidirect product of the Weyl group $W(\Delta)$ and the group $\mathrm{Aut}(\Pi)$.

Let $\sigma \in \mathrm{Aut}(\Pi) \subset \mathrm{Aut}(\Delta)$; the mapping $(\sigma^{-1})^t$ is an element of the group $\mathrm{Aut}(\Delta^\vee)$. It can be proved that there exists a unique automorphism φ_σ of $\mathfrak{g}$ such that $\varphi_\sigma(\alpha^\vee) = (\sigma^{-1})^t(\alpha^\vee)$ for any $\alpha \in \Pi$. The automorphism φ_σ is called the automorphism of $\mathfrak{g}$ induced by σ. Hence, the group $\mathrm{Aut}(\Pi)$ can be considered as a subgroup of $\mathrm{Aut}(\mathfrak{g})$. Furthermore, one can easily see that any automorphism of $\mathfrak{g}$ can be uniquely represented in the form $\psi \circ \varphi_\sigma$, where $\psi \in$

$\mathrm{Int}(\mathfrak{g})$ and $\sigma \in \mathrm{Aut}(\Pi)$. Therefore, for any complex semisimple Lie algebra, the group $\mathrm{Aut}(\mathfrak{g})$ has the representation

$$\mathrm{Aut}(\mathfrak{g}) = \mathrm{Int}(\mathfrak{g}) \rtimes \mathrm{Aut}(\Pi).$$

EXAMPLE 3.1 Considering the Dynkin diagrams of the complex simple Lie algebras (table 1.2), we conclude that the group $\mathrm{Aut}(\Pi)$ is $\mathbb{Z}_2$ for the Lie algebras of types A_r $(r > 1)$, D_r $(r > 4)$, E_6; and it is S_3 for the Lie algebras of type D_4. For the remaining simple Lie algebras, the group $\mathrm{Aut}(\Pi)$ is trivial; hence, for these Lie algebras the group of automorphisms coincides with the group of inner automorphisms.

Consider now the connection between automorphisms of Lie groups and Lie algebras. Let Σ be an automorphism of a real Lie group G; then the mapping

$$\sigma \equiv \Sigma_{*e} \tag{3.1}$$

is an automorphism of the corresponding Lie algebra $\mathfrak{g}$. On the other hand, if the group G is simply connected, then, for any automorphism σ of the Lie algebra $\mathfrak{g}$, there is a unique automorphism Σ of the group G such that $\Sigma_{*e} = \sigma$. It follows from the definition of the adjoint representation of a Lie group that

$$\sigma \circ \mathrm{Ad}(a) = \mathrm{Ad}(\Sigma(a)) \circ \sigma. \tag{3.2}$$

Furthermore, it can easily be shown that for the Maurer–Cartan form θ of G we have

$$\Sigma^*\theta = \sigma \circ \theta.$$

In the case of a complex Lie group G, it is natural to consider only holomorphic or antiholomorphic automorphisms of G. For a holomorphic automorphism Σ, relation (3.2) is valid. If Σ is an antiholomorphic automorphism of G, then

$$\Sigma_{*e}\left(T_e^{(1,0)}(G)\right) = T_e^{(0,1)}(G).$$

In this case we define the corresponding mapping $\sigma : \mathfrak{g} \to \mathfrak{g}$ by

$$\sigma(x) \equiv \overline{\Sigma_{*e}(x)}. \tag{3.3}$$

The mapping σ is now an antilinear automorphism of $\mathfrak{g}$. Definition (3.3) leads to the equality

$$\Sigma^*\theta(x) = \sigma \circ \theta(\bar{x}), \tag{3.4}$$

which is valid for any $x \in T_a(G)$, $a \in G$. Note that relation (3.2) is valid both for real and complex Lie groups with the mapping σ defined either by (3.1) or by (3.3).

3.1.2 Regular subalgebras and subgroups

In this subsection $\mathfrak{g}$ is a complex semisimple Lie algebra and $\mathfrak{h}$ is some fixed Cartan subalgebra of $\mathfrak{g}$.

A subalgebra $\mathfrak{f}$ of $\mathfrak{g}$ is called *regular* with respect to $\mathfrak{h}$, if $[\mathfrak{h}, \mathfrak{f}] \subset \mathfrak{f}$. In other words, $\mathfrak{f}$ is regular with respect to $\mathfrak{h}$ if $\mathfrak{h}$ is contained in the normaliser $N_{\mathfrak{g}}(\mathfrak{f})$. A Lie group G is called *semisimple* if its Lie algebra $\mathfrak{g}$ is semisimple. A subgroup F of a complex connected semisimple Lie group G is called *regular* with respect to a Cartan subgroup H if $N_G(F)$ contains H.

All regular subalgebras can be described in the following way. Let Δ be a root system of $\mathfrak{g}$ with respect to $\mathfrak{h}$. A subsystem $\Gamma \subset \Delta$ is called *closed* if for any $\alpha, \beta \in \Gamma$, such that $\alpha + \beta \in \Delta$, one has $\alpha + \beta \in \Gamma$. A subsystem $\Gamma \in \Delta$ is said to be *symmetric* if the inclusion $\alpha \in \Gamma$ implies that $-\alpha \in \Gamma$. Let Γ be a closed subsystem of Δ and let $\mathfrak{t}$ be a subalgebra of $\mathfrak{h}$ containing the element $\alpha^\vee$ for any $\alpha \in \Gamma \cap (-\Gamma)$. In this case

$$\mathfrak{f} \equiv \mathfrak{t} \oplus \bigoplus_{\alpha \in \Gamma} \mathfrak{g}^\alpha \tag{3.5}$$

is a subalgebra of $\mathfrak{g}$, which is regular with respect to $\mathfrak{h}$. We denote such a subalgebra by $\mathfrak{f}(\mathfrak{t}, \Gamma)$. On the other hand, any subalgebra $\mathfrak{f}$ of $\mathfrak{g}$ which is regular with respect to $\mathfrak{h}$ can be represented in the form of (3.5). A subalgebra $\mathfrak{f}(\mathfrak{t}, \Gamma)$ is semisimple if and only if Γ is symmetric and $\mathfrak{t}$ is generated by the elements $\alpha^\vee$, $\alpha \in \Gamma$. In such a case the subalgebra $\mathfrak{t}$ is a Cartan subalgebra of $\mathfrak{f}(\mathfrak{t}, \Gamma)$, and Γ is the root system of $\mathfrak{f}(\mathfrak{t}, \Gamma)$ with respect to $\mathfrak{t}$. Since a semisimple subalgebra of type $\mathfrak{f}(\mathfrak{t}, \Gamma)$ is determined only by the subsystem Γ, we use the notation $\mathfrak{f}(\Gamma)$ for it.

Consider the semisimple regular subalgebras in more detail. Let Ψ be a subsystem of Δ; denote by $[\Psi]$ the set of the elements of Δ which can be represented as linear combinations of elements of Ψ with integer coefficients. It is clear that if Π is a base of Δ, then $[\Pi] = \Delta$. For any subsystem Ψ of Δ, the set $[\Psi]$ is a closed symmetric subsystem of Δ which defines a semisimple regular subalgebra of $\mathfrak{g}$.

A subsystem Ψ of Δ is called a *π-system* if Ψ is linearly independent and $\alpha - \beta \notin \Delta$ for all $\alpha, \beta \in \Psi$. Any π-system Ψ is a base of the root system $\Gamma \equiv [\Psi]$ of the semisimple regular subalgebra $\mathfrak{f}([\Psi])$. On the other hand, any base of a closed symmetric subsystem of Δ is a π-system in Δ. Thus, the problem of the classification of the semisimple regular subalgebras is reduced to the problem of the enumeration of all π-systems in Δ. If two π-systems are connected by a transformation of $\mathrm{Aut}(\Delta)$, then the corresponding subalgebras are connected by an element of $\mathrm{Aut}(\mathfrak{g})$. If two π-systems are connected by a transformation of the Weyl group $W(\Delta)$, then the corresponding subalgebras are conjugated.

Note that any subsystem of a π-system is also a π-system. Moreover, if $\mathfrak{g}$ is of rank r, then any π-system is a subsystem of a π-system consisting of r elements. Hence, to find all π-systems, it suffices to find all π-systems consisting of r elements.

Let $\Pi = \{\alpha_1, \ldots, \alpha_r\}$ be a base of Δ; add to Π the minimal root α_0. The resulting subsystem of Δ is called the *extended system of simple roots* of $\mathfrak{g}$, and the corresponding Dynkin diagram is called the *extended Dynkin diagram* of $\mathfrak{g}$. The extended Dynkin diagrams for simple Lie algebras and their standard notations are listed in table 3.1. The vertices corresponding to the minimal roots are blackened. Note that the Dynkin diagrams of table 3.1 correspond to the so-called untwisted affine Lie algebras.

Now let Ψ be a π-system in Δ. As we noted above, Ψ is a base of the root system of the semisimple Lie algebra $\mathfrak{f}([\Psi])$. In particular, any irreducible component of Ψ is a base of the root system of the corresponding simple subalgebra of $\mathfrak{f}([\Psi])$. Complement Ψ by the minimal root corresponding to some irreducible component of Ψ, and then remove any root belonging to this component. As a result we obtain another π-system Ψ' in Δ, and we say that the π-system Ψ' is obtained from the π-system Ψ by an elementary transformation. It appears that any π-system in Δ consisting of r elements, with r being the rank of $\mathfrak{g}$, can be obtained from some base Π of Δ by a chain of elementary transformations.

There exists an important class of nonsemisimple regular subalgebras and subgroups. A Lie subalgebra $\mathfrak{b}$ of a Lie algebra $\mathfrak{g}$ is called a *Borel subalgebra* if $\mathfrak{b}$ is a maximal solvable subalgebra of $\mathfrak{g}$. A connected subgroup B of a real or complex Lie group G is said to be a *Borel subgroup* of G if the Lie algebra $\mathfrak{b}$ of B is a

Table 3.1.

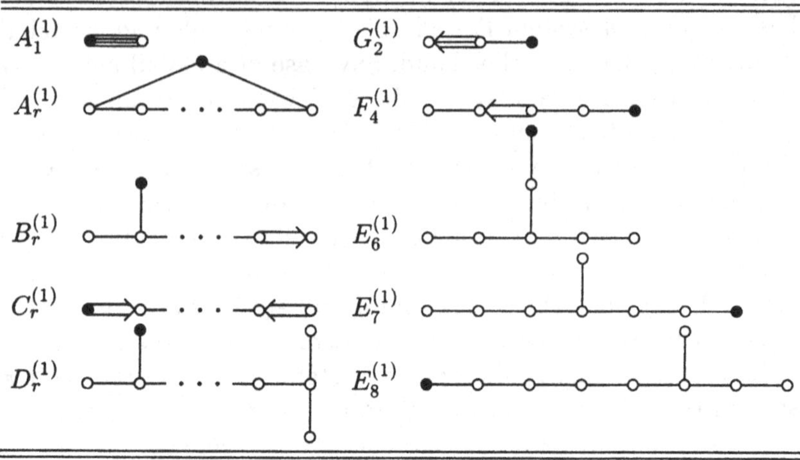

Borel subalgebra of the Lie algebra $\mathfrak{g}$ of G. According to the *Borel–Morozov theorem*, all Borel subalgebras of a complex Lie algebra and all Borel subgroups of a complex Lie group are conjugated to each other. The Borel subalgebras of a complex Lie algebra and Borel subgroups of a complex Lie group are regular subalgebras and regular subgroups, respectively. For any Borel subgroup B of a complex connected Lie group G, the homogeneous space G/B is a simply connected projective manifold.

The root decomposition (1.15) of $\mathfrak{g}$ implies that

$$\mathfrak{g} = \mathfrak{n}_- \oplus \mathfrak{h} \oplus \mathfrak{n}_+, \tag{3.6}$$

where

$$\mathfrak{n}_\pm \equiv \bigoplus_{\alpha \in \Delta_\pm} \mathfrak{g}^\alpha$$

are nilpotent subalgebras of $\mathfrak{g}$. The corresponding connected nilpotent Lie subgroups of G are denoted by $N_\pm$. For the connected subgroup generated by $\mathfrak{h}$ we use the notation H. It can be shown that the subalgebras

$$\mathfrak{b}_\pm \equiv \mathfrak{h} \oplus \mathfrak{n}_\pm \tag{3.7}$$

are Borel subalgebras of $\mathfrak{g}$. The Borel subalgebra $\mathfrak{b}_-$ is called *opposite* to the Borel subalgebra $\mathfrak{b}_+$. Using the notation introduced for regular subalgebras, we can write $\mathfrak{b}_\pm = \mathfrak{f}(\mathfrak{h}, \Delta_\pm)$. The subalgebras $\mathfrak{b}_\pm$ generate Borel subgroups $B_\pm$.

EXAMPLE 3.2 Let $\mathfrak{g}$ be one of the matrix complex simple Lie algebras $\mathfrak{sl}(m, \mathbb{C})$, $\widetilde{\mathfrak{o}}(m, \mathbb{C})$ or $\widetilde{\mathfrak{sp}}(m/2, \mathbb{C})$ considered in section 1.3. Recall that a Cartan subalgebra $\mathfrak{h}$ for all these Lie algebras can be chosen as $\mathfrak{h} = \mathfrak{g} \cap \mathfrak{d}(m, \mathbb{C})$, where $\mathfrak{d}(m, \mathbb{C})$ is the Lie algebra of all diagonal complex $m \times m$ matrices. In this case, using the bases of the corresponding root systems defined in section 1.3, we find that $\mathfrak{n}_\pm = \mathfrak{g} \cap \mathfrak{n}_\pm(m, \mathbb{C})$, where $\mathfrak{n}_+(m, \mathbb{C})$ and $\mathfrak{n}_-(m, \mathbb{C})$ are the Lie algebras of all strictly upper and lower triangular complex $m \times m$ matrices respectively. Therefore, the Borel subalgebras $\mathfrak{b}_\pm$ coincide with $\mathfrak{g} \cap \mathfrak{t}_\pm(m, \mathbb{C})$, where $\mathfrak{t}_+(m, \mathbb{C})$ and $\mathfrak{t}_-(m, \mathbb{C})$ are the Lie algebras of all upper and lower triangular complex $m \times m$ matrices respectively.

Let us consider the corresponding complex Lie groups. Denote by $\widetilde{\mathrm{SO}}(m, \mathbb{C})$ the complex Lie group consisting of complex $m \times m$ matrices which satisfy condition (2.47) with $b = \widetilde{I}_m$ and have unit determinant. Similarly, for an even m denote by $\widetilde{\mathrm{Sp}}(m/2, \mathbb{C})$ the complex Lie group formed by all complex $m \times m$ matrices satisfying (2.47) with $b = \widetilde{J}_{m/2}$. The Lie groups $\widetilde{\mathrm{SO}}(m, \mathbb{C})$ and $\widetilde{\mathrm{Sp}}(m/2, \mathbb{C})$ are isomorphic to the Lie groups $\mathrm{SO}(m, \mathbb{C})$ and $\mathrm{Sp}(m/2, \mathbb{C})$ respectively. The Lie algebras of $\widetilde{\mathrm{SO}}(m, \mathbb{C})$ and $\widetilde{\mathrm{Sp}}(m/2, \mathbb{C})$ are $\widetilde{\mathfrak{o}}(m, \mathbb{C})$ and $\widetilde{\mathfrak{sp}}(m/2, \mathbb{C})$.

Now let G be one of the complex Lie groups $\mathrm{SL}(m, \mathbb{C})$, $\widetilde{\mathrm{SO}}(m, \mathbb{C})$ or $\widetilde{\mathrm{Sp}}(m/2, \mathbb{C})$. The subgroup H in such a case coincides with $G \cap \mathrm{D}(m, \mathbb{C})$, where $\mathrm{D}(m, \mathbb{C})$ is the Lie group of all nonsingular diagonal complex $m \times m$ matrices. Further, we have $N_\pm = G \cap \mathrm{N}_\pm(m, \mathbb{C})$, where $\mathrm{N}_+(m, \mathbb{C})$ and $\mathrm{N}_-(m, \mathbb{C})$ are the complex Lie groups of all complex upper or lower triangular $m \times m$ matrices with unit diagonal elements. Finally, the Borel subgroups $B_\pm$ are given by the intersection $G \cap \mathrm{T}_\pm(m, \mathbb{C})$ with $\mathrm{T}_+(m, \mathbb{C})$ and $\mathrm{T}_-(m, \mathbb{C})$ being the complex Lie groups formed by upper and lower triangular nondegenerate complex $m \times m$ matrices respectively.

By definition, a *parabolic subalgebra* of $\mathfrak{g}$ is a subalgebra of $\mathfrak{g}$ which contains some Borel subalgebra of $\mathfrak{g}$. A subgroup P of a Lie group G is called a *parabolic subgroup* if it contains a Borel subgroup of G. The parabolic subalgebras and parabolic subgroups are regular subalgebras and regular subgroups respectively.

Let $\Pi = \{\alpha_1, \dots, \alpha_r\}$ be a system of simple roots of Δ and let Ψ be a subsystem of Π. Introduce the notation

$$\mathfrak{p}_{\pm\Psi} \equiv \mathfrak{b}_\pm \oplus \bigoplus_{\alpha \in [\Psi] \cap \Sigma_\mp} \mathfrak{g}^\alpha. \tag{3.8}$$

The subalgebras $\mathfrak{p}_{\pm\Psi}$ are parabolic subalgebras and any parabolic subalgebra of $\mathfrak{g}$, up to a transformation of the group $\mathrm{Aut}(\mathfrak{g})$, can be obtained in such a way. Thus, we have a transparent classification of the parabolic subalgebras of any complex semisimple Lie algebra. One usually writes $\mathfrak{p}_{\pm i_1, \dots, i_k}$ for the parabolic subalgebra corresponding to the subsystem $\Psi = \{\alpha_{i_1}, \dots, \alpha_{i_k}\}$. The parabolic subalgebras $\mathfrak{p}_{\pm(\Pi - \Psi)}$ are denoted by $\mathfrak{p}'_{\pm\Psi}$. Therefore, it is natural to write $\mathfrak{p}'_{\pm i_1, \dots, i_k}$ for the parabolic subalgebras corresponding to the subsystem of Π which contains all simple roots except the roots $\alpha_{i_1}, \dots, \alpha_{i_k}$. In particular, $\mathfrak{p}'_i$ is the parabolic subalgebra corresponding to a subsystem containing all simple roots except the root α_i.

Any maximal parabolic subalgebra of $\mathfrak{g}$ is not semisimple and, vice versa, any nonsemisimple maximal subalgebra of $\mathfrak{g}$ is a parabolic subalgebra.

For the parabolic subgroups we use the same notation as for the parabolic subalgebras with the change of $\mathfrak{p}$ to P. For example, the parabolic subgroups corresponding to the parabolic subalgebras $\mathfrak{p}_{\pm i_1, \dots, i_k}$, are denoted by $P_{\pm i_1, \dots, i_k}$.

Again let Ψ be a subsystem of the system of simple roots Π. Below we use the following notations:

$$\widetilde{\mathfrak{n}}_{\pm\Psi} \equiv \mathfrak{f}(0, \Delta^\pm - [\Psi]), \tag{3.9}$$

$$\widetilde{\mathfrak{h}}_\Psi \equiv \mathfrak{f}(\mathfrak{h}, [\Psi]). \tag{3.10}$$

When it is obvious which subsystem Ψ is under consideration, we write simply $\widetilde{\mathfrak{n}}_\pm$ and $\widetilde{\mathfrak{h}}$. It is clear that

$$\mathfrak{g} = \widetilde{\mathfrak{n}} \oplus \widetilde{\mathfrak{h}} \oplus \widetilde{\mathfrak{n}}. \tag{3.11}$$

Actually, $\widetilde{\mathfrak{n}}_{\pm\emptyset} = \mathfrak{n}_\pm$, $\widetilde{\mathfrak{h}}_\emptyset = \mathfrak{h}$, and we come to (3.6). In the general case $\widetilde{\mathfrak{n}}_{\pm\Psi} \subset \mathfrak{n}_\pm$ and $\widetilde{\mathfrak{h}}_\Psi \supset \mathfrak{h}$. From such point of view, it is also natural to introduce the notations

$$\widetilde{\mathfrak{b}}_{\pm\Psi} \equiv \widetilde{\mathfrak{h}}_\Psi \oplus \widetilde{\mathfrak{n}}_{\pm\Psi} = \mathfrak{p}_{\pm\Psi}. \tag{3.12}$$

For the Lie subgroups corresponding to the subalgebras $\widetilde{\mathfrak{n}}_\pm$, $\widetilde{\mathfrak{b}}_\pm$ and $\widetilde{\mathfrak{h}}$ we use the notations $\widetilde{N}_\pm$, $\widetilde{B}_\pm$ and $\widetilde{H}$ respectively.

EXAMPLE 3.3 Consider the Lie algebra $\mathfrak{sl}(m, \mathbb{C})$. Let $\{n_\alpha\}_{\alpha=1}^{k+1}$ be a fixed set of positive integers such that $\sum_{\alpha=1}^{k+1} n_\alpha = m$. It is obvious that the subalgebra of $\mathfrak{sl}(m, \mathbb{C})$, formed by the matrices a having the following block form:

$$a = \begin{pmatrix} x_1 & * & \cdots & * \\ 0 & x_2 & \cdots & * \\ \vdots & \vdots & \ddots & \vdots \\ 0 & 0 & \cdots & x_{k+1} \end{pmatrix},$$

where x_α are complex $n_\alpha \times n_\alpha$ matrices such that $\sum_{\alpha=1}^{k+1} \operatorname{tr} x_\alpha = 0$, is a parabolic subalgebra of $\mathfrak{sl}(m, \mathbb{C})$. Using the Cartan subalgebra of $\mathfrak{sl}(m, \mathbb{C})$ and the base of the corresponding root system introduced in section 1.3.1, we can easily see that one has here the parabolic subalgebra $\mathfrak{p}'_{i_1,\ldots,i_k}$ with $i_l = \sum_{\alpha=1}^{l} n_\alpha$. The corresponding parabolic subgroups of $\mathrm{SL}(m, \mathbb{C})$ consist of the matrices of the block form

$$a = \begin{pmatrix} X_1 & * & \cdots & * \\ 0 & X_2 & \cdots & * \\ \vdots & \vdots & \ddots & \vdots \\ 0 & 0 & \cdots & X_{k+1} \end{pmatrix}, \qquad (3.13)$$

where X_α are arbitrary complex $n_\alpha \times n_\alpha$ matrices such that $\prod_{\alpha=1}^{k+1} \det X_\alpha = 1$.

Any parabolic subgroup of a complex connected semisimple Lie group is connected and the homogeneous space G/P is a simply connected projective manifold. In particular, G/P is a compact manifold. The homogeneous space G/P is called a *flag manifold* or, quite rarely, a *parabolic space*. The flag manifolds corresponding to the parabolic subgroups $P_{\pm i_1,\ldots,i_k}$ are denoted by $F_{\mp i_1,\ldots,i_k}$; while for the flag manifolds corresponding to the parabolic subgroup $P'_{\pm i_1,\ldots,i_k}$ we use the notation $F'_{\mp i_1,\ldots,i_k}$.

EXAMPLE 3.4 The relation of the notion of a flag manifold to the notion of a flag is explained as follows. Let V be an m-dimensional vector space and let $i_1, \ldots, i_k$ be a set of integers such that $0 < i_1 < \ldots < i_k < m$. A family $\{V_l\}$ of subspaces of V such that

$\dim V_l = i_l$, $l = 1, \ldots, k$, and $V_1 \subset \ldots \subset V_k$ is called a *flag* of type $(i_1, \ldots, i_k)$ in V. A flag of type $(1, \ldots, m-1)$ is called the *full flag*. The set of all flags of type $(i_1, \ldots, i_k)$ in V is denoted by $F_{i_1,\ldots,i_k}(V)$. It is obvious that $F_i(V)$ is the Grassmann manifold $\mathbb{G}^i(V)$ defined in example 2.33.

Consider the case of the complex m-dimensional vector space V. It is clear that there is defined a natural left action of the Lie group $\mathrm{SL}(V)$ on the set $F_{i_1,\ldots,i_k}(V)$. It can be verified that this action is transitive. Let $\{e_i\}$ be a basis of V; consider the flag formed by the subspaces

$$V_l \equiv \bigoplus_{1 \le i \le i_l} \mathbb{C} e_i, \qquad l = 1, \ldots, k.$$

The corresponding isotropy subgroup consists of the elements $A \in \mathrm{SL}(V)$ whose matrix representation a with respect to the basis $\{e_i\}$ has the block form (3.13), where X_α, $\alpha = 1, \ldots, k+1$, are $n_\alpha \times n_\alpha$ matrices with $n_\alpha = i_\alpha - i_{\alpha-1}$ $(i_0 \equiv 0,\ i_{k+1} \equiv m)$. Thus, we see that the set $F_{i_1,\ldots,i_k}(V)$ can be identified with the homogeneous space $P\backslash\mathrm{SL}(m, \mathbb{C})$, where P is a parabolic subgroup of $\mathrm{SL}(m, \mathbb{C})$ formed by the matrices of form (3.13). This allows us to consider $F_{i_1,\ldots,i_k}(V)$ as a complex manifold. Note that the mapping $a \in \mathrm{SL}(m, \mathbb{C}) \mapsto a^{-1} \in \mathrm{SL}(m, \mathbb{C})$ provides identification of the homogeneous spaces $P\backslash\mathrm{SL}(m, \mathbb{C})$ and $\mathrm{SL}(m, \mathbb{C})/P$. Taking into account that $P = P'_{i_1,\ldots,i_k}$, we conclude that the manifold $F_{i_1,\ldots,i_k}(V)$ is diffeomorphic to the flag manifold $F'_{i_1,\ldots,i_k}$ of the Lie group $\mathrm{SL}(m, \mathbb{C})$. Note here that the flag manifold F'_i of $\mathrm{SL}(m, \mathbb{C})$ is diffeomorphic to the Grassmann manifold $\mathbb{C}\mathbb{G}^{i,m-i}$; in particular, the flag manifold F'_1 of $\mathrm{SL}(m, \mathbb{C})$ can be identified with the projective space $\mathbb{C}\mathbb{P}^{m-1}$.

3.1.3 $\mathbb{Z}$-graded Lie algebras

The notion of a gradation of an algebra is important both for the elucidation of its structure and for the classification of algebras. Let us begin with the definition of a graded vector space.

Let V be a vector space over a field $\mathbb{K}$ and let M be an abelian group. An *M-gradation* of V is a family $\{V_m\}_{m \in M}$ of subspaces of V such that

$$V = \bigoplus_{m \in M} V_m.$$

A vector space V is said to be *M-graded* if it is equipped with an M-gradation. An element of an M-graded vector space V is called *homogeneous of degree* $m \in M$ if it is an element of V_m.

An algebra A is called *M-graded* if it is an M-graded vector space and

$$A_m A_n \subset A_{m+n}, \quad m, n \in M.$$

In other words, the product of any element of the subspace A_m with any element of the subspace A_n belongs to the subspace A_{m+n}. Note that the subspace A_0 is a subalgebra of A.

Let an algebra A be endowed with an M-gradation

$$A = \bigoplus_{m \in M} A_m \tag{3.14}$$

and let φ be an automorphism of A. The representation

$$A = \bigoplus_{m \in M} \varphi(A_m) \tag{3.15}$$

defines another M-gradation of A. The M-gradations (3.14) and (3.15) are said to be connected by the automorphism φ. Two M-gradations of A, connected by an inner automorphism of A, are called *conjugated*.

EXAMPLE 3.5 Let Δ be a root system of a semisimple complex Lie algebra with respect to some Cartan subalgebra $\mathfrak{h}$ and let $\Pi = \{\alpha_1, \ldots, \alpha_r\}$ be a system of simple roots. Associate with a root $\alpha = \sum_{i=1}^r m_i \alpha_i \in \Delta$ the element $m = (m_1, \ldots, m_r) \in \mathbb{Z}^r$, and denote the root space $\mathfrak{g}^\alpha$ by $\mathfrak{g}_m$. If for $m = (m_1, \ldots, m_r) \in \mathbb{Z}^r$ one has $\sum_{i=1}^r m_i \alpha_i \notin \Delta$, put $\mathfrak{g}_m \equiv \{0\}$. Also denote the Cartan subalgebra $\mathfrak{h}$ by $\mathfrak{g}_{(0,\ldots,0)}$. It is easy to verify that the decomposition

$$\mathfrak{g} = \bigoplus_{m \in \mathbb{Z}^r} \mathfrak{g}_m$$

is a $\mathbb{Z}^r$-gradation of $\mathfrak{g}$.

According to a general definition, a $\mathbb{Z}$-gradation of a Lie algebra $\mathfrak{g}$ is a decomposition of $\mathfrak{g}$ into a direct sum of subspaces $\mathfrak{g}_m$

$$\mathfrak{g} = \bigoplus_{m \in \mathbb{Z}} \mathfrak{g}_m$$

such that

$$[\mathfrak{g}_m, \mathfrak{g}_k] \subset \mathfrak{g}_{m+k}.$$

For any $\mathbb{Z}$-gradation of $\mathfrak{g}$ we can define the derivation D of $\mathfrak{g}$ by

$$Dx \equiv \sum_{m \in \mathbb{Z}} m x_m.$$

We restrict ourselves to the case of semisimple Lie algebras. Since in this case any derivation of $\mathfrak{g}$ is an inner derivation, and the centre of $\mathfrak{g}$ is trivial, then there exists a unique element $q \in \mathfrak{g}$ such that

$$Dx = [q, x]$$

for all $x \in \mathfrak{g}$. In other words, for any $x \in \mathfrak{g}_m$ we have

$$[q, x] = mx.$$

The element q is called the *grading operator* of the $\mathbb{Z}$-gradation under consideration. Thus, for a semisimple Lie algebra, any $\mathbb{Z}$-gradation may be defined with the help of the corresponding grading operator.

Suppose now that $\mathfrak{g}$ is a complex semisimple Lie algebra. It is clear that for any grading operator q, the linear operator $\mathrm{ad}(q)$ is semisimple and satisfies the relation

$$\exp(2\pi\sqrt{-1}\,\mathrm{ad}(q)) = \mathrm{id}_{\mathfrak{g}}. \tag{3.16}$$

On the other hand, it is clear that any semisimple element q of $\mathfrak{g}$ satisfying (3.16) can be considered as the grading operator of some $\mathbb{Z}$-gradation of $\mathfrak{g}$.

Since any grading operator q is semisimple, we can suppose without loss of generality that q belongs to some Cartan subalgebra $\mathfrak{h}$ of $\mathfrak{g}$. Let $\Pi = \{\alpha_1, \ldots, \alpha_r\}$ be a base of the root system Δ of $\mathfrak{g}$ with respect to $\mathfrak{h}$. From (3.16) it follows that $\langle \alpha_i, q \rangle$, $i = 1, \ldots, r$, are integers. Furthermore, the element q belongs to the closure of some Weyl chamber C, and if Π is the base of Δ corresponding to this Weyl chamber, then all the numbers $n_i \equiv \alpha_i(q)$ are nonnegative. The numbers n_i do not depend on the choice of a Weyl chamber whose closure contains q. The corresponding Dynkin diagram, with the vertices labelled by the numbers n_i, is called the *characteristic* of q.

On the other hand, choosing an arbitrary set of nonnegative integers $\{n_i\}_{i=1}^r$, we can construct the element $q \in \mathfrak{h}$ as

$$q = \sum_{i,j=1}^{r} (k^{-1})_{ij} n_j h_i. \tag{3.17}$$

Here k^{-1} is the inverse of the Cartan matrix k of $\mathfrak{g}$. It is clear that the element q, defined by (3.17), is semisimple and satisfies (3.16). In other words, q is a grading operator. Moreover, for this element we have $\langle \alpha_i, q \rangle = n_i$. Thus, the Dynkin diagram, with the vertices labelled by arbitrary nonnegative integers, is the characteristic of some grading operator. Two grading operators are connected by an automorphism of $\mathfrak{g}$ if and only if they have the same characteristics. Two grading operators having the same characteristics, may not be conjugated. The number of the classes of conjugated grading operators coincides with the order of the symmetry group of the Dynkin diagram divided by the order of the symmetry group of the characteristic.

Note that the subspace $\mathfrak{g}_m$ is the sum of the root spaces $\mathfrak{g}^\alpha$ corresponding to the roots $\alpha = \sum_{1 \le i \le r} c_i \alpha_i$ with $\sum_{1 \le i \le r} c_i n_i = m$. The subspace $\mathfrak{g}_0$ also includes the Cartan subalgebra $\mathfrak{h}$. For any positive (negative) root α, the root space $\mathfrak{g}^\alpha$ belongs to some subspace $\mathfrak{g}_m$ with $m \ge 0$ ($m \le 0$). Further, if $\mathfrak{g}^\alpha \subset \mathfrak{g}_m$, then $\mathfrak{g}^{-\alpha} \subset \mathfrak{g}_{-m}$; hence, we have

$$\dim \mathfrak{g}_m = \dim \mathfrak{g}_{-m}.$$

Let a $\mathbb{Z}$-gradation of a complex semisimple Lie algebra $\mathfrak{g}$ be given. Find a Cartan subalgebra $\mathfrak{h}$ and a base $\Pi = \{\alpha_1, \ldots, \alpha_r\}$ of the root system Δ of $\mathfrak{g}$ with respect to $\mathfrak{h}$ such that for the corresponding grading operator q we have $\langle \alpha_i, q \rangle \ge 0$. Introduce the notation

$$\Psi \equiv \{\alpha_i \in \Pi \mid \langle \alpha_i, q \rangle = 0\}.$$

In accordance with (3.8), the subsystem Ψ defines a parabolic subalgebra of $\mathfrak{g}$. Here, for the subalgebras $\widetilde{\mathfrak{n}}_\pm(\Psi)$, $\widetilde{\mathfrak{b}}_\pm(\Psi)$ and $\widetilde{\mathfrak{h}}(\Psi)$ defined by (3.9), (3.12) and (3.10), we have

$$\widetilde{\mathfrak{n}}_- = \bigoplus_{m<0} \mathfrak{g}_m, \qquad \widetilde{\mathfrak{n}}_+ = \bigoplus_{m>0} \mathfrak{g}_m, \tag{3.18}$$

$$\widetilde{\mathfrak{b}}_- = \bigoplus_{m\le 0} \mathfrak{g}_m, \qquad \widetilde{\mathfrak{b}}_+ = \bigoplus_{m\ge 0} \mathfrak{g}_m, \tag{3.19}$$

$$\widetilde{\mathfrak{h}} = \mathfrak{g}_0. \tag{3.20}$$

For any complex semisimple Lie algebra one distinguished $\mathbb{Z}$-gradation arises when we choose all the numbers n_i equal to 1. In this case the corresponding grading operator has the form

$$q = \sum_{i=1}^{r} k_i h_i, \qquad k_i \equiv \sum_{j=1}^{r} (k^{-1})_{ij}. \tag{3.21}$$

Such a gradation is called the *canonical* or *principal gradation*. For the principal gradation $\widetilde{\mathfrak{n}}_\pm = \mathfrak{n}_\pm$, $\widetilde{\mathfrak{b}}_\pm = \mathfrak{b}_\pm$ and $\widetilde{\mathfrak{h}} = \mathfrak{h}$. Hence, the subalgebra $\mathfrak{g}_0$ is in this case abelian and the subspaces $\mathfrak{g}_{\pm 1}$ coincide with the linear spans of the Chevalley generators $x_{\pm i}$, $i = 1, \ldots, r$.

3.1.4 $\mathfrak{sl}(2, \mathbb{C})$-subalgebras

In this section we discuss the embeddings of the Lie algebra $\mathfrak{sl}(2, \mathbb{C})$ into a complex semisimple Lie algebra and the corresponding $\mathbb{Z}$-gradations. By an *embedding of* $\mathfrak{sl}(2, \mathbb{C})$ into $\mathfrak{g}$ we mean a nontrivial homomorphism from $\mathfrak{sl}(2, \mathbb{C})$ into $\mathfrak{g}$. Let h, $x_\pm$ be the elements of the standard basis of $\mathfrak{sl}(2, \mathbb{C})$ introduced in example 1.11. Note that h is a Cartan generator, while $x_\pm$ are Chevalley generators of $\mathfrak{sl}(2, \mathbb{C})$. The images of the elements h and $x_\pm$ under the homomorphism defining the embedding of $\mathfrak{sl}(2, \mathbb{C})$ under consideration are usually denoted by the same letters, optionally endowed with an index distinguishing different embeddings. The image of the whole $\mathfrak{sl}(2, \mathbb{C})$ is called an $\mathfrak{sl}(2, \mathbb{C})$-*subalgebra* of $\mathfrak{g}$. It can be shown that two $\mathfrak{sl}(2, \mathbb{C})$-subalgebras of $\mathfrak{g}$ are connected by an automorphism of $\mathfrak{g}$ if and only if the corresponding Cartan generators are connected by an automorphism of $\mathfrak{g}$. Here $\mathfrak{sl}(2, \mathbb{C})$-subalgebras are conjugated in $\mathfrak{g}$ if and only if their Cartan generators are conjugated in $\mathfrak{g}$.

EXAMPLE 3.6 Let $\mathfrak{h}$ be a Cartan subalgebra of a complex semisimple Lie algebra $\mathfrak{g}$ and let α be a root of $\mathfrak{g}$ with respect to $\mathfrak{h}$. It can be shown that one can choose the elements $x_{\pm\alpha} \in \mathfrak{g}^{\pm\alpha}$ in such a way that

$$K(x_\alpha, x_{-\alpha}) = 2/(\alpha, \alpha),$$

where K is the Killing form of $\mathfrak{g}$, and $(\cdot\,, \cdot)$ is the bilinear form on $\mathfrak{h}^*$ induced by the restriction of K to $\mathfrak{h}$. In this case we have

$$[x_\alpha, x_{-\alpha}] = \alpha^\vee, \qquad [\alpha^\vee, x_{\pm\alpha}] = \pm 2 x_{\pm\alpha}.$$

Hence, the elements $h \equiv \alpha^\vee$, $x_\pm \equiv x_{\pm\alpha}$ form a basis of some $\mathfrak{sl}(2, \mathbb{C})$-subalgebra of $\mathfrak{g}$.

For a given embedding of $\mathfrak{sl}(2, \mathbb{C})$ into $\mathfrak{g}$, the adjoint representation of $\mathfrak{g}$ defines the representation of the Lie algebra $\mathfrak{sl}(2, \mathbb{C})$ in $\mathfrak{g}$. From the properties of the finite-dimensional representations of $\mathfrak{sl}(2, \mathbb{C})$ it follows that the element h of $\mathfrak{g}$ must be semisimple, and the elements $x_\pm \in \mathfrak{g}$ must be nilpotent. Moreover, it is clear that

$$\exp(2\pi\sqrt{-1}\,\mathrm{ad}(h)) = \mathrm{id}_\mathfrak{g}. \tag{3.22}$$

Therefore, the element h can be used as the grading operator defining some $\mathbb{Z}$-gradation on $\mathfrak{g}$. As above, without any loss of generality, one can suppose that h belongs to some Cartan subalgebra $\mathfrak{h}$ of $\mathfrak{g}$, and we can choose a base $\Pi = \{\alpha_1, \ldots, \alpha_r\}$ of the root system of $\mathfrak{g}$ with respect to $\mathfrak{h}$ in such a way that the numbers $\langle \alpha_i, h \rangle$, $i = 1, \ldots, r$, are nonnegative integers. It can be shown that the numbers $\langle \alpha_i, h \rangle$ can be equal only to 0, 1 and 2. In fact, it is more convenient for our purposes to define the grading operator q, connected with the given embedding of $\mathfrak{sl}(2, \mathbb{C})$ into $\mathfrak{g}$, by the relation

$$h = 2q.$$

It is clear that this definition necessarily leads to $\mathbb{Z}/2$-gradations of $\mathfrak{g}$ as well. If, instead of (3.22), one has the relation

$$\exp(\pi\sqrt{-1}\,\mathrm{ad}(h)) = \mathrm{id}_\mathfrak{g},$$

we call the corresponding embedding *integral*; otherwise we deal with a *semi-integral embedding*. For an integral embedding the numbers $\langle \alpha_i, h \rangle$ are equal to 0 or 2. Note here that not every element $h \in \mathfrak{h}$, even one satisfying the requirement $\langle \alpha_i, h \rangle = 0, 1$ or 2, can be considered as the corresponding element of some embedding of $\mathfrak{sl}(2, \mathbb{C})$ into $\mathfrak{g}$, and we do not have here a direct relation to parabolic subalgebras, as for the case of a general $\mathbb{Z}$-gradation. Note also that the properties of the finite-dimensional representations of $\mathfrak{sl}(2, \mathbb{C})$ imply that if one considers a $\mathbb{Z}$-gradation or a $\mathbb{Z}/2$-gradation associated with an embedding of $\mathfrak{sl}(2, \mathbb{C})$ into $\mathfrak{g}$, then

$$\dim \mathfrak{g}_0 \geq \dim \mathfrak{g}_{\pm 1}.$$

Specifying in a certain way an $\mathfrak{sl}(2, \mathbb{C})$-subalgebra of a complex semisimple Lie algebra $\mathfrak{g}$, one can parametrise $\mathfrak{g}$ in accordance with the arising representation of $\mathfrak{sl}(2, \mathbb{C})$. Here all the elements of the algebra $\mathfrak{g}$ can be collected into multiplets corresponding to finite-dimensional irreducible representations of $\mathfrak{sl}(2, \mathbb{C})$. In order to parametrise the elements of $\mathfrak{g}$, one usually uses one of the two bases. The first is the root basis which is universal for all semisimple Lie algebras but is not very suitable for physical applications with tensor calculations, in particular for standard methods in atomic physics. The second basis uses rather cumbersome tensor notations which are restricted in their applications to the classical series and are quite tedious. Grouping of the elements of Lie algebras $\mathfrak{g}$ based on the consideration of some embedding of the Lie algebra $\mathfrak{sl}(2, \mathbb{C})$ takes, in a sense, some intermediate place since in the framework of this classification the generality of the root language is complemented by the visuality of the multiplet structure that physicists use and find convenient.

We now proceed to the discussion of concrete $\mathfrak{sl}(2, \mathbb{C})$-subalgebras of complex semisimple Lie algebras. First consider the case of complex matrix semisimple Lie algebras. Here, as well as the representation of $\mathfrak{sl}(2, \mathbb{C})$ induced by the adjoint representation, we have one more representation of $\mathfrak{sl}(2, \mathbb{C})$ realised by the corresponding matrices. The simplest case here is the special linear algebra $\mathfrak{sl}(m, \mathbb{C})$. Having an $\mathfrak{sl}(2, \mathbb{C})$ subalgebra of $\mathfrak{sl}(m, \mathbb{C})$, we actually have a faithful m-dimensional representation of $\mathfrak{sl}(2, \mathbb{C})$. Recall that any finite-dimensional representation of $\mathfrak{sl}(2, \mathbb{C})$ is a direct sum of irreducible representations, each of them being isomorphic to some of the representations considered in example 1.11. Any such representation is uniquely determined by a nonnegative integer n and has the dimension $d = n+1$. So we have the splitting

$$m = d_1 + \cdots + d_s, \quad d_1 \geq d_2 \cdots \geq d_s > 1, \qquad (3.23)$$

where $d_1, \ldots, d_s$ are the dimensions of the irreducible components of the representation of $\mathfrak{sl}(2, \mathbb{C})$. The case of $s = m$, when all d_is are equal to 1 must be excluded because it corresponds to the trivial representation. On the other hand, any splitting of m of form (3.23) corresponds to a possible m-dimensional representation which is realised by matrices belonging to $\mathfrak{sl}(m, \mathbb{C})$. Therefore, such a splitting corresponds to an embedding of $\mathfrak{sl}(2, \mathbb{C})$ into $\mathfrak{sl}(m, \mathbb{C})$. Two representations corresponding to the same splitting are connected by a transformation of $\mathrm{SL}(m, \mathbb{C})$, and any two representations of $\mathfrak{sl}(2, \mathbb{C})$ corresponding to different splittings cannot be connected by an automorphism of $\mathfrak{sl}(m, \mathbb{C})$. Thus, the splittings of the integer m of form (3.23) and nonconjugated embeddings of

$\mathfrak{sl}(2,\mathbb{C})$ into $\mathfrak{sl}(m,\mathbb{C})$ are in bijective correspondence.

A representation ρ of a Lie algebra $\mathfrak{g}$ in a vector space V is said to be *orthogonal* (*symplectic*) if there exists a symmetric (skew-symmetric) nondegenerate bilinear form B on V such that

$$B(v,\rho(x)u) + B(\rho(x)v,u) = 0$$

for all $v, u \in V$ and $x \in \mathfrak{g}$. An irreducible representation of $\mathfrak{sl}(2,\mathbb{C})$ is orthogonal (symplectic) if and only if its dimension is odd (even).

A representation of $\mathfrak{sl}(2,\mathbb{C})$ is orthogonal (symplectic) if and only if the multiplicities of its even-dimensional (odd-dimensional) irreducible components are even. These properties of the representations of $\mathfrak{sl}(2,\mathbb{C})$ imply that if one has an embedding of $\mathfrak{sl}(2,\mathbb{C})$ into $\mathfrak{o}(m,\mathbb{C})$ ($\mathfrak{sp}(m/2,\mathbb{C})$), then the corresponding splitting (3.23) contains any even (odd) summand an even number of times. On the other hand, any representation of $\mathfrak{sl}(2,\mathbb{C})$ defined by such a splitting can be realised by orthogonal (symplectic) matrices. Moreover, any two representations corresponding to the same splitting are connected by an element of the group $\mathrm{SO}(m,\mathbb{C})$ ($\mathrm{Sp}(m/2,\mathbb{C})$), except in the case of the orthogonal representations of $\mathfrak{sl}(2,\mathbb{C})$ with all the dimensions of the irreducible components being even. In the last case, realised only when m is a multiple of 4, the corresponding representations fall into two classes. The representations belonging to the same class are connected by elements of $\mathrm{SO}(m,\mathbb{C})$, while the representations of different classes are connected by elements of $\mathrm{O}(m,\mathbb{C})$. Thus, with only one exception, the nonconjugated embeddings of $\mathfrak{sl}(2,\mathbb{C})$ into $\mathfrak{o}(m,\mathbb{C})$ ($\mathfrak{sp}(m/2,\mathbb{C})$) are in bijective correspondence with the splittings (3.23) of m, where any even (odd) summand is contained an even number of times. The embeddings of $\mathfrak{sl}(2,\mathbb{C})$ into $\mathfrak{o}(m,\mathbb{C})$ corresponding to a splitting into even numbers only fall into two classes of conjugated embeddings connected by an 'external' automorphism of $\mathfrak{o}(m,\mathbb{C})$.

Thus, we have a complete classification of $\mathfrak{sl}(2,\mathbb{C})$-subalgebras of complex matrix simple Lie algebras. In table 3.2 we present the results of this classification for the algebras of rank less than or equal to 3. Underlined numbers denote the dimensions of the irreducible components of the corresponding representation of $\mathfrak{sl}(2,\mathbb{C})$, while the ordinary numbers denote the multiplicities.

Table 3.2.

$\mathfrak{sl}(2,\mathbb{C})$	A_1	$\underline{2}$
$\mathfrak{sl}(3,\mathbb{C})$	A_2	$\underline{3},\ \underline{2}+\underline{1}$
$\mathfrak{sl}(4,\mathbb{C})$	A_3	$\underline{4},\ \underline{3}+\underline{1},\ 2\cdot\underline{2},\ \underline{2}+2\cdot\underline{1}$
$\mathfrak{o}(3,\mathbb{C})$	A_1	$\underline{3}$
$\mathfrak{o}(4,\mathbb{C})$	$A_1\times A_1$	$\underline{3}+\underline{1},\ 2\cdot\underline{2}$
$\mathfrak{o}(5,\mathbb{C})$	C_2	$\underline{5},\ \underline{3}+2\cdot\underline{1},\ 2\cdot\underline{2}+\underline{1}$
$\mathfrak{o}(6,\mathbb{C})$	A_3	$\underline{5}+\underline{1},\ 2\cdot\underline{3},\ \underline{3}+3\cdot\underline{1},\ 2\cdot\underline{2}+2\cdot\underline{1}$
$\mathfrak{o}(7,\mathbb{C})$	B_3	$\underline{7},\ \underline{5}+2\cdot\underline{1},\ 2\cdot\underline{3}+\underline{1},\ \underline{3}+2\cdot\underline{2},\ \underline{3}+4\cdot\underline{1},\ 2\cdot\underline{2}+3\cdot\underline{1}$
$\mathfrak{sp}(1,\mathbb{C})$	A_1	$\underline{2}$
$\mathfrak{sp}(2,\mathbb{C})$	C_2	$\underline{4},\ 2\cdot\underline{2},\ \underline{2}+2\cdot\underline{1}$
$\mathfrak{sp}(3,\mathbb{C})$	C_3	$\underline{6},\ \underline{4}+\underline{2},\ \underline{4}+2\cdot\underline{1},\ 2\cdot\underline{3},\ 3\cdot\underline{2},\ 2\cdot\underline{2}+2\cdot\underline{1},\ \underline{2}+4\cdot\underline{1}$

Unfortunately, the above consideration is applicable only to the case of the classical series of Lie algebras; moreover, it does not give a constructive procedure for finding the concrete form of the elements h, $x_\pm$ specifying a concrete embedding. A general method, which can be used for all simple Lie algebras, was developed by Dynkin (1957a). This method is based on the fact that an embedding of $\mathfrak{sl}(2,\mathbb{C})$ into a complex semisimple Lie algebra $\mathfrak{g}$ is uniquely characterised by the characteristic of the corresponding element $h \in \mathfrak{g}$. Namely, two $\mathfrak{sl}(2,\mathbb{C})$-subalgebras of $\mathfrak{g}$ are connected by an automorphism of $\mathfrak{g}$ if and only if the corresponding Cartan generators have the same characteristics. The number of classes of conjugated $\mathfrak{sl}(2,\mathbb{C})$-subalgebras corresponding to the same characteristic coincides with the order of the symmetry group of the Dynkin diagram divided by the order of the symmetry group of the characteristic.

It is important for the method considered that for any complex semisimple Lie algebra $\mathfrak{g}$ there is an embedding of $\mathfrak{sl}(2,\mathbb{C})$ leading to the principal gradation. This embedding is defined by

$$x_\pm = \sum_{i=1}^{r}(2k_i)^{1/2}x_{\pm i}, \qquad h = 2\sum_{i=1}^{r} k_i h_i, \tag{3.24}$$

where h_i and $x_{\pm i}$ are Cartan and Chevalley generators of $\mathfrak{g}$ and the numbers k_i are defined by (3.21). This embedding is called the

Table 3.3.

	k	k^{-1}	$2k_i$
A_2	$\begin{pmatrix} 2 & -1 \\ -1 & 2 \end{pmatrix}$	$\frac{1}{3}\begin{pmatrix} 2 & 1 \\ 1 & 2 \end{pmatrix}$	2 2
A_3	$\begin{pmatrix} 2 & -1 & 0 \\ -1 & 2 & -1 \\ 0 & -1 & 2 \end{pmatrix}$	$\frac{1}{4}\begin{pmatrix} 3 & 2 & 1 \\ 2 & 4 & 2 \\ 1 & 2 & 3 \end{pmatrix}$	3 4 3
B_3	$\begin{pmatrix} 2 & -1 & 0 \\ 1 & 2 & -2 \\ 0 & -1 & 2 \end{pmatrix}$	$\frac{1}{2}\begin{pmatrix} 2 & 2 & 2 \\ 2 & 4 & 4 \\ 1 & 2 & 3 \end{pmatrix}$	6 10 6
C_2	$\begin{pmatrix} 2 & -1 \\ -2 & 2 \end{pmatrix}$	$\frac{1}{2}\begin{pmatrix} 2 & 1 \\ 2 & 2 \end{pmatrix}$	3 4
C_3	$\begin{pmatrix} 2 & -1 & 0 \\ 1 & 2 & -1 \\ 0 & -2 & 2 \end{pmatrix}$	$\frac{1}{2}\begin{pmatrix} 2 & 2 & 1 \\ 2 & 4 & 2 \\ 2 & 4 & 3 \end{pmatrix}$	5 8 9
G_2	$\begin{pmatrix} 2 & -1 \\ -3 & 2 \end{pmatrix}$	$\begin{pmatrix} 2 & 1 \\ 3 & 2 \end{pmatrix}$	6 10

principal embedding ; its exhaustive investigation was performed by Kostant (1959). For the principal embedding all the numbers $\langle \alpha_i, h \rangle$ are equal to 2. Hence, this embedding is defined up to conjugation. Note that for a principal embedding $\dim \mathfrak{g}_0 = \dim \mathfrak{g}_{\pm 1}$. In table 3.3 we give the necessary information about the principal embeddings of $\mathfrak{sl}(2, \mathbb{C})$ into simple Lie algebras of rank less or equal to 3.

For a principal $\mathfrak{sl}(2, \mathbb{C})$-subalgebra of a complex semisimple algebra $\mathfrak{g}$, the representation of $\mathfrak{sl}(2, \mathbb{C})$, obtained by reducing the adjoint representation of $\mathfrak{g}$ to $\mathfrak{sl}(2, \mathbb{C})$, has the number of irreducible components equal to the rank of $\mathfrak{g}$. The dimensions of the irreducible components can be calculated as follows. Let

$\Pi = \{\alpha_1, \dots, \alpha_r\}$ be a base of the root system Δ of $\mathfrak{g}$. Consider the element $s \equiv s_1 \cdots s_r$ of the Weyl group $W(\Delta)$, where, as usual, $s_i \equiv s_{\alpha_i}$, $i = 1, \dots, r$. The eigenvalues of s have the form $\exp(2\pi\sqrt{-1}l_i/c)$, where l_i, $i = 1, \dots r$, are positive integers and c is the Coxeter number of Δ. The numbers l_i are called the *exponents* of Δ. The dimensions n_i of the irreducible components of the considered representation of $\mathfrak{sl}(2, \mathbb{C})$ are $n_i = 2l_i + 1$. Furthermore, in the case of complex simple Lie algebras, the multiplicity of a given irreducible representation is always unity, except the case of the series D_r for an even r, where there are two representations of dimension $2r - 1$.

An $\mathfrak{sl}(2, \mathbb{C})$-subalgebra of $\mathfrak{g}$ is called *semiprincipal* if it is not contained in any proper regular subalgebra of $\mathfrak{g}$. For any semiprincipal embedding, the labels of the Dynkin diagram, specifying the corresponding characteristic, are equal to 0 or 2 only. Any principal $\mathfrak{sl}(2, \mathbb{C})$-subalgebra of a complex semisimple Lie algebra is also semiprincipal. For the Lie algebras of types A_r, B_r, C_r, G_2 and F_4, any semiprincipal $\mathfrak{sl}(2, \mathbb{C})$ subalgebra is a principal $\mathfrak{sl}(2, \mathbb{C})$-subalgebra. For a Lie algebra of type D_r there are $[(r-2)/2]$ nonconjugated semiprincipal $\mathfrak{sl}(2, \mathbb{C})$-subalgebras which are not principal. For the Lie algebras of types E_r, $r = 6, 7, 8$, there are $[(r-3)/2]$ such subalgebras.

Note that the classification of the semiprincipal $\mathfrak{sl}(2, \mathbb{C})$ subalgebras of complex semisimple Lie algebras is reduced to the classification of the semiprincipal subalgebras of complex simple Lie algebras in the following way. Let $\mathfrak{g}$ be a complex semisimple Lie algebra and let

$$\mathfrak{g} = \mathfrak{g}_1 \times \cdots \times \mathfrak{g}_k$$

be its representation as a direct product of simple ideals. Consider a set of homomorphisms $\iota_i : \mathfrak{sl}(2, \mathbb{C}) \to \mathfrak{g}_i$, $i = 1, \dots, k$, which specify semiprincipal subalgebras of Lie algebras $\mathfrak{g}_k$. The mapping $\iota : \mathfrak{sl}(2, \mathbb{C}) \to \mathfrak{g}$, defined by

$$\iota(x) \equiv \iota_1(x) + \cdots + \iota_k(x),$$

for all $x \in \mathfrak{sl}(2, \mathbb{C})$, specifies a semiprincipal $\mathfrak{sl}(2, \mathbb{C})$-subalgebra of $\mathfrak{g}$. It appears that any semiprincipal $\mathfrak{sl}(2, \mathbb{C})$-subalgebra of $\mathfrak{g}$ can be obtained in this way.

From the definition of semiprincipal $\mathfrak{sl}(2, \mathbb{C})$-subalgebras it follows that any $\mathfrak{sl}(2, \mathbb{C})$-subalgebra $\mathfrak{g}'$ of a complex semisimple Lie

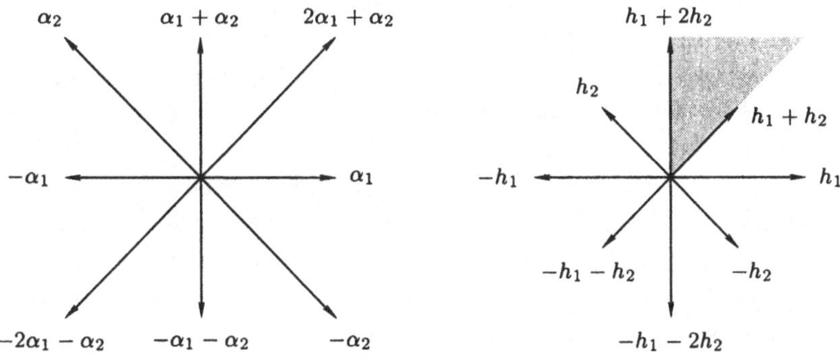

Fig. 3.1

algebra $\mathfrak{g}$ is a semiprincipal $\mathfrak{sl}(2,\mathbb{C})$-subalgebra of any minimal regular subalgebra of $\mathfrak{g}$ containing $\mathfrak{g}'$. This fact allows one to formulate the following constructive procedure for finding all $\mathfrak{sl}(2,\mathbb{C})$-subalgebras of a complex semisimple Lie algebra $\mathfrak{g}$. First, enumerate all semisimple regular subalgebras of $\mathfrak{g}$. This can be done using the method described in subsection 3.1.2. Then, for each of these subalgebras, consider all its semiprincipal $\mathfrak{sl}(2,\mathbb{C})$-subalgebras. For any such subalgebra find the characteristic of the corresponding Cartan generator. Finally, comparing the characteristics obtained, single out the nonconjugated subalgebras.

For a detailed and rather explicit consideration of $\mathfrak{sl}(2,\mathbb{C})$-subalgebras of complex simple Lie algebras we refer to Dynkin (1957a) and Lorente & Gruber (1972). Here we consider only one example.

EXAMPLE 3.7 Consider the case of the Lie algebra $\mathfrak{sp}(2,\mathbb{C})$. As follows from table 3.2, there are three nonconjugated $\mathfrak{sl}(2,\mathbb{C})$-subalgebras. The Lie algebra $\mathfrak{sp}(2,\mathbb{C})$ is of type C_2. The corresponding root system is $\Delta=\{\pm\alpha_1,\pm\alpha_2,\pm(\alpha_1+\alpha_2),\pm(2\alpha_1+\alpha_2)\}$. This root system and the corresponding dual root system are depicted in figure 3.1, where the notation $h_i\equiv(\alpha_i)^\vee$, $i=1,2$, is used and the Weyl chamber corresponding to the base $\Pi=\{\alpha_1,\alpha_2\}$ is coloured gray. The analysis based on the usage of the extended Dynkin diagram shows that one has four π-systems: $\Psi_1=\{\alpha_1,\alpha_2\}$, $\Psi_2=\{\alpha_1\}$, $\Psi_3=\{\alpha_2\}$ and $\Psi_4=\{\alpha_2,2\alpha_1+\alpha_2\}$, which are not connected by transformations of the Weyl group $W(\Delta)$. The first π-system corresponds to the principal embedding and, as follows from table 3.3, the corresponding Cartan generator here is $3h_1 + 4h_2$. For the next three cases, the Cartan generators are h_1, h_2 and $h_1 + 2h_2$. To find the characteristics, consider for each case a transformation of $W(\Delta)$ which brings the Cartan generator to the closure of the Weyl chamber corresponding to the base Π. It is clear that this procedure gives new Cartan generators $3h_1 + 4h_2$, $h_1 + 2h_2$, $h_1 + h_2$, and $h_1 + 2h_2$. Hence, the π-systems Ψ_2 and Ψ_4 give conjugated $\mathfrak{sl}(2, \mathbb{C})$-subalgebras; therefore, one can exclude the π-system Ψ_4 from the consideration. The labels of the characteristic are calculated using the Cartan matrix. The characteristics and the grading subspaces corresponding to the grading operator $q = h/2$ are

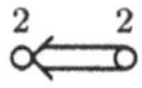

2 ⇐ 2

$$\mathfrak{g}_0 = \mathfrak{h}, \; \mathfrak{g}_{\pm 1} = \mathfrak{g}^{\pm\alpha_1} \oplus \mathfrak{g}^{\pm\alpha_2}, \; \mathfrak{g}_{\pm 2} = \mathfrak{g}^{\pm(\alpha_1+\alpha_2)}, \; \mathfrak{g}_{\pm 3} = \mathfrak{g}^{\pm(2\alpha_1+\alpha_2)},$$

0 ⇐ 2

$$\mathfrak{g}_0 = \mathfrak{h} \oplus \mathfrak{g}^{\alpha_1} \oplus \mathfrak{g}^{-\alpha_1}, \; \mathfrak{g}_{\pm 1} = \mathfrak{g}^{\pm\alpha_2} \oplus \mathfrak{g}^{\pm(\alpha_1+\alpha_2)} \oplus \mathfrak{g}^{\pm(2\alpha_1+\alpha_2)},$$

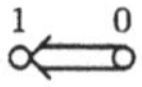

1 ⇐ 0

$$\mathfrak{g}_0 = \mathfrak{h} \oplus \mathfrak{g}^{\alpha_2} \oplus \mathfrak{g}^{-\alpha_2}, \; \mathfrak{g}_{\pm 1/2} = \mathfrak{g}^{\pm\alpha_1} \oplus \mathfrak{g}^{\pm(\alpha_1+\alpha_2)}, \; \mathfrak{g}_{\pm 1} = \mathfrak{g}^{\pm(2\alpha_1+\alpha_2)}.$$

As we have seen, any $\mathfrak{sl}(2, \mathbb{C})$-subalgebra of a complex semisimple Lie algebra $\mathfrak{g}$ gives an integral or a semi-integral gradation of $\mathfrak{g}$. The grading operator here is $h/2$. On the other hand, let a complex semisimple Lie algebra $\mathfrak{g}$ be equipped with an integral or semi-integral gradation and let q be the corresponding grading operator. For any element $x \in \mathfrak{g}_{+1}$ there is an $\mathfrak{sl}(2, \mathbb{C})$ subalgebra such that $x_+ = x$, $h \in \mathfrak{g}_0$ and $x_- \in \mathfrak{g}_{-1}$; see Delduc, Ragoucy & Sorba (1992); Fehér *et al.* (1992). Writing the grading operator in the form $q = h/2 + y$, one can prove that the element y belongs to the subspace

$$\mathfrak{g}_0^0 \equiv \widetilde{\mathfrak{h}}^0 \equiv \{x \in \mathfrak{g}_0 \mid [x_\pm, x] = 0\}. \qquad (3.25)$$

Thus, any integral or semi-integral gradation of $\mathfrak{g}$ gives a set of $\mathfrak{sl}(2, \mathbb{C}) \times \mathfrak{gl}(1, \mathbb{C})$-subalgebras of $\mathfrak{g}$ specified by the choice made for the element x_+.

Exercises

3.1 Construct the generalised Cartan matrices corresponding to the extended Dynkin diagrams for the simple Lie algebras of rank 2 and 3.

3.2 Verify that the Cartan generators of the principal embedding of $\mathfrak{sl}(2,\mathbb{C})$ into complex matrix simple Lie algebras are given by

$$\mathfrak{sl}(r+1,\mathbb{C}) : \; h = \sum_{i=1}^{r} i(r-i+1)h_i,$$

$$\tilde{\mathfrak{o}}(2r+1,\mathbb{C}) : \; h = \sum_{i=1}^{r-1} i(2r-i+1)h_i + \frac{r(r+1)}{2}h_r,$$

$$\widetilde{\mathfrak{sp}}(r,\mathbb{C}) : \; h = \sum_{i=1}^{r-1} i(2r-i)h_i + r^2 h_r,$$

$$\tilde{\mathfrak{o}}(2r,\mathbb{C}) : \; h = \sum_{i=1}^{r-2} i(2r-i-1)h_i + \frac{r(r-1)}{2}(h_{r-1}+h_r).$$

3.3 Study the embeddings of $\mathfrak{sl}(2,\mathbb{C})$ into complex matrix simple Lie algebras of rank 2 and 3; namely, find the characteristics and the grading subspaces.

3.4 Show the multiplet structure of the complex simple Lie algebras of rank 2 with respect to the integral embeddings of $\mathfrak{sl}(2,\mathbb{C})$.

3.5 Consider the $\mathfrak{sl}(2,\mathbb{C})$-subalgebra of $\mathfrak{sl}(m,\mathbb{C})$, such that the corresponding representation of $\mathfrak{sl}(2,\mathbb{C})$ has the splitting into irreducible components of the form $\underline{m} = k \cdot \underline{d}$, or of the form $\underline{m} = k \cdot \underline{d} + \underline{1}$. Find the structure of the grading subspaces.

3.2 Zero curvature representation of Toda-type systems

3.2.1 Gauge transformations

In the physical literature, principal bundle isomorphisms are often called *gauge transformations*. In other words, a gauge transformation of a principal fibre G-bundle $P \xrightarrow{\pi} M$ is a diffeomorphism

$\varphi : P \to P$ satisfying the relations

$$\pi \circ \varphi = \pi, \tag{3.26}$$

$$\varphi \circ R^P_a = R^P_a \circ \varphi \tag{3.27}$$

for any $a \in G$. The notion of a gauge transformation is very important in modern mathematical physics; therefore, we consider it in a more general framework than is really ideal for the problems considered in the book.

Let φ be a gauge transformation of a principal fibre G-bundle $P \xrightarrow{\pi} M$. Consider a set of local sections $s_\alpha : U_\alpha \to P$, $\alpha \in \mathcal{A}$, covering M. Recall that the set $\{s_\alpha\}$ generates a bundle atlas $\{U_\alpha, \psi_\alpha\}_{\alpha \in \mathcal{A}}$, where the mappings $\psi_\alpha : \pi^{-1}(U_\alpha) \to U_\alpha \times G$ are defined with the help of the relation

$$\psi_\alpha^{-1}(q, a) \equiv s_\alpha(q) \cdot a. \tag{3.28}$$

Note also that the mappings $g_\alpha : \pi^{-1}(U_\alpha) \to G$, defined by

$$g_\alpha \equiv \mathrm{pr}_G \circ \psi_\alpha,$$

satisfy the evident relation

$$g_\alpha(p \cdot a) = g_\alpha(p)a \tag{3.29}$$

for any $a \in G$. Using the mappings g_α, we can also write

$$\psi_\alpha(p) = (\pi(p), g_\alpha(p)), \quad p \in \pi^{-1}(U_\alpha). \tag{3.30}$$

Proposition 3.1 *For any $(q, a) \in U_\alpha \times G$ one has*

$$\psi_\alpha \circ \varphi \circ \psi_\alpha^{-1}(q, a) = (q, \varphi_\alpha(q)a), \tag{3.31}$$

where the mappings $\varphi_\alpha : U_\alpha \to G$ are defined by

$$\varphi_\alpha \equiv g_\alpha \circ \varphi \circ s_\alpha.$$

Proof Using relations (3.28) and (3.27), we obtain

$$\psi_\alpha \circ \varphi \circ \psi_\alpha^{-1}(q, a) = \psi_\alpha \circ \varphi(s_\alpha(q) \cdot a).$$

Now (3.30) and (3.29) give

$$\psi_\alpha \circ \varphi \circ \psi_\alpha^{-1}(q, a) = (\pi \circ \varphi \circ s_\alpha(q), g_\alpha \circ \varphi \circ s_\alpha(q)a).$$

Taking into account (3.26) and the relation

$$\pi \circ s_\alpha = \mathrm{id}_M,$$

we obtain (3.31). □

From (3.28), (3.31) and (3.30) it follows directly that

$$\varphi|_{U_\alpha}(p) = s_\alpha(\pi(p)) \cdot (\varphi_\alpha(\pi(p)) g_\alpha(p)). \tag{3.32}$$

Using this relation, we can recover the mapping φ from the mappings φ_α, $\alpha \in \mathcal{A}$.

We introduce some notations which will be used for mappings from a set to a group. Let φ be a mapping from a set S to a group G. Denote by φ^{-1} the mapping from S to G defined by

$$\varphi^{-1}(p) \equiv (\varphi(p))^{-1}.$$

Further, for any two mappings φ and ψ from S to G we use the notation $\varphi\psi$ for the mapping from S to G defined by

$$(\varphi\psi)(p) \equiv \varphi(p)\psi(p).$$

Proposition 3.2 *On $U_\alpha \cap U_\beta$ one has*

$$\varphi_\beta = g_{\beta\alpha} \varphi_\alpha g_{\alpha\beta} = g_{\beta\alpha} \varphi_\alpha g_{\beta\alpha}^{-1}, \tag{3.33}$$

where the $g_{\alpha\beta}$ are the transition functions of the bundle atlas on P, generated by the set $\{s_\alpha\}$.

Proof Formulas (3.28) and (3.30) imply

$$p = s_\alpha(\pi(p)) \cdot g_\alpha(p) \tag{3.34}$$

for any $p \in \pi^{-1}(U_\alpha)$. Recall that the transition functions $g_{\alpha\beta}$ are given by

$$g_{\alpha\beta} = g_\alpha \circ s_\beta. \tag{3.35}$$

Now, from the definition of the mappings g_α and from (3.34) and (3.35) we obtain

$$g_\beta(p) = g_{\beta\alpha}(\pi(p)) g_\alpha(p), \quad p \in \pi^{-1}(U_\alpha \cap U_\beta). \tag{3.36}$$

Putting $p = s_\beta(q)$, $q \in U_\alpha \cap U_\beta$ in (3.34) we obtain the following relation:

$$s_\beta(q) = s_\alpha(q) g_{\alpha\beta}(q). \tag{3.37}$$

Now writing

$$\varphi_\beta(q) = g_\beta \circ \varphi \circ s_\beta(q)$$

and taking (3.36) and (3.37) into account, we obtain (3.33). □

Proposition 3.3 *Any set of mappings φ_α, $\alpha \in \mathcal{A}$, satisfying (3.33) defines a gauge transformation φ of P via (3.32).*

Proof First, we should show that (3.32) defines the mapping φ in a correct way. In other words, for any $p \in \pi^{-1}(U_\alpha \cap U_\beta)$ we must have

$$s_\alpha(\pi(p)) \cdot (\varphi_\alpha(\pi(p)) g_\alpha(p)) = s_\beta(\pi(p)) \cdot (\varphi_\beta(\pi(p)) g_\beta(p)).$$

It is easy to show that, due to (3.33), (3.36) and (3.37), this equality is valid.

Now we should verify the validity of relations (3.26) and (3.27). In fact, (3.26) is obvious, while (3.27) follows from (3.29). □

Proposition 3.4 *Let ω be the connection form of some connection on P. The $\mathfrak{g}$-valued 1-form $\varphi^*\omega$ defines some new connection on P.*

Proof Show that for the form $\varphi^*\omega$ the corresponding analogues of relations (2.70) and (2.71) are valid.

From (2.12), for any $v \in \mathfrak{g}$ and any $f \in \mathfrak{F}(P)$ we obtain

$$(\varphi_* X_v^P)_{\varphi(p)}(f) = X_{vp}^P(\varphi^* f).$$

Hence, thanks to (2.64) and (3.27),

$$X_{vp}^P(\varphi^* f) = \frac{d}{dt} f(\varphi(p) \cdot \exp(tv))\bigg|_{t=0} = X_{v\varphi(p)}^P(f).$$

Therefore, one sees that

$$\varphi^* X_v^P = X_v^P.$$

Relation (2.70) now implies that

$$\varphi^*\omega(X_v) = v.$$

Using (3.27) and (2.71), for any $a \in G$ we obtain

$$R_a^{P*}\varphi^*\omega = \varphi^* R_a^{P*}\omega = \mathrm{Ad}(a^{-1}) \circ \varphi^*\omega.$$

Thus, the $\mathfrak{g}$-valued 1-form $\varphi^*\omega$ defines some connection on P. □

Recall that the set $\{\omega_\alpha\}$ of 1-forms

$$\omega_\alpha \equiv s_\alpha^*\omega$$

completely determines the connection form ω. The connection form $\varphi^*\omega$ generates the set $\{(\varphi^*\omega)_\alpha\}$ of 1-forms, defined as

$$(\varphi^*\omega)_\alpha \equiv s_\alpha^*\varphi^*\omega. \tag{3.38}$$

Proposition 3.5 *The following relation:*

$$(\varphi^* \omega)_\alpha = \mathrm{Ad}(\varphi_\alpha^{-1}) \circ \omega_\alpha + \varphi_\alpha^* \theta,$$

with θ being the Maurer–Cartan form of G, is valid.

Proof The statement of the propositions follows from (3.38) and (2.72). □

Proposition 3.6 *Let Ω be the curvature form associated with the connection form ω. Then the curvature form corresponding to the connection form $\varphi^*\omega$, is $\varphi^*\Omega$.*

Proof From relation (2.74), with account of (2.50) and (2.51), one obtains

$$\varphi^*\Omega = d\varphi^*\omega + \frac{1}{2}[\varphi^*\omega, \varphi^*\omega].$$

Hence, $\varphi^*\Omega$ is the curvature form of the connection determined by the connection form $\varphi^*\omega$. □

The curvature form Ω is completely determined by the set $\{\Omega_\alpha\}$ of the 2-forms

$$\Omega_\alpha \equiv s_\alpha^* \Omega.$$

The corresponding forms determining the curvature form $\varphi^*\Omega$ are defined by

$$(\varphi^*\Omega)_\alpha \equiv s_\alpha^* \varphi^* \Omega. \tag{3.39}$$

Proposition 3.7 *The relation*

$$(\varphi^*\Omega)_\alpha = \mathrm{Ad}(\varphi_\alpha^{-1}) \circ \Omega_\alpha$$

is valid.

Proof The relation in question follows directly from (3.39) and (2.75). □

3.2.2 Zero curvature condition

Also in this chapter we consider flat connections on a trivial principal fibre bundle. Let M be a manifold and let G be a Lie group. Consider the trivial principal G-bundle $M \times G \to M$.

Any such bundle has a bundle atlas consisting of only one chart $(M \times G, \mathrm{id}_{M\times G})$. Hence, as follows from the discussion given in section 2.8.2, we have in the case under consideration a bijective correspondence between connection forms and $\mathfrak{g}$-valued 1-forms on M. Bearing this correspondence in mind, we call a $\mathfrak{g}$-valued 1-form on M a connection form, or simply a connection. The curvature 2-form of a connection ω is determined by the 2-form Ω on M, related to ω by the formula

$$\Omega = d\omega + \frac{1}{2}[\omega, \omega].$$

From (2.75) we conclude that the connection ω is flat if and only if

$$d\omega + \frac{1}{2}[\omega, \omega] = 0. \tag{3.40}$$

We call relation (3.40) the *zero curvature condition.*

For any smooth mapping $\varphi : M \to G$, define a $\mathfrak{g}$-valued 1-form ${}^{\varphi}\omega$ as

$${}^{\varphi}\omega \equiv \varphi^*\theta, \tag{3.41}$$

where θ is the Maurer–Cartan form of G. From (2.57) it follows that ${}^{\varphi}\omega$ satisfies the zero curvature condition. Note that if G is a matrix group, then φ is a matrix valued function and one can write

$${}^{\varphi}\omega = \varphi^{-1}d\varphi.$$

It appears that if the manifold M is simply connected, then any connection satisfying the zero curvature condition has the form ${}^{\varphi}\omega$ for some smooth mapping $\varphi : M \to G$. Note that ${}^{\varphi}\omega = 0$ if and only if φ is a constant mapping.

Proposition 3.8 *For any two mappings $\varphi, \psi : M \to G$, the following relation:*

$${}^{\varphi\psi}\omega = \mathrm{Ad}(\psi^{-1}) \circ {}^{\varphi}\omega + {}^{\psi}\omega \tag{3.42}$$

is valid. In particular, one has

$${}^{\varphi^{-1}}\omega = -\mathrm{Ad}(\varphi) \circ {}^{\varphi}\omega.$$

Proof For any $p \in M$ and $v \in T_p^{\mathbb{C}}(M)$ we have

$${}^{\varphi\psi}\omega(v) = (\varphi\psi)^*\theta(v) = \theta((\varphi\psi)_{*p}(v)).$$

It is not difficult to show that

$$(\varphi\psi)_{*p} = R_{\psi(p)*\varphi(p)} \circ \varphi_{*p} + L_{\varphi(p)*\psi(p)} \circ \psi_{*p}.$$

Now, using (2.54) and (2.55), we arrive at (3.42). □

The gauge transformations in the case under consideration are described by smooth mappings from M to G. Let ψ be such a mapping and let ω be a connection form. Proposition 3.5 implies that the gauge transformed connection form ω^ψ is given by the relation

$$\omega^\psi = \mathrm{Ad}(\psi^{-1}) \circ \omega + \psi^*\theta. \tag{3.43}$$

For the case of a matrix group G, ω is a matrix valued 1-form, ψ is a matrix valued function, and (3.43) takes the form

$$\omega^\psi = \psi^{-1}\omega\psi + \psi^{-1}d\psi.$$

From proposition 3.7 it follows that the zero curvature condition is invariant with respect to the gauge transformations (3.43). In other words, if a connection ω satisfies this condition, then the connection ω^ψ also satisfies it. It is convenient to call the gauge transformations defined by (3.43) G-gauge transformations. In fact, using proposition 3.8, one can easily show that

$$({}^\varphi\omega)^\psi = {}^{\varphi\psi}\omega$$

for any smooth mappings φ and ψ.

Proposition 3.9 *The equality*

$${}^\varphi\omega = {}^{\varphi'}\omega, \tag{3.44}$$

is valid if and only if $\varphi'\varphi^{-1}$ is a constant mapping.

Proof Performing the gauge transformation corresponding to the mapping φ^{-1}, from (3.44) we obtain the equality

$${}^{\varphi'\varphi^{-1}}\omega = 0;$$

hence, $\varphi'\varphi^{-1}$ is a constant mapping. □

Actually, we will consider the zero curvature condition for the case where M is a complex one-dimensional manifold and G is a complex semisimple Lie group. It is convenient to use the notations z^- and z^+, respectively, for a local coordinate z on M and its conjugate $\bar{z}$. Write for ω the representation

$$\omega = \omega_- dz^- + \omega_+ dz^+,$$

where $\omega_\pm$ are some mappings from M to $\mathfrak{g}$. In what follows the superscripts minus and plus mean for 1-forms on M the corresponding components in the expansion over the local basis formed by dz^- and dz^+. In terms of $\omega_\pm$ the zero curvature condition takes the form

$$\partial_- \omega_+ - \partial_+ \omega_- + [\omega_-, \omega_+] = 0. \tag{3.45}$$

Here and in what follows we use the notation

$$\partial_- \equiv \partial/\partial z^-, \quad \partial_+ \equiv \partial/\partial z^+.$$

Choosing a basis in $\mathfrak{g}$ and treating the components of the expansion of $\omega_\pm$ over this basis as fields, we can consider the zero curvature condition as a nonlinear system of partial differential equations for the fields.

Suppose also that the manifold M is simply connected. In this case any flat connection can be gauge transformed to zero. In this sense system (3.45) is trivial. On the other hand, the majority of two-dimensional integrable equations can be obtained from system (3.45) by imposing some gauge noninvariant constraints on the connection form ω. Note that, in general, for the case of infinite-dimensional Lie algebras and Lie groups one needs a generalisation of the scheme, see, for example, Leznov & Saveliev (1992), but in the present book we restrict ourselves to the finite-dimensional case. Consider one of the methods of imposing the conditions in question, giving, in fact, a differential geometric formulation of the group-algebraic approach for integrating nonlinear systems in the spirit of Leznov & Saveliev (1992).

3.2.3 Grading condition

Suppose that the Lie algebra $\mathfrak{g}$ is a $\mathbb{Z}$-graded Lie algebra. The first condition we impose, in accordance with Leznov & Saveliev (1992), on the connection ω is the following. Let $\widetilde{\mathfrak{b}}_\pm$ be the subalgebras of $\mathfrak{g}$ given by (3.19). Require that the (1,0)-component of the form ω takes values in $\widetilde{\mathfrak{b}}_-$, and that its (0,1)-component takes values in $\widetilde{\mathfrak{b}}_+$. We call this condition the *general grading condition*. Any connection ω satisfying the general grading condition is certainly of the form ${}^\varphi\omega$ for some mapping $\varphi : M \to G$; however not each mapping φ leads to the connection ${}^\varphi\omega$ satisfying this condition.

Let us formulate the requirements which should be imposed on the mapping φ to guarantee the validity of the general grading condition.

Note first that the connected subgroups $\widetilde{B}_\pm$ of G corresponding to the parabolic subalgebras $\widetilde{\mathfrak{b}}_\pm$ are parabolic subgroups. Hence, the homogeneous spaces $F_\pm = G/\widetilde{B}_\mp$ are flag manifolds. Let $\pi_\pm : G \to F_\pm$ be the canonical projections. Define the mappings $\varphi_\pm : M \to F_\pm$ by

$$\varphi_\pm = \pi_\pm \circ \varphi.$$

The mappings $\pi_\pm$ are holomorphic. Hence, for any $a \in G$ we have

$$\pi_{\pm * a} \circ J^G_a = J^{F_\pm}_{\pi_\pm(a)} \circ \pi_{\pm * a}. \tag{3.46}$$

Further, there are defined the natural left actions $L^{F_\pm}$ of the Lie group G on $F_\pm$, satisfying the condition

$$\pi_\pm \circ L_a = L^{F_\pm}_a \circ \pi_\pm \tag{3.47}$$

for any $a \in G$.

Theorem 3.1 *The connection ${}^\varphi\omega$ satisfies the general grading conditions if and only if the mapping φ_- is holomorphic and the mapping φ_+ is antiholomorphic.*

Proof Suppose that the 1-form $({}^\varphi\omega)^{(0,1)}$ takes values in $\widetilde{\mathfrak{b}}_+$. Thus, for any $v \in T^{\mathbb{C}}_p(M)$ we have

$$\pi_{-*e}[{}^\varphi\omega(\bar{P}^M_p(v))] = 0, \tag{3.48}$$

where the linear operator $\bar{P}^M_p$ projects the tangent vector v to its $(0,1)$-component, see (2.18). Using (3.41) we obtain from (3.48) the equality

$$\pi_{-*e}[\theta(\varphi_{*p} \circ \overline{P}^M_p(v))] = 0.$$

Taking (2.61) into account, we obtain the relation

$$\pi_{*e} \circ L_{\varphi^{-1}(p)*\varphi(p)} \circ P^G_{\varphi(p)} \circ \varphi_{*p} \circ \overline{P}^M_p = 0. \tag{3.49}$$

From (3.47) it follows that

$$\pi_{\pm * ab} \circ L_{a*b} = L^{F_\pm}_{a*\pi_\pm(b)} \circ \pi_{\pm * b}$$

for all $a, b \in G$. Hence,

$$\pi_{-*e} \circ L_{\varphi^{-1}(p)*\varphi(p)} = L^{F_-}_{\varphi^{-1}*\pi_-(\varphi(p))} \circ \pi_{-*\varphi(p)}. \tag{3.50}$$

Further, relation (3.46) implies that

$$\pi_{\pm * a} \circ P_a^G = P_{\pi_\pm(a)}^{F_\pm} \circ \pi_{\pm * a}$$

for any $a \in G$. Therefore,

$$\pi_{- * \varphi(p)} \circ P_{\varphi(p)}^G = P_{\varphi_-(p)}^{F_-} \circ \pi_{- * \varphi(p)}. \tag{3.51}$$

Using (3.50) and (3.51) we obtain the following equality from (3.49):

$$L_{\varphi^{-1}(p) * \varphi_-(p)}^{F_-} \circ P_{\varphi_-(p)}^{F_-} \circ \varphi_{- * p} \circ \overline{P}_p^M = 0.$$

Since, for any $a \in G$ the mapping $L_a^{F_-}$ is a diffeomorphism, we conclude from this equality that

$$P_{\varphi_-(p)}^{F_-} \circ \varphi_{- * p} \circ \overline{P}_p^M = 0. \tag{3.52}$$

The mapping $\varphi_{- * p}$ is real; therefore, after the complex conjugation of (3.52), we obtain

$$\overline{P}_{\varphi_-(p)}^{F_-} \circ \varphi_{- * p} \circ P_p^M = 0. \tag{3.53}$$

It follows from (3.52) and (3.53) that

$$J_{\varphi_-(p)}^{F_-} \circ \varphi_{- * p} = \varphi_{- * p} \circ J_p^M,$$

and so the mapping φ_- is holomorphic.

Suppose now that the mapping φ_- is holomorphic. Reversing the arguments given above, we conclude that the form $\omega^{(0,1)}$ takes values in $\widetilde{\mathfrak{b}}_+$.

The case of the mapping φ_+ can be considered in the same way. □

We call a mapping φ generating a connection which satisfies the general grading condition a mapping satisfying the general grading condition.

EXAMPLE 3.8 Consider the case of $G = \mathrm{SL}(2, \mathbb{C})$ and endow the Lie algebra $\mathfrak{sl}(2, \mathbb{C})$ with the principal gradation corresponding to the choice of the Cartan and Chevalley generators described in section 1.3.1. In this case the flag manifolds $F_\pm$ are diffeomorphic to the projective space $\mathbb{CP}^1$ and we will identify them. It is not difficult to show that the projections $\pi_\pm$ can be defined as

$$\pi_+(a) \equiv (a_{12} : a_{22}), \qquad \pi_-(a) \equiv (a_{11} : a_{21}),$$

where

$$a = \begin{pmatrix} a_{11} & a_{12} \\ a_{21} & a_{22} \end{pmatrix}.$$

Define the functions $\varphi_{ij} : M \to \mathbb{C}$, $i, j = 1, 2$, by

$$\varphi_{ij} \equiv g_{ij} \circ \varphi,$$

where the mappings $g_{ij} : \mathrm{SL}(2, \mathbb{C}) \to \mathbb{C}$, $i, j = 1, 2$, are given by

$$g_{ij}(a) \equiv a_{ij}.$$

Note that for any $p \in M$ either $\varphi_{12}(p)$ or $\varphi_{22}(p)$ is different from zero. So in some neighbourhood of p at least one of the functions $\varphi_{12}/\varphi_{22}$ and $\varphi_{22}/\varphi_{12}$ is well defined. From theorem 3.26 it follows now that if the mapping φ satisfies the general grading condition, then these functions are holomorphic. Similarly, in some neighbourhood of p at least one of the functions $\varphi_{11}/\varphi_{21}$ and $\varphi_{21}/\varphi_{11}$ is well defined, and if φ satisfies the general grading condition, then they are antiholomorphic.

Now, again following Leznov & Saveliev (1992), perform a further specification of the grading condition. Define the subspaces $\widetilde{\mathfrak{m}}_\pm$ of $\mathfrak{g}$ by

$$\widetilde{\mathfrak{m}}_+ \equiv \bigoplus_{1 \le m \le l_+} \mathfrak{g}_m, \quad \widetilde{\mathfrak{m}}_- \equiv \bigoplus_{-l_- \le m \le -1} \mathfrak{g}_m, \tag{3.54}$$

where $l_\pm$ are some positive integers. Let us require that the (1,0)-component of the connection ω takes values in the linear space $\widetilde{\mathfrak{m}}_- \oplus \widetilde{\mathfrak{h}}$, and that the (0,1)-component of it takes values in $\widetilde{\mathfrak{h}} \oplus \widetilde{\mathfrak{m}}_+$. We will call such a requirement the *specified grading condition*. To reformulate this grading condition as a condition imposed on the corresponding mapping φ, introduce some holomorphic distributions on the manifolds F_+ and F_-.

Note that the subspace $\widetilde{\mathfrak{b}}_- \oplus \widetilde{\mathfrak{m}}_+$ is invariant with respect to the adjoint action of the subgroup $\widetilde{B}_-$ in $\mathfrak{g}$. Let $p \in F_+$ and let a be any element of G such that $\pi_+(a) = p$. Define the subspace $\mathcal{M}_{+p} \subset T_p^{(1,0)}(F_+)$ by

$$\mathcal{M}_{+p} \equiv \pi_{+*a}(\widetilde{\mathfrak{b}}_{-a} \oplus \widetilde{\mathfrak{m}}_{+a}) = \pi_{+*a}(\widetilde{\mathfrak{m}}_{+a}),$$

where

$$\widetilde{\mathfrak{b}}_{-a} \equiv L_{a*e}(\widetilde{\mathfrak{b}}_-), \quad \widetilde{\mathfrak{m}}_{+a} \equiv L_{a*e}(\widetilde{\mathfrak{m}}_+).$$

The subspaces $\mathcal{M}_{+p}$, $p \in F_+$, generate a distribution on F_+ which will be denoted by $\mathcal{M}_+$. In the same way we can define an analogous distribution $\mathcal{M}_-$ on F_-.

Theorem 3.2 *The connection ${}^{\varphi}\omega$ satisfies the specified grading condition if and only if the mapping φ_- is holomorphic and tangent to the distribution $\mathcal{M}_-$ while the mapping φ_+ is antiholomorphic and tangent to the distribution $\mathcal{M}_+$.*

Proof Suppose that the connection ${}^{\varphi}\omega$ satisfies the specified grading condition. Let $p \in M$ and $v \in T_p^{\mathbb{C}}(M)$. Following the proof of theorem 3.1, we see that

$$ {}^{\varphi}\omega(\overline{P}_p^M v) = L_{\varphi^{-1}(p)*\varphi(p)}(P_{\varphi(p)}^G \circ \varphi_{*p} \circ \overline{P}_p^M(v)). $$

From this relation it follows that $P_{\varphi(p)}^G \circ \varphi_{*p} \circ \overline{P}_p^M(v) \in \widetilde{\mathfrak{h}} \oplus \widetilde{\mathfrak{m}}_{+p}$. Therefore, $\pi_{+*\varphi(p)} \circ P_{\varphi(p)}^G \circ \varphi_{*p} \circ \overline{P}_p^M(v) \in \mathcal{M}_{+\varphi_+(p)}$. Since the mapping π_+ is holomorphic, while φ_+ is antiholomorphic, we obtain

$$ \pi_{+*\varphi(p)} \circ P_{\varphi(p)}^G \circ \varphi_{*p} \circ \overline{P}_p^M = P_{\varphi_+(p)}^{F_+} \circ \varphi_{+*p}. $$

Hence, $P_{\varphi_+(p)}^{F_+} \circ \varphi_{+*p}(x) \in \mathcal{M}_{+\varphi_+(p)}$; and the mapping φ_+ is tangent to the distribution $\mathcal{M}_+$. □

Theorem 3.2 shows that the specified grading condition is directly related to the notion of a superhorisontal mapping, see Burstall & Rawnsley (1990).

It is clear that the specified grading condition is not invariant under the action of an arbitrary G-gauge transformation, but it is invariant under the action of gauge transformations (3.43) generated by the mappings taking values in the subgroup $\widetilde{H}$ corresponding to the subalgebra $\widetilde{\mathfrak{h}} = \mathfrak{g}_0$. Such transformations form a subgroup of the group of G-gauge transformations. We call a gauge transformation from this subgroup an $\widetilde{H}$-gauge transformation. To obtain a system of equations having no $\widetilde{H}$-gauge invariance, we should impose further restrictions on the connection form. In fact, following Leznov & Saveliev (1992) we choose another way leading to the system in question. It consists of constructing some $\widetilde{H}$-gauge invariant mappings and rewriting the zero curvature conditions in terms of them. To this end, let us first consider the structure of the holomorphic principal fibre bundles $G \to F_\pm$.

3.2.4 Modified Gauss decomposition

Consider first the Gauss decomposition for matrices. Taking into account future applications, we deal with matrices over an algebra. In other words, we suppose that the matrix elements of the considered matrices are elements of an associative unital algebra. In this case it is convenient to use the theory of quasideterminants in the form given by Gelfand & Retakh (1991, 1992).

Let $\mathcal{I}$ and $\mathcal{J}$ be two ordered sets, each consisting of m elements. Consider a matrix $a = (a_{ij})$, $i \in \mathcal{I}$, $j \in \mathcal{J}$, with the matrix elements belonging to some associative unital algebra A. Define the family of m^2 elements $|a|_{kl}$, $k \in \mathcal{I}, l \in \mathcal{J}$ of the algebra A which are called the *quasideterminants* of a. In the case of $m = 1$ there is only one quasideterminant defined by $|a|_{kk} \equiv a_{kk}$. Denote by $a^{(i;j)}$ the matrix which is obtained from the matrix a by eliminating the ith row and jth column. The quasideterminant $|a|_{kl}$ of the matrix a is defined by

$$|a|_{kl} \equiv a_{kl} - \sum_{i \neq k, j \neq l} a_{ki} |a^{(k;l)}|_{ji}^{-1} a_{jl}.$$

It is clear that the quasideterminant $|a|_{kl}$ exists only if the quasideterminants $|a^{(k;l)}|_{ij}$ for any $i \neq k$ and $j \neq l$ are invertible elements of A.

For the matrix

$$a = \begin{pmatrix} a_{11} & a_{12} \\ a_{21} & a_{22} \end{pmatrix}$$

one obtains four quasideterminants:

$$\begin{aligned} |a|_{11} &= a_{11} - a_{12} a_{22}^{-1} a_{21}, \quad & |a|_{12} &= a_{12} - a_{11} a_{21}^{-1} a_{22}, \\ |a|_{21} &= a_{21} - a_{22} a_{12}^{-1} a_{11}, \quad & |a|_{22} &= a_{22} - a_{21} a_{11}^{-1} a_{12}. \end{aligned}$$

Put $b_{ij} = |a|_{ji}^{-1}$; it appears that

$$\sum_{k \in \mathcal{J}} a_{ik} b_{kj} = \sum_{k \in \mathcal{I}} b_{ik} a_{kj} = \delta_{ij}.$$

Hence, the matrix $b \equiv (b_{ij})$ is the inverse of the matrix a. If the algebra A is commutative, then

$$|a|_{kl} = (-1)^{k+l} \frac{\det a}{\det a^{(k;l)}}, \tag{3.55}$$

and one arrives at the usual expression for the inverse matrix.

We now introduce some more notations. Let $\mathcal{K} \subset \mathcal{I}$ and $\mathcal{L} \subset \mathcal{J}$. Denote by $a^{(\mathcal{K};\mathcal{L})}$ the matrix obtained from a by eliminating the rows with the indices from $\mathcal{K}$ and the columns with the indices from $\mathcal{L}$. The notation $a_{(\mathcal{K};\mathcal{L})}$ will be used for the submatrix of a

composed of the matrix elements a_{kl} with $k \in \mathcal{K}$ and $l \in \mathcal{L}$. In other words, $a_{(\mathcal{K};\mathcal{L})} = a^{(\mathcal{I}-\mathcal{K};\mathcal{J}-\mathcal{L})}$.

Consider now the Gauss decomposition of matrices over an associative algebra with unit. Suppose that for the matrix $a = (a_{ij})$, $i, j = 1, \ldots, m$, there exist the quasideterminants

$$h_{-kk} \equiv |a_{(k,\ldots,m;k,\ldots,m)}|_{kk}.$$

Define

$$m_{+kl} \equiv |a_{(k,l+1,\ldots,m;l,\ldots,m)}|_{kl} h_{-ll}^{-1}, \quad 1 \le k < l \le m,$$
$$n_{-kl} \equiv |a_{(k,\ldots,m;l,k+1,\ldots,m)}|_{kl} h_{-ll}^{-1}, \quad 1 \le l < k \le m,$$

and put $h_{-kl} = 0$ for $k \neq l$; $m_{+kl} = 0$ for $l < k$, $m_{+kk} = 1$; $n_{-kl} = 0$ for $k < l$, $n_{-kk} = 1$. One can now prove that

$$a = m_+ n_- h_-. \tag{3.56}$$

This decomposition is called the *Gauss decomposition* of the matrix a. Such a decomposition, if it exists, is unique. For the case where $m = 2$ one obtains

$$h_- = \begin{pmatrix} a_{11} - a_{12}a_{22}^{-1}a_{21} & 0 \\ 0 & a_{22} \end{pmatrix}, \quad m_+ = \begin{pmatrix} 1 & a_{12}a_{22}^{-1} \\ 0 & 1 \end{pmatrix},$$

$$n_- = \begin{pmatrix} 1 & 0 \\ a_{21}(a_{11} - a_{12}a_{22}^{-1}a_{21})^{-1} & 1 \end{pmatrix}.$$

Consider the case where the algebra A is commutative. Denote

$$d_{-k} \equiv \det a_{(k,\ldots,m;k,\ldots,m)}, \qquad 1 \le k \le m,$$

and put $d_{-(m+1)} \equiv 1$. Now, using (3.55), one obtains

$$h_{-kk} = \frac{d_{-k}}{d_{-(k+1)}}, \qquad 1 \le k \le m,$$

$$m_{+kl} = \frac{\det a_{(k,l+1,\ldots,m;l,\ldots,m)}}{d_{-l}}, \qquad 1 < k < l \le m,$$

$$n_{-kl} = \frac{\det a_{(k,\ldots,m;l,k+1,\ldots,m)} d_{-(l+1)}}{d_{-(k+1)} d_{-l}}, \qquad 1 < l < k \le m.$$

These are standard expressions for the Gauss decomposition of matrices.

There is another form of the Gauss decomposition. It is constructed as follows. Let us suppose that for the matrix $a = (a_{ij})$, $i, j = 1, \ldots, m$, there exist the quasideterminants

$$h_{+kk} \equiv |a_{(1,\ldots,k;1,\ldots,k)}|_{kk}.$$

Define

$$m_{-kl} \equiv |a_{(1,\ldots,l-1,k;1,\ldots,l)}|_{kl} h_{+ll}^{-1}, \quad 1 \le l < k \le m,$$
$$n_{+kl} \equiv |a_{(1,\ldots,k;1,\ldots,k-1,l)}|_{kl} h_{+ll}^{-1}, \quad 1 \le k < l \le m,$$

and put $h_{+kl} = 0$ for $k \neq l$; $m_{-kl} = 0$ for $k < l$, $m_{-kk} = 1$; $n_{+kl} = 0$ for $l < k$, $n_{+kk} = 1$. In this case one has

$$a = m_- n_+ h_+. \tag{3.57}$$

For the case where $m = 2$ one obtains

$$h_+ = \begin{pmatrix} a_{11} & 0 \\ 0 & a_{22} - a_{21}a_{11}^{-1}a_{12} \end{pmatrix}, \quad m_- = \begin{pmatrix} 1 & 0 \\ a_{21}a_{11}^{-1} & 1 \end{pmatrix},$$
$$n_+ = \begin{pmatrix} 1 & a_{12}(a_{22} - a_{21}a_{11}^{-1}a_{12})^{-1} \\ 0 & 1 \end{pmatrix}.$$

Consider again the case where the algebra A is commutative. Here we denote

$$d_{+k} \equiv \det a_{(1,\ldots,k;1,\ldots,k)}, \qquad 1 \le k \le m,$$

and put $d_{+0} \equiv 1$. Again using (3.55), one obtains

$$h_{+kk} = \frac{d_{+k}}{d_{+(k-1)}}, \qquad 1 \le k \le m,$$
$$m_{-kl} = \frac{\det a_{(1,\ldots,l-1,k;1,\ldots,l)}}{d_{+l}}, \qquad 1 < l < k \le m,$$
$$n_{+kl} = \frac{\det a_{(1,\ldots,k;1,\ldots,k-1,l)} d_{+(l-1)}}{d_{+(k-1)} d_{+l}}, \qquad 1 < k < l \le m$$

One can use (3.56) and (3.57) to rewrite the Gauss decomposition in various different forms. For example, defining $k_- \equiv m_-$, $h \equiv h_+$ and $k_+ \equiv h_+^{-1} n_+ h_+$, one obtains from (3.57) the equality

$$a = k_- h k_+.$$

In particular, for the case where $m = 2$ one obtains

$$h = \begin{pmatrix} a_{11} & 0 \\ 0 & a_{22} - a_{21}a_{11}^{-1}a_{12} \end{pmatrix}, \quad k_- = \begin{pmatrix} 1 & 0 \\ a_{21}a_{11}^{-1} & 1 \end{pmatrix},$$
$$k_+ = \begin{pmatrix} 1 & a_{11}^{-1}a_{12} \\ 0 & 1 \end{pmatrix}.$$

For more information about quasideterminants we refer the reader to the papers by Gelfand & Retakh (1991, 1992).

We now proceed to the case of Lie groups. Let G be a complex connected semisimple Lie group. Decomposition (3.11) of the corresponding Lie algebra $\mathfrak{g}$ leads to the following decomposition of G. Let $\widetilde{N}_\pm$ and $\widetilde{H}$ be the subgroups of G corresponding to the subalgebras $\widetilde{\mathfrak{n}}_\pm$ and $\widetilde{\mathfrak{h}}$. It can be shown that

$$G = \overline{\widetilde{N}_+\widetilde{N}_-\widetilde{H}} = \overline{\widetilde{N}_-\widetilde{N}_+\widetilde{H}} = \overline{\widetilde{N}_-\widetilde{H}\widetilde{N}_+},$$

where the bar means the closure. Thus, any element a belonging to some dense subset of G can be presented in one of the forms

$$a = m_+n_-h_-, \qquad a = m_-n_+h_+, \qquad a = k_-hk_+,$$

where $m_\pm, n_\pm, k_\pm \in \widetilde{N}_\pm$ and $h_\pm, h \in \widetilde{H}$. The representation of an element of G in any such a form is called the *Gauss decomposition*. It is clear that if an element of G possesses the Gauss decomposition of one or another form, then such a decomposition is unique.

EXAMPLE 3.9 Consider the case of the Lie group $\mathrm{SL}(2,\mathbb{C})$. This group consists of all complex 2×2 matrices

$$a = \begin{pmatrix} a_{11} & a_{12} \\ a_{21} & a_{22} \end{pmatrix}$$

satisfying the condition

$$a_{11}a_{22} - a_{21}a_{12} = 1.$$

The subgroups N_- and N_+ are formed by the matrices of the form

$$n_- = \begin{pmatrix} 1 & 0 \\ a & 1 \end{pmatrix}, \qquad n_+ = \begin{pmatrix} 1 & b \\ 0 & 1 \end{pmatrix}$$

respectively. The subgroup H consists of the diagonal matrices

$$h = \begin{pmatrix} c & 0 \\ 0 & 1/c \end{pmatrix}$$

with $c \neq 0$. The first form of the Gauss decomposition exists for all elements of $\mathrm{SL}(2,\mathbb{C})$ with $a_{22} \neq 0$. A concrete expression here is

$$a = m_+n_-h_- = \begin{pmatrix} 1 & a_{12}/a_{22} \\ 0 & 1 \end{pmatrix} \begin{pmatrix} 1 & 0 \\ a_{21}a_{22} & 1 \end{pmatrix} \begin{pmatrix} 1/a_{22} & 0 \\ 0 & a_{22} \end{pmatrix}.$$

An element of $\mathrm{SL}(2,\mathbb{C})$ can be represented with the help of the second form of the Gauss decomposition if $a_{11} \neq 0$. Here we have

$$a = m_-n_+h_+ = \begin{pmatrix} 1 & 0 \\ a_{21}/a_{11} & 1 \end{pmatrix} \begin{pmatrix} 1 & a_{12}a_{11} \\ 0 & 1 \end{pmatrix} \begin{pmatrix} a_{11} & 0 \\ 0 & 1/a_{11} \end{pmatrix}.$$

The third form of the Gauss decomposition exists simultaneously with the second form. This decomposition is given by

$$a = k_- h k_+ = \begin{pmatrix} 1 & 0 \\ a_{21}/a_{11} & 1 \end{pmatrix} \begin{pmatrix} a_{11} & 0 \\ 0 & 1/a_{11} \end{pmatrix} \begin{pmatrix} 1 & a_{12}/a_{11} \\ 0 & 1 \end{pmatrix}.$$

As we can see, the Gauss decomposition is not a global one and not each element of a complex semisimple Lie group possesses it. Below we need a decomposition applicable for an arbitrary element. To construct such a decomposition we start by proving three useful lemmas, the second of which belongs to Harish-Chandra (1953).

Lemma 3.1 *Let b be an arbitrary element of G. The set $\widetilde{N}_+ \cap b\widetilde{B}_-$ consists of at most one element of G.*

Proof Suppose that $n_1, n_2 \in \widetilde{N}_+ \cap b\widetilde{B}_-$; then there exist the elements $b_1, b_2 \in \widetilde{B}_-$ such that

$$n_1 = bb_1, \qquad n_2 = bb_2.$$

These relations imply the equality

$$n_1^{-1} n_2 = b_1^{-1} b_2.$$

Note that $\widetilde{N}_+ \cap \widetilde{B}_- = \{e\}$. Hence, we have $n_1^{-1} n_2 = b_1^{-1} b_2 = e$. □

Lemma 3.2 *Let G be a Lie group; and let H, K be such Lie subgroups of G that*

$$\mathfrak{g} = \mathfrak{h} \oplus \mathfrak{k}. \qquad (3.58)$$

The mapping $\chi : (h, k) \in H \times K \mapsto hk \in G$ is regular at any point $(h, k) \in H \times K$.

Proof Actually, due to (3.58) we should prove only that $\ker \chi_{*(h,k)} = \{0\}$ for any point $(h, k) \in H \times K$. Suppose that for some $u \in T_{(h,k)}(H \times K)$ we have

$$\chi_{*(h,k)}(u) = 0. \qquad (3.59)$$

The discussion given in example 2.16 shows that there are unique tangent vectors $v \in T_h(H)$ and $w \in T_k(K)$ such that

$$u = \iota^H_{k*h}(v) + \iota^K_{h*k}(w),$$

where the mappings $\iota_k^H : H \to H \times K$ and $\iota_h^K : K \to H \times K$ are given by

$$\iota_k^H(h) \equiv (h, k), \qquad \iota_h^K(k) \equiv (h, k). \tag{3.60}$$

On the other hand, there are unique tangent vectors $v' \in T_e(H) = \mathfrak{h}$ and $w' \in T_e(K) = \mathfrak{k}$ such that

$$v = L_{h*e}^H(v'), \qquad w = L_{k*e}^K(w'). \tag{3.61}$$

Using relations (3.60) and (3.61), we write equality (3.59) in the form

$$\chi_{*(h,k)}(\iota_{k*h}^H(L_{h*e}^H(v')) + \iota_{h*k}^K(L_{k*e}^K(w'))) = 0. \tag{3.62}$$

It is not difficult to verify that

$$\chi \circ \iota_k^H \circ L_h^H = L_{hk}^G \circ L_{k^{-1}}^G \circ R_k^G \circ \iota^H,$$
$$\chi \circ \iota_h^K \circ L_k^K = L_{hk}^G \circ \iota^K,$$

where ι^H and ι^K are the inclusion mappings of H and K into G. These relations show that equality (3.62) is valid if and only if

$$L_{hk*e}^G(\mathrm{Ad}(k^{-1})v' + w') = 0.$$

Since for any $a \in G$ the mapping L_a^G is a diffeomorphism, we have $\mathrm{Ad}(k^{-1})v' + w' = 0$, that is, equivalent to $v' + \mathrm{Ad}(k)w' = 0$. This is possible if and only if $v' = 0$ and $w' = 0$. Therefore, $u = 0$, and $\ker \chi_{*(h,k)} = \{0\}$. □

Lemma 3.3 *The set $\pi_+(\widetilde{N}_+)$ is an open set.*

Proof From lemma 3.2 it follows that the mapping $\chi : (n, b) \in \widetilde{N}_+ \times \widetilde{B}_- \mapsto nb \in G$ is regular for any $(n, b) \in \widetilde{N}_+ \times \widetilde{B}_-$. Then the inverse mapping theorem implies that for any $(n, b) \in \widetilde{N}_+ \times \widetilde{B}_-$ there is an open set $U_{(n,b)}$ such that $(n, b) \in U_{(n,b)}$ and the set $\chi(U_{(n,b)})$ is open. Hence, the set $\widetilde{N}_+\widetilde{B}_-$ is open. On the other hand, the mapping π_+ is an open mapping; thus $\pi_+(\widetilde{N}_+)$ is an open set. □

Using the assertions of lemmas 3.1 and 3.3, we can define a local section s_+ of the fibre bundle $G \to F_+$, assuming that

$$s_+(p) \equiv \widetilde{N}_+ \cap (\pi_+)^{-1}(p), \qquad p \in \pi_+(\widetilde{N}_+).$$

Since $\widetilde{N}_+$ is a submanifold of a complex manifold, the section s_+ is holomorphic.

The subspaces $\widetilde{\mathfrak{n}}_{+a} \equiv L_{a*e}(\widetilde{\mathfrak{n}}_+)$, $a \in G$, generate an involutive holomorphic distribution $\mathcal{N}_+$ on G. The image $\widetilde{N}_+$ of the section s_+ is an integral manifold of this distribution. We can also say that the mapping s_+ is tangent to the distribution $\mathcal{N}_+$. Note that for any $a \in G$, the set $a\widetilde{N}_+$ is a maximal integral manifold of the distribution $\mathcal{N}_+$, and any maximal integral manifold of $\mathcal{N}_+$ has such a form. The following proposition is now almost evident.

Proposition 3.10 *There exists an open covering $\{U_{+\alpha}\}_{\alpha \in \mathcal{A}}$ of the manifold F_+ and a family of local holomorphic sections $s_{+\alpha}$: $U_{+\alpha} \to G$, $\alpha \in \mathcal{A}$, of the fibre bundle $G \to F_+$ such that for any $\alpha \in \mathcal{A}$ the section $s_{+\alpha}$ is tangent to the distribution $\mathcal{N}_+$.*

Proof As the first element of the required covering and the corresponding section we can take the set $U_+ = \pi_+(\widetilde{N}_+)$ and the section s_+ defined above. Let $p \notin U_+$, and $a \in (\pi_+)^{-1}(p)$. The set $a\widetilde{N}_+$ possesses properties similar to the properties of the set $\widetilde{N}_+$. Namely, if $a\widetilde{N}_+ \cap b\widetilde{B}_- \neq \emptyset$, then the set $a\widetilde{N}_+ \cap b\widetilde{B}_-$ contains just one point, and the set $\pi_+(a\widetilde{N}_+)$ is open. Therefore, we can define a local holomorphic section $s'_+ : U'_+ \to G$, where $U'_+ = \pi^+(a\widetilde{N}^+)$. Here the set U'_+ contains the point p. Repeating this procedure, we obtain a family of local holomorphic sections of the fibre bundle $G \to F^+$ with the required properties. □

Actually, we shall consider families of local sections constructed with the help of the procedure used in the proof of proposition 3.29. In this case, for any $\alpha \in \mathcal{A}$ we have

$$s_{+\alpha}(U_{+\alpha}) = a_{+\alpha}\widetilde{N}_+ \equiv \widetilde{N}_{+\alpha}$$

for some $a_{+\alpha} \in G$. It is clear that a similar family of local sections can also be constructed for the fibre bundle $G \to F_-$. If the Lie algebra $\mathfrak{g}$ is endowed with an involutive antilinear automorphism consistent with the $\mathbb{Z}$-gradation, such a family of sections of the fibre bundle $G \to F_-$ can be constructed on the basis of the given family of sections of the fibre bundle $G \to F_+$. The corresponding method for this is considered in section 3.3.4.

As was discussed in section 2.8.2, any family of holomorphic local sections of a holomorphic principal fibre bundle covering the base space allows one to define a bundle atlas. The corresponding procedure in our case looks as follows.

For any fixed $\alpha \in \mathcal{A}$, consider a holomorphic mapping $m_{+\alpha} : (\pi_+)^{-1}(U_{+\alpha}) \to G$ defined as

$$m_{+\alpha}(a) \equiv s_{+\alpha}(\pi_+(a)), \qquad a \in (\pi_+)^{-1}(U_{+\alpha}).$$

This mapping allows one to introduce another holomorphic mapping $b_{-\alpha}$ defined on $(\pi_+)^{-1}(U_{+\alpha})$ by

$$b_{-\alpha}(a) \equiv m_{+\alpha}^{-1}(a)a.$$

Thus, for any $a \in (\pi_+)^{-1}(U_{+\alpha})$ we can write

$$a = m_{+\alpha}(a)b_{-\alpha}(a). \tag{3.63}$$

Since $s_{+\alpha}$ is a section of the fibre bundle $G \to F_+$, i.e.,

$$\pi_+ \circ s_{+\alpha} = \mathrm{id}_{U_{+\alpha}},$$

we have

$$\pi_+(m_{+\alpha}(a)) = \pi_+(a),$$

and from (3.63) it can readily be seen that the mapping $b_{-\alpha}$ takes values in the subgroup $\widetilde{B}_-$. Note here that the mappings $m_{+\alpha}$ and $b_{-\alpha}$ have the following properties:

$$m_{+\alpha}(ab) = m_{+\alpha}(a), \qquad b_{-\alpha}(ab) = b_{-\alpha}(a)b \tag{3.64}$$

for any $b \in B_-$.

It is clear that the mapping $\psi_{+\alpha} : (\pi_+)^{-1}(U_{+\alpha}) \to U_{+\alpha} \times \widetilde{B}_-$, defined as

$$\psi_{+\alpha}(a) \equiv (\pi_+(a), b_{-\alpha}(a)),$$

provides a bundle chart on the fibre bundle $G \to F_+$. Considering all possible values of the index α, we obtain an atlas of this fibre bundle.

Let a be an element of G such that $\pi_+(a) \in U_{+\alpha} \cap U_{+\beta}$. In this case, using (3.64), we can write

$$b_{-\alpha}(a) = b_{-\alpha}(s_{+\beta}(p)b_{-\beta}(a)) = b_{-\alpha}(s_{+\beta}(p))b_{-\beta}(a),$$

where $p = \pi_+(a)$. Hence, we have

$$b_{-\alpha}(a) = b_{-\alpha\beta}(\pi_+(a))b_{-\beta}(a), \tag{3.65}$$

where

$$b_{-\alpha\beta} \equiv b_{-\alpha} \circ s_{+\beta}.$$

It is clear that the mappings $b_{-\alpha\beta}$, $\alpha, \beta \in \mathcal{A}$, are the transition functions of the atlas we have defined. These transition functions are obviously holomorphic.

Again let $\pi_+(a) \in U_{+\alpha} \cap U_{+\beta}$. In this case we have

$$a = m_{+\alpha}(a) b_{-\alpha} = m_{+\beta}(a) b_{-\beta}(a).$$

This relation, taking account of (3.65), gives

$$m_{+\beta}(a) = m_{+\alpha}(a) b_{-\alpha\beta}(\pi_+(a)). \tag{3.66}$$

Proposition 3.11 *The groups $\widetilde{B}_\pm$ have the holomorphic decomposition*

$$\widetilde{B}_\pm = \widetilde{N}_\pm \widetilde{H}.$$

Proof Since $\widetilde{\mathfrak{b}}_+ = \widetilde{\mathfrak{n}}_+ \oplus \widetilde{\mathfrak{h}}$, then, as was done in the proof of lemma 3.3, one can show that $\widetilde{N}_+ \widetilde{H}$ is an open subset of $\widetilde{B}_+$. The subspace $\widetilde{\mathfrak{n}}_+$ is an ideal of $\widetilde{\mathfrak{b}}_+$. Therefore, $\widetilde{N}_+$ is a normal subgroup of $\widetilde{B}_+$. Hence $\widetilde{N}_+ \widetilde{H}$ is a subgroup of $\widetilde{B}_+$. The Lie algebras of $\widetilde{N}_+ \widetilde{H}$ and $\widetilde{B}_+$ coincide. Thus, these Lie groups also coincide. The case of B_- can be considered in a similar way. □

The last proposition implies that we can uniquely represent the mapping $b_{-\alpha}$ in the form

$$b_{-\alpha} = n_{-\alpha} h_{-\alpha}, \tag{3.67}$$

where $n_{-\alpha}$ and $h_{-\alpha}$ are holomorphic mappings from $(\pi_+)^{-1}(U_{+\alpha})$ to the subgroups $\widetilde{N}_-$ and $\widetilde{H}$ respectively. Analogously, for the transition functions $b_{-\alpha\beta}$ we have

$$b_{-\alpha\beta} = n_{-\alpha\beta} h_{-\alpha\beta},$$

where $n_{-\alpha\beta}$ and $h_{-\alpha\beta}$ are holomorphic mappings from $U_{+\alpha} \cap U_{+\beta}$ to $\widetilde{N}_-$ and $\widetilde{H}$ respectively.

From (3.64) we obtain the following relations:

$$m_{+\alpha}(an) = m_{+\alpha}(a), \quad m_{+\alpha}(ah) = m_{+\alpha}(a), \tag{3.68}$$

$$n_{-\alpha}(an) = n_{-\alpha}(a) h_{-\alpha}(a) n h_{-\alpha}^{-1}(a), \tag{3.69}$$

$$n_{-\alpha}(ah) = n_{-\alpha}(a), \tag{3.70}$$

$$h_{-\alpha}(an) = h_{-\alpha}(a), \quad h_{-\alpha}(ah) = h_{-\alpha}(a) h, \tag{3.71}$$

which are valid for any $n \in \widetilde{N}_-$ and $h \in \widetilde{H}$.

Using (3.65), for any $a \in G$ such that $p = \pi_+(a) \in U_{+\alpha} \cap U_{+\beta}$, we also obtain

$$n_{-\alpha}(a) = n_{-\alpha\beta}(p) h_{-\alpha\beta}(p) n_{-\beta}(a) h_{-\alpha\beta}^{-1}(p), \tag{3.72}$$

$$h_{-\alpha}(a) = h_{-\alpha\beta}(p) h_{-\beta}(a). \tag{3.73}$$

Proposition 3.12 *Any element $a \in (\pi_+)^{-1}(U_{+\alpha}) = \widetilde{N}_{+\alpha}\widetilde{B}_-$ can be uniquely represented in the form*

$$a = m_+ n_- h_-, \tag{3.74}$$

where $m_+ \in \widetilde{N}_{+\alpha}$, $n_- \in \widetilde{N}_-$, $h_- \in \widetilde{H}$. The elements m_+, n_-, and h_- holomorphically depend on a.

Proof We come to representation (3.74) putting

$$m_+ = m_{+\alpha}(a), \qquad n_- = n_{-\alpha}(a), \qquad h_- = h_{-\alpha}(a).$$

The uniqueness of decomposition (3.74) follows directly from the fact that $\widetilde{N}_+ \cap \widetilde{N}_- = \widetilde{N}_+ \cap \widetilde{H} = \widetilde{N}_- \cap \widetilde{H} = \{e\}$. □

We now have a proposition similar to the previous one.

Proposition 3.13 *Any element $a \in (\pi_-)^{-1}(U_{-\alpha}) = \widetilde{N}_{-\alpha}\widetilde{B}_+$ can be uniquely represented in the form*

$$a = m_- n_+ h_+, \tag{3.75}$$

where $m_- \in \widetilde{N}_{-\alpha}$, $n_+ \in \widetilde{N}_+$, $h_+ \in \widetilde{H}$. The elements m_-, n_+, and h_+ holomorphically depend on a.

We call decompositions (3.74) and (3.75) the *modified Gauss decompositions.*

3.2.5 Toda-type systems

Now we will use the modified Gauss decomposition to define the required $\widetilde{H}$-gauge invariant mappings and to derive the equations that they satisfy.

Proposition 3.14 *Let $\varphi : M \to G$ be an arbitrary mapping and let $p \in M$.*

(i) *There exists an open neighbourhood V_+ of the point p such that the mapping φ, being restricted to V_+, has the unique decomposition*

$$\varphi = \mu_+ \nu_- \eta_-, \tag{3.76}$$

where the mapping μ_+ takes values in $\widetilde{N}_{+\alpha}$ for some $\alpha \in \mathcal{A}$, while the mappings ν_- and η_- take values in $\widetilde{N}_-$ and $\widetilde{H}$ respectively.

(ii) *There exists an open neighbourhood V_- of the point p such that the mapping φ, being restricted to V_-, has the unique decomposition*

$$\varphi = \mu_- \nu_+ \eta_+, \tag{3.77}$$

where the mapping μ_- takes values in $\widetilde{N}_{-\alpha}$ for some $\alpha \in \mathcal{A}$ while the mappings ν_+ and η_+ take values in $\widetilde{N}_+$ and $\widetilde{H}$ respectively.

Proof The proof is based on the modified Gauss decomposition (3.74). It is clear that we can find $\alpha \in \mathcal{A}$ such that $\varphi(p) \in (\pi_+)^{-1}(U_{+\alpha})$. Define a mapping μ_+ by

$$\mu_+ \equiv m_{+\alpha} \circ \varphi = s_{+\alpha} \circ \pi_+ \circ \varphi = s_{+\alpha} \circ \varphi_+. \tag{3.78}$$

The domain of the mapping μ_+ is the open set $V_+ \equiv \varphi_+^{-1}(U_{+\alpha})$. We now introduce the mappings

$$\nu_- \equiv n_{-\alpha} \circ \varphi, \qquad \eta_- \equiv h_{-\alpha} \circ \varphi \tag{3.79}$$

with the same domain and arrive at the required decomposition (3.76).

The second part of the proposition can be proved in a similar way. □

Corollary 3.1 *For any $p \in M$, there exists an open set V such that $p \in V$, and the mapping φ, being restricted to V, possesses decompositions* (3.76) *and* (3.77) *simultaneously.*

Proposition 3.15 *If the mapping $\varphi : M \to G$ satisfies the specified grading condition, then the mapping μ_- is holomorphic and the holomorphic 1-form ${}^{\mu_-}\omega$ takes values in $\widetilde{\mathfrak{m}}_-$, while the mapping μ_+ is antiholomorphic and the antiholomorphic 1-form ${}^{\mu_+}\omega$ takes values in $\widetilde{\mathfrak{m}}_+$.*

Proof Since, by definition,

$$\mu_- = s_{-\alpha} \circ \varphi_- \tag{3.80}$$

and the mappings $s_{-\alpha}$ and φ_- are holomorphic, then μ_- is holomorphic.

From (3.80) we see that the mapping μ_- is tangent to the distribution $\mathcal{N}_-$; therefore, $P^G_{\mu_-(p)} \circ \mu_{-*p}(v) \in \widetilde{\mathfrak{n}}_{-\mu_-(p)}$ for any $v \in T_p^{\mathbb{C}}(M)$. On the other hand, from (3.80) we obtain

$$\pi_- \circ \mu_- = \varphi_-.$$

Recall that the mapping φ_- is tangent to the distribution $\mathcal{M}_-$; hence, $P^G_{\mu_-(p)} \circ \mu_{-*p}(v) \in \mathfrak{b}_{+\mu_-(p)} \oplus \widetilde{\mathfrak{m}}_{-\mu_-(p)}$. Thus, $P^G_{\mu_-(p)} \circ \mu_{-*p}(v) \in \widetilde{\mathfrak{m}}_{-\mu_-(p)}$. Writing the equality

$$^{\mu_-}\omega(v) = L_{\mu_-^{-1}(p)*\mu_-(p)} \circ P^G_{\mu_-(p)} \circ \mu_{-*p}(v),$$

we conclude that the 1-form $^{\mu_-}\omega$ takes values in $\widetilde{\mathfrak{m}}_-$. □

Let $x_\pm$ be some fixed nonzero elements of $\mathfrak{g}_{\pm l_\pm}$. For the case of $l_+ = l_- = 1$, when we consider a $\mathbb{Z}$-gradation of $\mathfrak{g}$ associated with an integral embedding of $\mathfrak{sl}(2, \mathbb{C})$ into $\mathfrak{g}$, we can, in particular, take as $x_\pm$ the corresponding elements defined by this embedding. Let $\mathcal{O}_\pm$ be the orbits of the elements $x_\pm$ generated by the restriction of the adjoint action of the group G to the subgroup $\widetilde{H}$. Note that the orbits $\mathcal{O}_\pm$ have the following property. If the element x belongs to $\mathcal{O}_+$ ($\mathcal{O}_-$), then, for any nonzero $c \in \mathbb{C}$, the element cx also belongs to $\mathcal{O}_+$ ($\mathcal{O}_-$). This statement follows from the fact that the grading operator q generates similarity transformations of the subspaces $\mathfrak{g}_{\pm l_\pm}$.

Denoting by $\widetilde{H}_\pm$ the isotropy subgroups of the elements $x_\pm$, we identify the orbits $\mathcal{O}_\pm$ with the homogeneous spaces $\widetilde{H}/\widetilde{H}_\pm$. More precisely, we establish such identification putting into correspondence to a coset $h\widetilde{H}_\pm$, $h \in \widetilde{H}$, the element $x(h\widetilde{H}_\pm) \in \mathcal{O}_\pm$ given by

$$x(h\widetilde{H}_\pm) = \mathrm{Ad}(h)x_\pm.$$

Denote by $\widetilde{\mathfrak{m}}'_+$ the subset of $\widetilde{\mathfrak{m}}_+$ which consists of the elements with the l_+th grading component belonging to $\mathcal{O}_+$. Similarly, $\widetilde{\mathfrak{m}}'_-$ consists of the elements of $\widetilde{\mathfrak{m}}_-$, which l_-th grading component belongs to $\mathcal{O}_-$. The corresponding subsets of $\mathcal{M}_{\pm p}$ will be denoted $\mathcal{M}'_{\pm p}$. We call a mapping $\varphi : M \to G$ an *admissible mapping*, if it satisfies the specified grading condition, and, moreover, $\varphi_{-*p}(\partial_{-p}) \in \mathcal{M}'_{-p}$ and $\varphi_{+*p}(\partial_{+p}) \in \mathcal{M}'_{+p}$ for any $p \in M$. Due to the properties of the orbits $\mathcal{O}_\pm$ discussed above, this definition does not depend on the choice of the local coordinate $z = z^-$. If the mapping φ is admissible, then we can write

$$^{\mu_\pm}\omega = \sqrt{-1}\lambda_\pm dz^\pm = \sqrt{-1}\sum_{m=1}^{l_\pm} \lambda_{\pm m} dz^\pm, \tag{3.81}$$

where $\mu_\pm$ are the mappings arising from the local decompositions (3.76), (3.77); and $\lambda_{\pm m}$, $1 \le m \le l_\pm - 1$, are the mappings taking values in $\mathfrak{g}_{\pm m}$, while the mapping $\lambda_{\pm l_\pm}$ takes values in $\mathcal{O}_\pm$. The factor $\sqrt{-1}$ is introduced in this relation for the future convenience. The mappings $\lambda_{\pm m}$ are defined in the open set V from corollary 3.1. It follows from proposition 3.15, that the mappings λ_{-m} are holomorphic, and the mapping λ_{+m} are antiholomorphic.

Let $\gamma_\pm$ be local lifts of the mappings $\lambda_{\pm l_\pm}$ to the group $\widetilde{H}$. These mappings are defined in some open set $W \subset V$, and satisfy the relations

$$\lambda_{\pm l_\pm} = \mathrm{Ad}(\gamma_\pm) x_\pm. \tag{3.82}$$

Note that in the case when the groups $\widetilde{H}_\pm$ are nontrivial, the mappings $\gamma_\pm$ are defined ambiguously, but in any case they can be chosen in such a way that the mapping γ_- would be holomorphic, and the mapping γ_+ would be antiholomorphic. In what follows we use in our consideration such a choice.

Theorem 3.3 *Let $\varphi : M \to G$ be an admissible mapping. There exists a local $\widetilde{H}$-gauge transformation that transforms a connection ${}^\varphi\omega$ to the connection ω of the form*

$$\omega = \sqrt{-1}\Big(x_- + \sum_{m=1}^{l_- -1} \upsilon_{-m} - \sqrt{-1}({}^\gamma\omega)_-\Big)dz^- + \sqrt{-1}\,\mathrm{Ad}(\gamma^{-1})\Big(\sum_{m=1}^{l_+ -1} \upsilon_{+m} + x_+\Big)dz^+, \tag{3.83}$$

where γ is a mapping taking values in $\widetilde{H}$ and $\upsilon_{\pm m}$ are mappings taking values in $\mathfrak{g}_{\pm m}$.

Proof Using representation (3.77) and proposition 3.8 we can write

$${}^\varphi\omega = {}^{\mu_-\nu_+\eta_+}\omega = \mathrm{Ad}(\eta_+^{-1}\nu_+^{-1})({}^{\mu_-}\omega) + \mathrm{Ad}(\eta_+^{-1})({}^{\nu_+}\omega) + {}^{\eta_+}\omega. \tag{3.84}$$

In the same way, representation (3.76) gives

$${}^\varphi\omega = {}^{\mu_+\nu_-\eta_-}\omega = \mathrm{Ad}(\eta_-^{-1}\nu_-^{-1})({}^{\mu_+}\omega) + \mathrm{Ad}(\eta_-^{-1})({}^{\nu_-}\omega) + {}^{\eta_-}\omega. \tag{3.85}$$

The form ${}^{\mu_-}\omega$ is holomorphic, and the form ${}^{\mu_+}\omega$ is antiholomorphic, therefore,

$$({}^{\mu_\mp}\omega)_\pm = 0.$$

Taking these relations into account, we arrive at the following consequences of (3.84) and (3.85):

$$({}^{\varphi}\omega)_{\pm} = \mathrm{Ad}(\eta_{\pm}^{-1})({}^{\nu_{\pm}}\omega)_{\pm} + ({}^{\eta_{\pm}}\omega)_{\pm}. \tag{3.86}$$

Consider now the mapping

$$\kappa \equiv \mu_{+}^{-1}\mu_{-}. \tag{3.87}$$

Proposition 3.8 provides the relation

$${}^{\kappa}\omega = {}^{\mu_{-}}\omega - \mathrm{Ad}(\kappa^{-1})({}^{\mu_{+}}\omega).$$

Using (3.81), we obtain from this relation the equalities

$$({}^{\kappa}\omega)_{+} = -\sqrt{-1}\,\mathrm{Ad}(\kappa^{-1})\lambda_{+}, \qquad ({}^{\kappa}\omega)_{-} = \sqrt{-1}\lambda_{-}.$$

Thanks to decompositions (3.76) and (3.77), we conclude that the mapping κ can also be represented as

$$\kappa = \nu_{-}\eta\nu_{+}^{-1}. \tag{3.88}$$

where

$$\eta \equiv \eta_{-}\eta_{+}^{-1}. \tag{3.89}$$

Representation (3.88) leads to the equality

$${}^{\kappa}\omega = \mathrm{Ad}(\nu_{+}\eta^{-1})({}^{\nu_{-}}\omega - {}^{\eta^{-1}}\omega - \mathrm{Ad}(\eta)({}^{\nu_{+}}\omega)), \tag{3.90}$$

which results in the formula

$$\sqrt{-1}\,\mathrm{Ad}(\eta\nu_{+}^{-1})\lambda_{-} = ({}^{\nu_{-}}\omega)_{-} - ({}^{\eta^{-1}}\omega)_{-} - \mathrm{Ad}(\eta)({}^{\nu_{+}}\omega)_{-}. \tag{3.91}$$

Taking the $\widetilde{\mathfrak{n}}_{-}$-component of (3.91), we obtain the relation

$$({}^{\nu_{-}}\omega)_{-} = \sqrt{-1}(\mathrm{Ad}(\eta\nu_{+}^{-1})\lambda_{-})_{\widetilde{\mathfrak{n}}_{-}}. \tag{3.92}$$

Therefore, the mapping $({}^{\nu_{-}}\omega)_{-}$ takes values in $\widetilde{\mathfrak{m}}'_{-}$ and we can write

$$\mathrm{Ad}((\eta\gamma_{-})^{-1})({}^{\nu_{-}}\omega)_{-} = \sqrt{-1}\sum_{m=1}^{l_{-}} \upsilon_{-m}. \tag{3.93}$$

It is not difficult to show that $\upsilon_{-l_{-}} = x_{-}$. Using (3.93), we obtain from (3.86) the following equality:

$$({}^{\varphi}\omega)_{-} = \sqrt{-1}\,\mathrm{Ad}(\eta_{+}^{-1}\gamma_{-})\sum_{m=1}^{l_{-}} \upsilon_{-m} + ({}^{\eta_{-}}\omega)_{-}. \tag{3.94}$$

Similarly, it follows from (3.90) that

$$-\sqrt{-1}\,\mathrm{Ad}(\eta^{-1}\nu_{-}^{-1})\lambda_{+} = \mathrm{Ad}(\eta^{-1})({}^{\nu_{-}}\omega)_{+} + ({}^{\eta}\omega)_{+} - ({}^{\nu_{+}}\omega)_{+},$$

and the $\widetilde{\mathfrak{n}}_{+}$-component of this relation is

$$({}^{\nu_{+}}\omega)_{+} = \sqrt{-1}(\mathrm{Ad}(\eta^{-1}\nu_{-}^{-1})\lambda_{+})_{\widetilde{\mathfrak{n}}_{+}}. \tag{3.95}$$

Hence, the mapping $({}^{\nu_+}\omega)_+$ takes values in $\widetilde{\mathfrak{m}}'_+$ and we can write

$$\mathrm{Ad}(\gamma_+^{-1}\eta)({}^{\nu_+}\omega)_- = \sqrt{-1}\sum_{m=1}^{l_+} \upsilon_{+m}. \qquad (3.96)$$

Here $\upsilon_{+l_+} = x_+$ and the analogue of (3.94) is

$$({}^{\varphi}\omega)_+ = \sqrt{-1}\,\mathrm{Ad}(\eta_-^{-1}\gamma_+)\sum_{m=1}^{l_+} \upsilon_{+m} + ({}^{\eta_+}\omega)_+. \qquad (3.97)$$

We now perform the gauge transformation defined by the mapping $\eta_+^{-1}\gamma_-$. Taking (3.94) into account and (3.97), we conclude that this gauge transformation brings the connection ${}^{\varphi}\omega$ to the connection of the form (3.83) with the mapping γ given by

$$\gamma \equiv \gamma_+^{-1}\eta\gamma_-. \qquad (3.98)$$

This is to be proved. □

Thus we see that, up to a local $\widetilde{H}$-gauge transformation, we consider the connections of form (3.83). In what follows we deal with the matrix Lie groups only. In this case we have

$$\omega_- = \sqrt{-1}\sum_{m=1}^{l_-} \upsilon_{-m} + \gamma^{-1}\partial_-\gamma, \quad \omega_+ = \sqrt{-1}\sum_{m=1}^{l_+} \gamma^{-1}\upsilon_{+m}\gamma. \quad (3.99)$$

Let us write explicitly the equations for γ and $\upsilon_{\pm m}$ which follow from the zero curvature condition (3.45). We restrict ourselves to the case of $l_- = l_+ = l$; the generalisation to the case of $l_+ \neq l_-$ is straightforward. Using (3.99), we obtain

$$\begin{aligned}\partial_-\omega_+ - \partial_+\omega_- &= \sqrt{-1}\sum_{m=1}^{l} \partial_-(\gamma^{-1}\upsilon_{+m}\gamma) \\ &\quad - \sqrt{-1}\sum_{m=1}^{l} \partial_+\upsilon_{-m} - \partial_+(\gamma^{-1}\partial_-\gamma). \quad (3.100)\end{aligned}$$

The same relations give

$$[\omega_-, \omega_+] = -\sum_{m,n=1}^{l} [\upsilon_{-m}, \gamma^{-1}\upsilon_{+n}\gamma] + \sqrt{-1}\sum_{m=1}^{l}[\gamma^{-1}\partial_-\gamma, \gamma^{-1}\upsilon_{+m}\gamma].$$

Taking into account the identity

$$\partial_-(\gamma^{-1})\gamma = -\gamma^{-1}\partial_-\gamma,$$

we obtain the relation

$$[\gamma^{-1}\partial_-\gamma, \gamma^{-1}\upsilon_{+m}\gamma] = -\partial_-(\gamma^{-1}\upsilon_{+m}\gamma) + \gamma^{-1}(\partial_-\upsilon_{+m})\gamma.$$

Therefore, we have

$$[\omega_-, \omega_+] = -\sum_{m,n=1}^{l} [\upsilon_{-m}, \gamma^{-1}\upsilon_{+n}\gamma]$$

$$-\sqrt{-1}\sum_{m=1}^{l} \partial_-(\gamma^{-1}\upsilon_{+m}\gamma) + \sqrt{-1}\sum_{m=1}^{l} \gamma^{-1}(\partial_-\upsilon_{+m})\gamma. \quad (3.101)$$

Finally, relations (3.100) and (3.101) result in the following system of equations:

$$\sqrt{-1}\partial_+\upsilon_{-m} = \sum_{n=1}^{l-m}[\gamma^{-1}\upsilon_{+n}\gamma, \upsilon_{-(m+n)}], \quad (3.102)$$

$$\partial_+(\gamma^{-1}\partial_-\gamma) = \sum_{m=1}^{l}[\gamma^{-1}\upsilon_{+m}\gamma, \upsilon_{-m}], \quad (3.103)$$

$$\sqrt{-1}\partial_-\upsilon_{+m} = \sum_{n=1}^{l-m}[\gamma\upsilon_{-n}\gamma^{-1}, \upsilon_{+(m+n)}], \quad (3.104)$$

where $\upsilon_{\pm l} = x_\pm$. This form of writing the equations is equivalent to that given in Gervais & Saveliev (1995).

For the case where $l = 1$ one obtains the following equation:

$$\partial_+ (\gamma^{-1}\partial_-\gamma) = [\gamma^{-1}x_+\gamma, x_-]. \quad (3.105)$$

The equations for parameters of the group $\widetilde{H}$ which follow from (3.105), are called the *Toda equations*. In the case of a principal gradation, the subgroup $\widetilde{H}$ coincides with some Cartan subgroup H of G and is, by this reason, an abelian subgroup. The corresponding equations in this case are called the *abelian Toda* equations. In the case of a $\mathbb{Z}$-gradation associated with an arbitrary integral embedding of $\mathfrak{sl}(2, \mathbb{C})$ into $\mathfrak{g}$, the subgroup $\widetilde{H}$ is not necessarily abelian and we deal either with the abelian Toda equations or with some of their nonabelian versions.

EXAMPLE 3.10 Let us derive the concrete form of the abelian Toda equations. In the case under consideration, we can locally parametrise the mapping γ by the set of complex functions f_i as

$$\gamma = \exp\left(\sum_{i=1}^{r} f_i h_i\right), \quad (3.106)$$

where r is the rank of $\mathfrak{g}$ and the elements $h_i \in \mathfrak{h}$ are the Cartan generators. Choose as the elements $x_\pm \in \mathfrak{g}_{\pm 1}$ the elements describing the corresponding principal embedding of $\mathfrak{sl}(2, \mathbb{C})$ in $\mathfrak{g}$. The concrete form of such elements is given by (3.24). Using (1.18), we obtain the relation

$$\gamma^{-1} x_+ \gamma = \sum_{i=1}^{r} (2k_i)^{1/2} \exp[-(kf)_i] x_{+i},$$

where

$$(kf)_i \equiv \sum_{j=1}^{r} k_{ij} f_j.$$

From this relation one immediately obtains

$$[\gamma^{-1} x_+ \gamma, x_-] = \sum_{i=1}^{r} 2k_i \exp[-(kf)_i] h_i.$$

On the other hand, it is clear that

$$\partial_+ \left(\gamma^{-1} \partial_- \gamma\right) = \sum_{i=1}^{r} (\partial_+ \partial_- f_i) h_i.$$

Thus, in the case under consideration, equations (3.105) can be reduced to the system

$$\partial_+ \partial_- f_i = 2k_i \exp[-(kf)_i]. \tag{3.107}$$

Introducing the functions

$$v_i \equiv (kf)_i - \ln(2k_i),$$

we rewrite equations (3.107) in the form

$$\partial_+ \partial_- v_i = \sum_{j=1}^{r} k_{ij} \exp(-v_j),$$

which is standard for the abelian Toda equations.

3.2.6 Gauge invariance and dependence on lifts

Consider now the behaviour of the mappings γ and $\upsilon_{\pm m}$ under the $\widetilde{H}$-gauge transformation. Let $\varphi' = \varphi\psi$, where the mapping ψ takes values in $\widetilde{H}$. It is clear that to define the mappings γ' and $\upsilon'_{\pm m}$ corresponding to the mapping φ', we can use the same modified Gauss decompositions that we have used for the construction of the mapping γ. From relations (3.78) and (3.68) we have $\mu'_+ = \mu_+$. Using the same lift from $\widetilde{H}/\widetilde{H}_+$ to $\widetilde{H}$ as in transition from the mapping μ_+ to the mapping γ_+, we obtain $\gamma'_+ = \gamma_+$. On the other hand, relations (3.79) and (3.71) give $\eta'_- = \eta_-\psi$. In a similar way we have $\gamma'_- = \gamma_-$, and $\eta'_+ = \eta_+\psi$. Thus, it follows from (3.98) that $\gamma' = \gamma$. Further, equality (3.70) gives $\nu'_- = \nu_-$; therefore, from (3.93) it follows that $\upsilon'_{-m} = \upsilon_{-m}$. Similarly, we obtain $\upsilon'_{+m} = \upsilon_{+m}$. Just in this sense the mappings γ and $\upsilon_{\pm m}$ are $\widetilde{H}$-gauge invariant.

One more question that we are going to consider in this section is the dependence of the mappings γ and $\upsilon_{\pm m}$ on the choice of modified Gauss decompositions and local lifts from $\widetilde{H}/\widetilde{H}_\pm$ to $\widetilde{H}$. Suppose that we have two local decompositions of the mapping φ:

$$\varphi = \mu_+\nu_-\eta_-, \qquad \varphi = \mu'_+\nu'_-\eta'_-,$$

which are obtained with the help of the modified Gauss decompositions corresponding to the indices α and β respectively. Using (3.66) and (3.67), we obtain

$$\mu'_+ = m_{+\beta} \circ \varphi = \mu_+\nu_{-\alpha\beta}\eta_{-\alpha\beta}, \tag{3.108}$$

where

$$\nu_{-\alpha\beta} \equiv n_{-\alpha\beta} \circ \varphi_+, \qquad \eta_{-\alpha\beta} \equiv h_{-\alpha\beta} \circ \varphi_+.$$

It is obvious that the mappings $\nu_{-\alpha\beta}$ and $\eta_{-\alpha\beta}$ are antiholomorphic and take values in $\widetilde{N}_-$ and $\widetilde{H}$ respectively. From (3.108) it follows that

$${}^{\mu'_+}\omega = \mathrm{Ad}(\eta_{-\alpha\beta}^{-1})(\mathrm{Ad}(\nu_{-\alpha\beta}^{-1})({}^{\mu_+}\omega) + {}^{\nu_{-\alpha\beta}}\omega) + {}^{\eta_{-\alpha\beta}}\omega.$$

Taking (3.81) and (3.82) into account, one obtains

$$\mathrm{Ad}(\gamma'_+)x_+ = \mathrm{Ad}(\eta_{-\alpha\beta}\gamma_+)x_+.$$

It follows from this relation that

$$\gamma'_+ = \eta_{-\alpha\beta}^{-1}\gamma_+\xi_+, \tag{3.109}$$

where the mapping ξ_+ takes values in $\widetilde{H}_+$. It is clear that the mapping ξ_+ is antiholomorphic.

Further, (3.73) allows one to write

$$\eta_- = \eta_{-\alpha\beta}\eta'_-. \tag{3.110}$$

Combining (3.109) and (3.110), we obtain

$$\gamma_+^{\prime -1}\eta'_- = \xi_+^{-1}\gamma_+^{-1}\eta_-;$$

and, in a similar way,

$$\eta_+^{\prime -1}\gamma'_- = \eta_+^{-1}\gamma_-\xi_-,$$

where ξ_- is a holomorphic mapping taking values in $\widetilde{H}_-$. Finally, we obtain the relation

$$\gamma' = \xi_+^{-1}\gamma\xi_-. \tag{3.111}$$

For the mappings $\upsilon_{\pm m}$ one obtains

$$\upsilon'_{\pm m} = \xi_\pm^{-1}\upsilon_{\pm m}\xi_\pm. \tag{3.112}$$

Resuming our discussion, we can say that any admissible mapping φ leads to a set $\{\gamma_i, (\upsilon_{\pm m})_i\}_{i\in\mathcal{I}}$ of local solutions of equations (3.102)–(3.104). These solutions, in the overlaps of their domains, are connected by the relations

$$\gamma_i = \xi_{+ij}^{-1}\gamma_j\xi_{-ij}, \quad (\upsilon_{\pm m})_i = \xi_{\pm ij}^{-1}(\upsilon_{\pm m})_j\xi_{\pm ij},$$

where the mappings ξ_{-ij} are holomorphic and take values in $\widetilde{H}_-$, while the mappings ξ_{+ij} are antiholomorphic and take values in $\widetilde{H}_+$. Note here that (3.111) and (3.112) describe symmetry transformations of equations (3.102)–(3.104). For the case $l = 1$ and the principal gradation, the subgroups $\widetilde{H}_\pm$ are discrete.

Exercises

3.6 Find explicit expressions for the quasideterminants of a 3×3 matrix over an associative unital algebra.

3.7 Let $a = (a_{ij})$, $i, j = 1, 2, 3$, be a matrix over an associative algebra with unit. Prove the following Silvester identities:

$$|a|_{33} = |a_{(1,3;1,3)}|_{33} - |a_{(1,3;1,2)}|_{32}|a_{(1,2;1,2)}|_{22}^{-1}|a_{(1,2;1,3)}|_{23},$$

$$|a|_{33} = |a_{(2,3;2,3)}|_{33} - |a_{(2,3;1,2)}|_{31}|a_{(1,2;1,2)}|_{11}^{-1}|a_{(1,2;2,3)}|_{13}.$$

3.8 Construct explicit expressions for the matrices entering the Gauss decomposition of a 3×3 matrix over an associative unital algebra.

3.9 For the Lie groups $\mathrm{SL}(m,\mathbb{C})$, $\widetilde{\mathrm{SO}}(m,\mathbb{C})$ and $\widetilde{\mathrm{Sp}}(2n,\mathbb{C})$ find the subgroups $\widetilde{H}_\pm$ for the case where $l = 1$ and the principal gradation.

3.3 Construction of solutions and reality condition

3.3.1 General solution of Toda-type systems

In the preceding section we have shown that any admissible mapping $\varphi : M \to G$ allows one to construct a set of local solutions of equations (3.102)–(3.104). At first glance, the problem of constructing admissible mappings is rather complicated. On the other hand, to construct a solution we do not need to know the mapping φ itself. It is sufficient to deal with the mappings $\mu_\pm$ entering the Gauss decompositions (3.76) and (3.77); see Leznov & Saveliev (1992). Indeed, once we know the mappings $\mu_\pm$ corresponding to some admissible mapping φ, then using (3.76), (3.77) and (3.89), we obtain

$$\mu_+^{-1}\mu_- = \nu_-\eta\nu_+^{-1}. \tag{3.113}$$

This relation implies that we can find the mappings η and $\nu_\pm$ considering the corresponding Gauss decomposition of the mapping $\mu_+^{-1}\mu_-$. The mappings $\gamma_\pm$ are determined by relation (3.82). Knowing the mappings η, $\nu_\pm$ and $\gamma_\pm$, we construct the mappings γ, v_{-m} and v_{+m} using (3.98), (3.94) and (3.97) respectively.

Thus, to construct solutions of equations (3.102)–(3.104) we should specify the mappings $\mu_\pm : M \to G$. It appears that not each pair of such mappings gives a solution. Actually we must use only the mappings $\mu_\pm$ arising in the generalised Gauss decompositions of some admissible mapping $\varphi : M \to G$. To clarify the situation arising let us return to consideration of the mappings $\varphi_\pm$. We call a pair of mappings $\varphi_\pm : M \to F_\pm$ *consistent* if there exists a mapping $\varphi : M \to G$ such that

$$\varphi_\pm = \pi_\pm \circ \varphi.$$

If the mappings $\varphi_\pm$ are consistent, then the corresponding mapping φ is defined up to an $\widetilde{H}$-gauge transformation. To show this, let us prove the following simple lemma.

Lemma 3.4 *Let a and a' be two arbitrary elements of G. The equalities*

$$\pi_\pm(a) = \pi_\pm(a') \tag{3.114}$$

are valid if and only if $a' = ah$ for some element $h \in \widetilde{H}$.

Proof Suppose that the equalities (3.114) are valid; then

$$a' = ab_-, \qquad a' = ab_+$$

for some elements $b_\mp \in \widetilde{B}_\mp$. Since $\widetilde{B}_- \cap \widetilde{B}_+ = \widetilde{H}$, therefore $b_- = b_+ \equiv h \in \widetilde{H}$. The inverse statement is obvious. □

Using this lemma, it is easy to show the validity of the following proposition.

Proposition 3.16 *Let φ and φ' be two mappings from M to G. The equalities*

$$\varphi'_+ = \varphi_+, \qquad \varphi'_- = \varphi_-$$

are valid if and only if $\varphi' = \varphi\psi$, where the mapping ψ takes values in $\widetilde{H}$.

Now formulate a useful criterion which allows one to check the consistency of the mappings $\varphi_\pm$. Let us start with one more lemma.

Lemma 3.5 *Let p_- and p_+ be two arbitrary point of the flag manifolds F_- and F_+ respectively. Consider arbitrary elements $m_\pm \in G$ satisfying the relation $\pi_\pm(m_\pm) = p_\pm$. An element $a \in G$ such that*

$$\pi_\pm(a) = p_\pm \tag{3.115}$$

exists if and only if the element $m_+^{-1}m_-$ has the following Gauss decomposition:

$$m_+^{-1}m_- = n_- h n_+^{-1}, \tag{3.116}$$

where $h \in \widetilde{H}$ and $n_\pm \in \widetilde{N}_\pm$.

Proof Suppose that the Gauss decomposition (3.116) exists and define

$$a \equiv m_+ n_- h = m_- n_+.$$

It is clear that $\pi_\pm(a) = \pi_\pm(m_\pm) = p_\pm$. On the other hand, suppose that the element $a \in G$, satisfying relations (3.115), exists. Then

$$a = m_+ n_- h_- = m_- n_+ h_+$$

for some $m_+, n_+ \in \widetilde{N}_+$, $m_-, n_- \in \widetilde{N}_-$ and $h_\pm \in \widetilde{H}$. Therefore, we obtain the Gauss decomposition (3.116) with $h = h_- h_+^{-1}$. □

Now it is easy to prove the following proposition.

Proposition 3.17 *Let φ_- and φ_+ be two mappings from M to F_- and F_+, respectively, and let $\mu_\pm : M \to G$ be arbitrary mappings* *from M to G, satisfying the condition*

$$\pi_\pm \circ \mu_\pm = \varphi_\pm.$$

The mappings $\varphi_\pm$ are consistent if and only if the mapping $\mu_+^{-1}\mu_-$ has the following Gauss decomposition:

$$\mu_+^{-1}\mu_- = \nu_-\eta\nu_+^{-1}, \tag{3.117}$$

where the mappings ν_- and ν_+ take values in $\widetilde{N}_-$ and $\widetilde{N}_+$, respectively; and the mapping η takes values in $\widetilde{H}$.

Recall now that the mappings $\mu_\pm$ arising from the Gauss decompositions of an admissible mapping φ satisfy the following conditions. First of all, the mapping μ_- is holomorphic, and the mapping μ_+ is antiholomorphic. Further, the mapping $\mu_-^{-1}\partial_-\mu_-$ takes values in $\widetilde{\mathfrak{m}}'_-$, and the mapping $\mu_+^{-1}\partial_+\mu_+$ takes values in $\widetilde{\mathfrak{m}}'_+$. We call a pair of the mappings $\mu_\pm : M \to G$ satisfying these conditions and having the Gauss decomposition (3.117), *regular*.

Proposition 3.18 *Any regular pair of mappings $\mu_\pm : M \to G$ arises from the corresponding Gauss decompositions of some admissible mapping $\varphi : M \to G$.*

Proof Let $\mu_\pm$ be a regular pair of mappings. Using (3.117), define the mapping $\varphi : M \to G$ by

$$\varphi \equiv \mu_+\nu_-\eta = \mu_-\nu_+.$$

It is not difficult to show that the mapping φ is admissible. Moreover, it is clear that the mappings $\mu_\pm$ are the mappings arising when we perform the corresponding Gauss decompositions of the mapping φ. □

Thus, we obtain the general solution of equations (3.102)–(3.104) using all regular pairs of mappings $\mu_\pm$.

Proposition 3.19 *Let $\mu'_\pm$ and $\mu_\pm$ be two regular pairs of mappings. If*

$$\pi_\pm \circ \mu'_\pm = \pi_\pm \circ \mu_\pm, \tag{3.118}$$

then the corresponding solutions of equations (3.102)–(3.104) *are connected by a symmetry transformation of the form* (3.111), (3.112).

Proof The proof of the proposition follows the lines of section 3.2.6. □

The mappings $\mu_\pm$ entering a regular pair are, by definition, tangent to the distributions $\mathcal{N}_\pm$. Since these distributions are involutive, the mappings $\mu_\pm$ take values in maximal integral manifolds of $\mathcal{N}_\pm$. As we have already noted in section 3.2.4, the maximal integral manifolds of the distributions $\mathcal{N}_\pm$ have the form $a_\pm \widetilde{N}_\pm$ for some elements $a_\pm \in G$. Suppose that the mappings $\mu_\pm$ have the following Gauss decomposition:

$$\mu_\pm = \mu'_\pm \nu'_\mp \eta'_\mp, \tag{3.119}$$

where the mappings $\mu'_\pm$ take values in $\widetilde{N}_\pm$, the mappings $\nu'_\mp$ take values in $\widetilde{N}_\mp$, and the mappings $\eta'_\mp$ take values in $\widetilde{H}$. If the pair $\mu_\pm$ is regular, then the pair $\mu'_\pm$ is also regular. Furthermore, the mappings $\mu'_\pm$ and $\mu_\pm$ satisfy relation (3.118). Therefore, the corresponding solutions of equations (3.102)–(3.104) are connected by a symmetry transformation of the form (3.111), (3.112). Actually, such symmetry transformations are connected with the ambiguity arising in constructing the corresponding lifts from $\widetilde{H}/\widetilde{H}_\pm$ to $\widetilde{H}$. Thus, almost any solution of equations (3.102)–(3.104) can be constructed from the mappings $\mu_\pm$ taking values in $\widetilde{N}_\pm$. Such a construction fails only when the mappings $\mu_\pm$ do not possess the Gauss decomposition (3.119). In fact, we can obtain almost all solutions by choosing the mappings $\mu_\pm$ taking values in $a_\pm \widetilde{N}_\pm$ for some fixed elements $a_\pm \in G$.

In constructing a regular pair of mappings, one should satisfy a number of requirements. It is not difficult to satisfy the requirement of holomorphicity or antiholomorphicity. A more difficult problem is to construct mappings $\mu_\pm$ for which $\mu_-^{-1}\partial_-\mu_-$ takes values in $\widetilde{\mathfrak{m}}'_-$ and $\mu_+^{-1}\partial_+\mu_+$ takes values in $\widetilde{\mathfrak{m}}'_+$. To solve this problem one can take a set of arbitrary mappings $\lambda_{\pm m} : M \to \mathfrak{g}_m$, $m = 1, \ldots, l-1$, and $\gamma_\pm : M \to \widetilde{H}$, where λ_{-m} and γ_- are holomorphic, while λ_{+m} and γ_+ are antiholomorphic mappings. Then the mappings $\mu_\pm$ can be determined by the integration of the relations

$$\mu_\pm^{-1}\partial_\pm\mu_\pm = \sqrt{-1}\left(\sum_{m=1}^{l-1} \lambda_{\pm m} + \gamma_\pm x_\pm \gamma_\pm^{-1}\right).$$

EXAMPLE 3.11 Consider the abelian Toda system corresponding to the Lie group $\mathrm{SL}(2,\mathbb{C})$. The Lie algebra of $\mathrm{SL}(2,\mathbb{C})$ is $\mathfrak{sl}(2,\mathbb{C})$, and we can take the identity mapping as a principal embedding of $\mathfrak{sl}(2,\mathbb{C})$ into $\mathfrak{sl}(2,\mathbb{C})$. In this, the generators h and $x_\pm$ are of the standard form given in example 1.8. The corresponding Toda system consists of just one equation:

$$\partial_+\partial_- f = e^{-2f}. \tag{3.120}$$

This is the famous *Liouville equation*. Recall that the function f parametrises the mapping γ taking values in the subgroup H. Actually, it is a local parametrisation. Indeed, in the case under consideration the subgroup H consists of all 2×2 diagonal matrices of $\mathrm{SL}(2,\mathbb{C})$. Therefore, we can globally parametrise the mapping γ as

$$\gamma = \begin{pmatrix} \beta & 0 \\ 0 & \beta^{-1} \end{pmatrix},$$

where β is a function taking values in $\mathbb{C}^\times$. The functions f and β are connected by the relation

$$\beta = e^f, \tag{3.121}$$

and it is clear that the function f provides only a local parametrisation of the mapping γ. In terms of the function β the Liouville equation is written as

$$\partial_+(\beta^{-1}\partial_-\beta) = \beta^{-2}. \tag{3.122}$$

Note that the subgroups $H_\pm$ in the case under consideration are formed by the matrices $\pm I_2$, and the symmetry transformations (3.111) look as $\beta' = \pm\beta$. We now obtain the general solution of equation (3.122).

Parametrise the mappings $\gamma_\pm$ as

$$\gamma_\pm = \begin{pmatrix} \beta_\pm & 0 \\ 0 & \beta_\pm^{-1} \end{pmatrix}.$$

This parametrisation leads to the following expression for the mapping μ_-:

$$\mu_- = \begin{pmatrix} a_{-11} & a_{-12} \\ a_{-21} & a_{-22} \end{pmatrix} \begin{pmatrix} 1 & 0 \\ \psi_- & 1 \end{pmatrix}, \tag{3.123}$$

where $a_- \equiv (a_{-ij})$ is an arbitrary element of $\mathrm{SL}(2,\mathbb{C})$ and

$$\psi_-(z^-) = \sqrt{-1}\int_{c^-}^{z^-} dy^- \beta_-^{-2}(y^-),$$

where c^- is a fixed complex number. Similarly, for μ_+ we obtain

$$\mu_+ = \begin{pmatrix} a_{+11} & a_{+12} \\ a_{+21} & a_{+22} \end{pmatrix} \begin{pmatrix} 1 & \psi_+ \\ 0 & 1 \end{pmatrix}, \tag{3.124}$$

where $a_+ \equiv (a_{+ij})$ is one more arbitrary element of $\mathrm{SL}(2, \mathbb{C})$ and

$$\psi_+(z^+) = \sqrt{-1} \int_{c^+}^{z^+} dy^+ \beta_+^2(y^+),$$

with c^+ being another fixed complex number. Thus, we obtain the following expression:

$$\mu_+^{-1}\mu_- = \begin{pmatrix} a_{11} - a_{21}\psi_+ + a_{12}\psi_- - a_{22}\psi_+\psi_- & a_{12} - a_{21}\psi_+ \\ a_{21} + a_{22}\psi_- & a_{22} \end{pmatrix},$$

where $a = (a_{ij}) \equiv a_+^{-1}a_-$. Finally, using the Gauss decomposition of form (3.113) and relation (3.98), we arrive at

$$\beta = \beta_+\beta_-^{-1}(a_{11} - a_{21}\psi_+ + a_{12}\psi_- - a_{22}\psi_+\psi_-).$$

This is the general solution of equation (3.122).

Let us demonstrate the fact that to obtain almost all solutions of (3.122) it is enough to use the mappings $\mu_\pm$ taking values in the subgroups $N_\pm$. Let the mappings μ_- and μ_+ be of form (3.123) and (3.124) respectively. Find the mappings $\mu'_\pm$ entering decomposition (3.119). The mapping μ'_- has the form

$$\mu'_- = \begin{pmatrix} 1 & 0 \\ \dfrac{a_{-21} + a_{-22}\psi_-}{a_{-11} + a_{-12}\psi_-} & 1 \end{pmatrix}.$$

It is easy to verify that this mapping satisfies the equation

$$\mu'^{-1}_- \partial_- \mu'_- = \sqrt{-1}\gamma'_- x_- \gamma'^{-1}_-,$$

with the mapping γ'_- parametrised by the function

$$\beta'_- = \pm\beta_-(a_{-11} + a_{-12}\psi_-).$$

For the mapping μ'_+ we obtain

$$\mu'_+ = \begin{pmatrix} 1 & \dfrac{a_{+11}\psi_+ + a_{+12}}{a_{+21}\psi_+ + a_{+22}} \\ 0 & 1 \end{pmatrix},$$

and the corresponding function β'_+ is given by

$$\beta'_+ = \frac{\pm\beta_+}{a_{+21}\psi_+ + a_{+22}}.$$

A direct calculation shows that the mappings $\mu'_\pm$ and $\beta'_\pm$ give the solution which differs from the solution obtained with the use of the mappings $\mu_\pm$ and $\beta_\pm$ by a possible symmetry transformation of type (3.111), which, in our case, is simply a change of the sign of the function β.

As we noted above, almost all solutions of Toda-type systems can be obtained with the use of the mappings $\mu_\pm$ taking values in the subsets $a_\pm \widetilde{N}_\pm$ with fixed elements $a_\pm \in G$. Actually, the solutions depend only on the element $a \equiv a_+^{-1} a_-$, and we obtain them using the equality

$$\mu_+^{-1}\mu_- = \mu_+'^{-1} a \mu_-',$$

where the mappings $\mu'_\pm$ take values in $N_\pm$. Choosing different elements a, we obtain different representation of solutions.

In the case under consideration we can write the mappings $\mu'_\pm$ as

$$\mu'_- = \begin{pmatrix} 1 & 0 \\ m_{-21} & 1 \end{pmatrix} \begin{pmatrix} 1 & 0 \\ \psi_- & 1 \end{pmatrix} = \begin{pmatrix} 1 & 0 \\ m_{-21} + \psi_- & 1 \end{pmatrix},$$

$$\mu'_+ = \begin{pmatrix} 1 & m_{+12} \\ 0 & 1 \end{pmatrix} \begin{pmatrix} 1 & \psi_+ \\ 0 & 1 \end{pmatrix} = \begin{pmatrix} 1 & m_{+12} + \psi_+ \\ 0 & 1 \end{pmatrix},$$

where m_{-21} and m_{+12} are arbitrary complex numbers parametrising general elements of the subgroups N_- and N_+. Introducing the notations

$$\zeta_- = -\sqrt{-1}(m_{-21} + \psi_-), \qquad \zeta_+ = -\sqrt{-1}(m_{+12} + \psi_+),$$

we obtain the following relation:

$$\beta = \beta_+^{-1}\beta_-(a_{11} - \sqrt{-1}a_{21}\zeta_+ + \sqrt{-1}a_{12}\zeta_- + a_{22}\zeta_+\zeta_-).$$

Consider two different choices for the element a:

$$a = \begin{pmatrix} 1 & 0 \\ 0 & 1 \end{pmatrix}, \qquad a = \begin{pmatrix} 0 & -1 \\ 1 & 0 \end{pmatrix}.$$

These choices give the following formulas:

$$\beta = \beta_+^{-1}\beta_-(1 + \zeta_+\zeta_-),$$

$$\beta = -\sqrt{-1}\beta_+^{-1}\beta_-(\zeta_- - \zeta_+).$$

which describe almost all the solutions of equation (3.122).

Return now to the original Liouville equation. Taking (3.121) and the relations $\partial_-\zeta_- = \beta_-^{-2}$ and $\partial_+\zeta_+ = \beta_+^2$ into account, we can formally write the solutions of the Liouville equation (3.120) as

$$e^f = (\partial_+\zeta_+\partial_-\zeta_-)^{-1/2}(1 + \zeta_+\zeta_-),$$

or as

$$e^f = -\sqrt{-1}(\partial_+\zeta_+\partial_-\zeta_-)^{-1/2}(\zeta_- - \zeta_+).$$

These are usual the forms for writing the solution of the Liouville equation.

As we can see, the procedure of the construction of the general solution for the Toda-type systems is based mainly on Gauss decomposition. In some cases it is more convenient not to use the explicit form of Gauss decomposition but, rather, to appeal to the algebraic structure of the construction. The relevant object here is a representation of the corresponding Lie algebra, or, in other words, a module over this Lie algebra.

3.3.2 Modules over semisimple Lie algebras

In this section $\mathfrak{g}$ is a complex semisimple Lie algebra of rank r, $\mathfrak{h}$ is some fixed Cartan subalgebra of $\mathfrak{g}$, Δ is a root system of $\mathfrak{g}$ with respect to $\mathfrak{h}$, and $\Pi = \{\alpha_1, \ldots, \alpha_r\}$ is a base of Δ. The corresponding Cartan and Chevalley generators are denoted by h_i and $x_{\pm i}$ respectively.

Let V be a $\mathfrak{g}$-module and let $\lambda \in \mathfrak{h}^*$. Introduce the following notation:

$$V^\lambda \equiv \{v \in V \mid hv = \langle \lambda, h \rangle v \text{ for all } h \in \mathfrak{h}\}.$$

If $V^\lambda \neq \{0\}$, we say that V^λ is a *weight space* and λ is a *weight* of V. Elements of V_λ are called elements of weight λ, and the dimension of V^λ is called the *multiplicity* of λ. It is easy to demonstrate that for any $\lambda \in \mathfrak{h}^*$ and $\alpha \in \Delta$ we have

$$\mathfrak{g}^\alpha V^\lambda \subset V^{\lambda+\alpha}.$$

Since eigenvectors which correspond to different eigenvalues are linearly independent, then, for any two different weights λ and μ of V, we have $V^\lambda \cap V^\mu = \{0\}$. Therefore, we can consider the following direct sum:

$$V' \equiv \bigoplus_\lambda V^\lambda.$$

It is clear that V' is a submodule of V. If $V' = V$, the $\mathfrak{g}$-module is called *diagonalisable* with respect to $\mathfrak{h}$. It can be shown that any finite-dimensional $\mathfrak{g}$-module is diagonalisable.

Let V be a $\mathfrak{g}$-module. A nonzero element $v \in V$ is called a *singular vector* of weight λ if $v \in V_\lambda$ and $\mathfrak{n}_+ v = 0$. It is clear that $\mathfrak{n}_+ v = 0$ if and only if $x_{+i} v = 0$ for all $i = 1, \dots, r$. Let $v \in V$ be a singular vector. The subspace $W \equiv U(\mathfrak{g})v$ is a submodule of V.

A $\mathfrak{g}$-module V is called an *extremal module* of highest weight $\lambda \in \mathfrak{h}^*$ if v_λ is a singular vector of weight λ and $U(\mathfrak{g})v_\lambda = V$. Here λ is called the *highest weight* and the vector v_λ the *highest weight vector* of the module V. For any weight μ of an extremal module V of highest weight λ we have $\mu \prec \lambda$. This fact explains why the weight λ is called the highest weight. It is clear that any extremal module can be thought of as the result of the construction described above.

Let V be an extremal $\mathfrak{g}$-module of highest weight λ. From representation (3.6) it follows that $U(\mathfrak{g}) = U(\mathfrak{n}_-)U(\mathfrak{h})U(\mathfrak{n}_+)$. This relation implies that $U(\mathfrak{g})v_\lambda = U(\mathfrak{n}_-)v_\lambda$. Having enumerated the positive roots, we can write $\Delta_+ = \{\beta_1, \dots, \beta_s\}$ and $\Delta_- = \{-\beta_1, \dots, -\beta_s\}$. For any $\beta_a \in \Delta_+$ choose some nonzero vector $x_{-\beta_a}$ from $\mathfrak{g}^{-\beta_a}$. From the Poincaré–Birkhoff–Witt theorem it follows that the monomials $(x_{-\beta_1})^{k_1} \cdots (x_{-\beta_s})^{k_s}$ form a basis of $U(\mathfrak{n}_-)$. Therefore, the elements $(x_{-\beta_1})^{k_1} \cdots (x_{-\beta_s})^{k_s} v_\lambda$ span V, and the weights of V have the form

$$\lambda - \sum_{i=1}^{r} m_i \alpha_i,$$

where m_i are nonnegative integers. Recall that $\mathfrak{n}_-$ is generated by the Chevalley generators x_{-i}, $i = 1, \dots, r$, where r is the rank of $\mathfrak{g}$. By this reason, it is clear that the vectors of the form $x_{-i_1}^{k_{i_1}} \cdots x_{-i_s}^{k_{i_s}} v_\lambda$ also span $M(\lambda)$. A basis obtained by selecting from these vectors linearly independent ones is called a *Verma basis.* It is clear that the module V is diagonalisable, the multiplicity of λ is equal to 1 and the multiplicities of all weights are finite.

Any extremal $\mathfrak{g}$-module is indecomposable. If an extremal $\mathfrak{g}$-module V of highest weight λ is irreducible, then any singular vector of V is proportional to the highest weight vector v_λ.

It appears that for any $\lambda \in \mathfrak{h}^*$ there exists an irreducible extremal $\mathfrak{g}$-module of highest weight λ. Such a module is unique up to isomorphism. To demonstrate the existence of the module in question, note first that $U(\mathfrak{g})$ has the natural structure of a $\mathfrak{g}$-module, induced by the multiplication of the elements of $U(\mathfrak{g})$ by the elements of $\mathfrak{g}$ from the left. Consider the left ideal $J(\lambda)$ of $U(\mathfrak{g})$ generated by $\mathfrak{n}_+$ and by the elements $h - \langle \lambda, h \rangle$, $h \in \mathfrak{h}$ and denote $M(\lambda) \equiv U(\mathfrak{g})/J(\lambda)$. The $\mathfrak{g}$-module $M(\lambda)$ is an extremal module of highest weight λ which is called the *Verma module* of highest weight λ. Here the image of $1 \in U(\mathfrak{g})$ is the highest weight vector. It can be shown that any extremal module of highest weight λ is isomorphic to a quotient module of $M(\lambda)$.

Any singular vector of $M(\lambda)$ that is not proportional to v_λ generates a proper submodule of $M(\lambda)$. Any proper submodule of $M(\lambda)$ can be represented as a direct sum of such submodules. The direct sum of proper submodules of $M(\lambda)$ is a proper submodule of $M(\lambda)$. Hence $M(\lambda)$ has a unique maximal proper submodule $M'(\lambda)$. The quotient module $L(\lambda) \equiv M(\lambda)/M'(\lambda)$ is an irreducible extremal $\mathfrak{g}$-module of highest weight λ. It can be shown that any irreducible extremal $\mathfrak{g}$-module of highest weight λ is isomorphic to $L(\lambda)$.

Since any element of $\mathfrak{h}^*$ can be a weight of some $\mathfrak{g}$-module, we shall often call elements of $\mathfrak{h}^*$ weights. A weight λ is said to be *integral* if $\langle \lambda, h_i \rangle$ is an integer for $i = 1, \ldots, r$. The set of all integral weights is called the *weight lattice* and is denoted by Λ. Note that $\Delta \subset \Lambda$. An integral weight λ is called *dominant* (*regular dominant*) if $\langle \lambda, h_i \rangle \geq 0$ ($\langle \lambda, h_i \rangle > 0$) for $i = 1, \ldots, r$. The sets of dominant and regular dominant weights are denoted by Λ_+ and Λ_{++} respectively.

The module $L(\lambda)$ is finite-dimensional if and only if λ is a dominant weight. It can be shown that any finite-dimensional $\mathfrak{g}$-module has a singular vector and that the submodule generated by this vector is irreducible. Hence, any finite-dimensional irreducible $\mathfrak{g}$-module is extremal and, therefore, it is isomorphic to some $\mathfrak{g}$-module $L(\lambda)$ with λ being a dominant weight. The dominant weights ϵ_i, $i = 1, \ldots, r$, defined by

$$\langle \epsilon_i, h_j \rangle \equiv \delta_{ij},$$

are called the *fundamental weights*. The corresponding $\mathfrak{g}$-modules (representations of $\mathfrak{g}$) are called the *fundamental $\mathfrak{g}$-modules* (*fundamental representations of* $\mathfrak{g}$). Here we use the notation $L_i \equiv L(\epsilon_i)$.

Recall that any representation of a Lie group generates a representation of the corresponding Lie algebra. If a Lie group G is simply connected, then any representation of its Lie algebra $\mathfrak{g}$ can be 'integrated' up to the corresponding representation of G.

Let σ be a conjugation of $\mathfrak{g}$ and let $x \in \mathfrak{g} \mapsto x^\dagger \equiv -\sigma(x) \in \mathfrak{g}$ be the corresponding hermitian involution of $\mathfrak{g}$. The mapping σ defines a real form $\mathfrak{g}_\sigma$ of the Lie algebra $\mathfrak{g}$ by

$$\mathfrak{g}_\sigma \equiv \{x \in \mathfrak{g} \mid \sigma(x) = x\}.$$

An element $x \in \mathfrak{g}$ is said to be σ-hermitian if $x^\dagger = x$, and it is said to be σ-antihermitian if $x^\dagger = -x$. The subalgebra $\mathfrak{g}_\sigma$ is formed by all σ-antihermitian elements of $\mathfrak{g}$.

Suppose that σ can be extended to an antiholomorphic automorphism Σ of the group G. It is always possible when the group G is simply connected. In that case we have $\Sigma^2 = \mathrm{id}_G$. Define the mapping $a \in G \mapsto a^\dagger \in G$, where

$$a^\dagger \equiv \Sigma(a^{-1}) = (\Sigma(a))^{-1}.$$

It is obvious that this mapping is an antiholomorphic antiautomorphism of G satisfying the condition

$$(a^\dagger)^\dagger = a$$

for any $a \in G$. An element $a \in G$ is called Σ-*hermitian* if $a^\dagger = a$, and it is called Σ-*unitary* if $a^\dagger = a^{-1}$. In the case where the mapping Σ is determined by the mapping σ defined by (1.29), we simply say 'hermitian' and 'unitary'. The real Lie group G_σ corresponding to the real form $\mathfrak{g}_\sigma$ is formed by all Σ-unitary elements of G.

Suppose also that the mapping σ has the property

$$\sigma(\mathfrak{g}_m) = \mathfrak{g}_{-m}, \qquad m \in \mathbb{Z}. \tag{3.125}$$

In this case one has

$$(\widetilde{H})^\dagger = \widetilde{H}, \qquad (\widetilde{N}_\pm)^\dagger = \widetilde{N}_\mp.$$

A representation $\rho : G \to GL(V)$ of the group G in the linear space V over the field $\mathbb{C}$ is called Σ-*unitary* if the space V is equipped with a nondegenerate hermitian form $(\cdot, \cdot)$ such that

$$\rho(a)^\dagger = \rho(a^\dagger),$$

where $\dagger$ on the left-hand side means the hermitian conjugation with respect to the hermitian form $(\cdot, \cdot)$. It can be shown that the representation ρ is Σ-unitary if and only if the restriction of ρ to the real Lie group G_σ is unitary.

3.3.3 From representations to solutions

Below we will use the Dirac notation for complex vector spaces, see Dirac (1958). According to this notation, the elements of a complex vector space V are denoted by the symbol $|\,\rangle$. To distinguish different elements of V, the symbol $|\,\rangle$ is supplied with labels, for example, $|v\rangle$, $|u\rangle$. The elements of the dual space V^* are denoted by the symbol $\langle\,|$, also supplied with labels. The action of an element $\langle\alpha| \in V^*$ on an element $|v\rangle \in V$ is denoted by $\langle\alpha|v\rangle$. If a nondegenerate hermitian form $(\ ,\ )$ is defined on V, then one constructs the mapping from V to V^* which associates with an element $|v\rangle$ of V the element $\langle v|$ of V^*, such that

$$\langle v|u\rangle \equiv (|v\rangle, |u\rangle)$$

for all $|u\rangle \in V$. For more details we refer the reader to the book by Dirac (1958).

Let us now consider an arbitrary Σ-unitary representation ρ of the group G in a linear space V. Denote by V_+ the subspace of V formed by all elements $|v\rangle \in V$ such that

$$\rho(a)|v\rangle = |v\rangle$$

for all $a \in \widetilde{N}_+$. For any mapping $\varphi : M \to G$ and any vectors $|u\rangle, |v\rangle \in V$ we denote by $\langle u|\varphi|v\rangle$ the mapping from M to $\mathbb{C}$, defined as

$$\langle u|\varphi|v\rangle(p) \equiv \langle u|\rho(\varphi(p))|v\rangle.$$

Theorem 3.4 *For any $|u\rangle, |v\rangle \in V_+$ the following relation is valid:*

$$\langle u|\gamma|v\rangle = \langle u|(\mu_+\gamma_+)^{-1}(\mu_-\gamma_-)|v\rangle. \tag{3.126}$$

Proof Using definition (3.98) of the mapping γ and equality (3.113), we obtain the relation

$$\nu'_-\gamma\nu'^{-1}_+ = (\mu_+\gamma_+)^{-1}(\mu_-\gamma_-), \tag{3.127}$$

where

$$\nu'_- \equiv \gamma_+^{-1}\nu_-\gamma_+, \qquad \nu'_+ \equiv \gamma_-^{-1}\nu_+\gamma_-.$$

Then the validity of (3.126) follows from (3.127) and the definition of V_+. □

Thus, we can find some matrix elements of the linear operators corresponding to the mapping γ in a Σ-unitary representation of the group G by using only the mappings $\mu_\pm$, which in their turn are determined only by the mappings $\varphi_\pm$. It is natural to suppose here that in using a large enough set of representations, one will be able to recover the mapping γ from the mappings $\varphi_\pm$.

In more or less the same way one can consider the mappings $\upsilon_{\pm m}$. Here, however, one deals with matrix elements taken between the vectors annihilated by the elements of subspaces $\bigoplus_{n=1}^{m=1} \mathfrak{g}_{\pm n}$, see Gervais & Saveliev (1995). Note also that for $l = 1$ form (3.126) of writing the general solution of a Toda system coincides with that given in Leznov & Saveliev (1992).

3.3.4 Real solutions

Let us now consider a special class of solutions which are called *real solutions.* To this end, we introduce two antiholomorphic mappings $\Sigma_\pm : F_\pm \to F_\mp$ defined by

$$\Sigma_\pm(a\widetilde{B}_\mp) \equiv \Sigma(a)\widetilde{B}_\pm.$$

It is easy to show that the mappings $\Sigma_\pm$ are defined correctly. Directly from the definition of these mappings, we obtain the equalities

$$\Sigma_+ \circ \pi_+ = \pi_- \circ \Sigma, \qquad \Sigma_- \circ \pi_- = \pi_+ \circ \Sigma. \tag{3.128}$$

Moreover, these mappings are mutually inverse:

$$\Sigma_+ \circ \Sigma_- = \mathrm{id}_{F_-}, \qquad \Sigma_- \circ \Sigma_+ = \mathrm{id}_{F_+}. \tag{3.129}$$

A mapping $\varphi : M \to G$ is said to satisfy the *reality condition* if

$$\Sigma_+ \circ \varphi_+ = \varphi_-, \tag{3.130}$$

which can be also written as

$$\Sigma_- \circ \varphi_- = \varphi_+. \tag{3.131}$$

Proposition 3.20 *A mapping $\varphi : M \to G$ satisfies the reality condition if and only if*

$$\Sigma \circ \varphi = \varphi\psi, \tag{3.132}$$

where the mapping ψ takes values in $\widetilde{H}$.

Proof Let φ satisfy the reality condition. Using (3.128) in (3.130), we obtain

$$\pi_- \circ \Sigma \circ \varphi = \pi_- \circ \varphi.$$

Hence, $\Sigma \circ \varphi = \varphi\psi$, where ψ takes values in $\widetilde{B}_+$. In a similar way, from (3.131) one sees that ψ takes values in $\widetilde{B}_-$. Since $\widetilde{B}_+ \cap \widetilde{B}_- = \widetilde{H}$, we conclude that ψ takes values in $\widetilde{H}$.

The inverse statement of the proposition is obvious. □

A mapping $\psi : M \to G$ is called Σ-*hermitian* if for any $p \in M$ the mapping $\psi(p)$ is a Σ-hermitian element of G.

Proposition 3.21 *The mapping* ψ *entering proposition* 3.20 *is* Σ*-hermitian.*

Proof Since $\Sigma^2 = \mathrm{id}_G$, from (3.132) one obtains

$$\varphi = (\Sigma \circ \varphi)(\Sigma \circ \psi),$$

which can be written as

$$\Sigma \circ \varphi = \varphi(\Sigma \circ \psi)^{-1}. \tag{3.133}$$

Comparing (3.133) with (3.132), we have

$$(\Sigma \circ \psi)^{-1} = \psi.$$

Hence, the mapping ψ is Σ-hermitian. □

Proposition 3.22 *The reality condition is* $\widetilde{H}$*-gauge invariant.*

Proof Let a mapping φ satisfy the reality condition, and $\varphi' = \varphi\xi$, with ξ taking values in $\widetilde{H}$, be a gauge transformed mapping. Using (3.132) we have

$$\Sigma \circ \varphi' = \varphi' \circ \psi',$$

where

$$\psi' = \xi^{-1}\psi(\Sigma \circ \xi).$$

Since $\Sigma(\widetilde{H}) = \widetilde{H}$, the mapping ψ' satisfies the reality condition. □

Let $s_{+\alpha}$, $\alpha \in \mathcal{A}$, be a family of local holomorphic sections of the fibre bundle $G \to F_+$ with the properties described in proposition 3.29. The mapping Σ allows us to construct the corresponding family of local holomorphic sections of the fibre bundle $G \to F_-$. For each open set $U_{+\alpha}$, we define the open set $U_{-\alpha}$ by

$$U_{-\alpha} \equiv \Sigma_+(U_{+\alpha}).$$

Using (3.128) and (3.129), it is easy to show that for any $\alpha \in \mathcal{A}$ the mapping $s_{-\alpha} : U_{-\alpha} \to G$ given by

$$s_{-\alpha} \equiv \Sigma \circ s_{+\alpha} \circ \Sigma_-$$

is a local holomorphic section of the fibre bundle $G \to F_-$. Since $\Sigma(\widetilde{N}_+) = \widetilde{N}_-$, we also have

$$\Sigma_* \mathcal{N}_+ = \mathcal{N}_-;$$

hence, the section $s_{-\alpha}$ is tangent to the distribution $\mathcal{N}_-$. Thus, we obtain a family of holomorphic sections of the fibre bundle $G \to F_-$ with the required properties. Now, to construct the mappings needed to define the mapping γ, one uses for any section $s_{+\alpha}$ the corresponding section $s_{-\alpha}$ defined with the help of the procedure described above.

Proposition 3.23 *If a mapping φ satisfies the reality condition, then the mappings μ_+ and μ_- entering proposition* 3.14 *are connected by the relation*

$$\mu_\mp = \Sigma \circ \mu_\pm. \tag{3.134}$$

Proof Recall that the mappings $\mu_\pm$ are given by

$$\mu_\pm = s_{\pm\alpha} \circ \varphi_\pm.$$

From this relation we have

$$\mu_\mp = \Sigma \circ s_{\pm\alpha} \circ \Sigma_\mp \circ \varphi_\mp = \Sigma \circ s_{\pm\alpha} \circ \varphi_\pm = \Sigma \circ \mu_\pm.$$

This chain of equalities provides the assertion of the proposition. □

Proposition 3.24 *If a mapping φ satisfies the reality condition, then the mappings $\gamma_\pm$ satisfying* (3.82) *can be chosen in such a way that*

$$\Sigma \circ \gamma_\pm = \gamma_\mp. \tag{3.135}$$

Proof First, let us show that the mappings λ_+ and λ_- are connected by the relation

$$\lambda_- = -\sigma \circ \lambda_+. \tag{3.136}$$

From (3.134) we have

$$^{\mu_-}\omega = (\mu_+^* \circ \Sigma^*)\theta.$$

Now, using equality (3.4), one shows that

$$^{\mu_-}\omega(v) = \Sigma^*\theta(\mu_{+*}(v)) = \sigma\left(^{\mu_+}\omega(\bar{v})\right)$$

for any $v \in T_p^{\mathbb{C}}(M)$, $p \in M$. In particular, there is the following equality:

$$^{\mu_-}\omega(\partial_-) = \sigma\left(^{\mu_+}\omega(\partial_+)\right),$$

which leads directly to (3.136).

From (3.136) it follows that

$$\mathrm{Ad}(\gamma_-)x_- = (\sigma \circ \mathrm{Ad}(\gamma_+))x_+.$$

Suppose now that the elements $x_\pm$ entering (3.82) are chosen in such a way that

$$\sigma(x_+) = -x_-.$$

Taking relation (3.2) into account, we conclude that one can choose the mappings γ_+ and γ_- satisfying (3.135). □

Theorem 3.5 *If a mapping φ satisfies the reality condition, then the mapping γ can be chosen Σ-hermitian; while for the mappings $v_{\pm m}$ one has*

$$(v_{+m})^\dagger = v_{-m}. \tag{3.137}$$

Proof From proposition 3.20, using (3.76) and (3.77), we obtain

$$\Sigma \circ \varphi = (\Sigma \circ \mu_+)(\Sigma \circ \nu_-)(\Sigma \circ \eta_-) = \mu_-\nu_+\eta_+\psi. \tag{3.138}$$

Since decomposition (3.77) is unique, then

$$\Sigma \circ \eta_- = \eta_+\psi,$$

and hence, one can write the equality

$$\eta_- = (\eta_+^{-1})^\dagger\psi^{-1},$$

which leads to the relation

$$\eta = (\eta_+^{-1})^\dagger\psi^{-1}\eta_+^{-1}.$$

As follows from proposition 3.21, the mapping ψ is Σ-hermitian, thus the mapping η is also Σ-hermitian. Taking proposition 3.23 and the definition of the mapping γ into account, we conclude that it can be chosen to be Σ-hermitian.

From (3.138) it follows that

$$\nu_+^\dagger = \nu_-^{-1}.$$

Using relations (3.93) and (3.96) now one obtains (3.137). □

Thus, we can say that using the mappings φ which satisfy the reality condition, one can construct hermitian, in a sense real, solutions of equations (3.102)–(3.104). Since the solutions are actually determined by the mappings $\varphi_\pm$, then to obtain real solutions of the equations under consideration, we should choose the mappings $\varphi_\pm$ satisfying relation (3.130).

Suppose now that the involutive antilinear automorphism σ is defined by (1.29), so that the corresponding real form $\mathfrak{u}$ of $\mathfrak{g}$ is compact. In this case the grading operator is σ-hermitian, and (3.125) is valid. Consider the realification $\mathfrak{g}_\mathbb{R}$ of the Lie algebra $\mathfrak{g}$. Let J be the linear operator in $\mathfrak{g}_\mathbb{R}$ corresponding to multiplication by $\sqrt{-1}$ in $\mathfrak{g}$. It is clear that we have the expansion

$$\mathfrak{g}_\mathbb{R} = \mathfrak{u} \oplus J\mathfrak{a} \oplus \mathfrak{n}_{+\mathbb{R}},$$

where $\mathfrak{a}$ is a maximal abelian subalgebra of $\mathfrak{u}$. This expansion is called the *Iwasawa decomposition* of $\mathfrak{g}_\mathbb{R}$. Note that

$$\mathfrak{h}_\mathbb{R} = \mathfrak{a} \oplus J\mathfrak{a}.$$

There is the corresponding decomposition of the Lie group G considered as a real Lie group, see Helgason (1978). It has the form

$$G_\mathbb{R} = UN_{+\mathbb{R}}A^*, \tag{3.139}$$

where A^* is the real connected Lie group corresponding to the subalgebra $J\mathfrak{a}$.

Theorem 3.6 *If mappings $\varphi_\pm : M \to F_\pm$ satisfy relation* (3.130), *then they are consistent.*

Proof It is enough to show that if two points $p_+ \in F_+$ and $p_- \in F_-$ are connected by the relation

$$p_- = \Sigma_+(p_+),$$

then there exists an element $a \in G$, such that

$$p_+ = \pi_+(a), \qquad p_- = \pi_-(a).$$

This fact can be proved using the Iwasawa decomposition (3.139). Let a' be any element of G, such that $\pi_+(a') = p_+$. This element can be written as

$$a' = un_+a^*,$$

where $u \in U$, $n_+ \in N_+$ and $a^* \in A^*$, and the subgroups U, N_+ and A^* are defined above. We have $N_+A \subset B_+ \subset \tilde{B}_+$, hence $\pi_+(u) = p_+$. Using (3.128) one now obtains

$$(\Sigma_+ \circ \pi_+)(u) = (\pi_- \circ \Sigma)(u) = \pi_-(u) = \Sigma_+(p_+).$$

Thus, the element u can be taken as the element a that we are looking for. □

The generalisation of the results proved in this section and in the previous one, to the case of the semi-integral embeddings of $\mathfrak{sl}(2, \mathbb{C})$ into $\mathfrak{g}$ is straightforward, and can be performed following completely similar reasons. Recall that in that case we deal with $\mathbb{Z}/2$-gradations of $\mathfrak{g}$.

Exercises

3.10 Construct explicitly Verma bases of the first fundamental representation for the classical series of simple Lie algebras $\mathfrak{sl}(r, \mathbb{C})$, $\widetilde{\mathfrak{so}}(2r + 1, \mathbb{C})$, $\widetilde{\mathfrak{so}}(2r, \mathbb{C})$, $\widetilde{\mathfrak{sp}}(r, \mathbb{C})$, and for the exceptional Lie algebra G_2.

3.11 Let $|i\rangle$ be the highest weight vector of the ith fundamental representation of a complex semisimple Lie algebra $\mathfrak{g}$ normalised by $\langle i|i\rangle = 1$. Using the defining relations (1.16)–(1.18), find an explicit expression for the matrix elements $\langle i|x_{+j_1} \cdots x_{+j_m} x_{-i_m} \cdots x_{-i_1}|i\rangle$ in terms of the corresponding Cartan matrix.

3.4 Toda fields and generalised Plücker relations

In this section we give a derivation of the generalised Plücker relations.

3.4.1 Riemannian and Kähler manifolds

A smooth tensor field g of type $\binom{0}{2}$ on a real manifold M is called a *Riemannian metric* on M if for any $p \in M$ one has

(RM1) $g_p(v, u) = g_p(u, v)$ for all $v, u \in T_p(M)$;

(RM2) $g_p(v, v) > 0$ for all nonzero $v \in T_p(M)$.

In other words, g is a Riemannian metric on M if for any $p \in M$ the bilinear form g_p on $T_p(M)$ is symmetric and positive definite. It can be shown that any real manifold possesses a Riemannian metric. Here it is important that we consider only the manifolds which are second countable topological spaces. A manifold endowed with a Riemannian metric is called a *Riemannian manifold*.

Let M and N be two manifolds and let $\varphi : M \to N$ be an immersion. Suppose that N is endowed with a Riemannian metric g^N. Define a tensor field g^M on M by

$$g_p^M(v, u) \equiv g_{\varphi(p)}^N(\varphi_{*p}(v), \varphi_{*p}(u))$$

for any $p \in M$ and $v, u \in T_p(M)$. The tensor field g^M is a Riemannian metric on M, which is called the *Riemannian metric induced from g^N by φ*. In particular, any submanifold of a Riemannian manifold can be provided with the Riemannian metric induced by the inclusion mapping.

As it is for an arbitrary tensor field, a Riemannian metric g on a manifold M can be extended to a complex tensor field on M. Denote this tensor field also by g. Here, for any $p \in M$ one has

(CRM1) $g_p(v, u) = g_p(u, v)$ for all $v, u \in T_p^{\mathbb{C}}(M)$;

(CRM2) $g_p(v, \bar{v}) > 0$ for all nonzero $v \in T_p^{\mathbb{C}}(M)$;

(CRM3) $\overline{g_p(v, u)} = g_p(\bar{v}, \bar{u})$ for all $v, u \in T_p^{\mathbb{C}}(M)$.

On the other hand, any complex tensor field g on M having the properties (CRM1)–(CRM3) is the natural extension of a Riemannian metric on M. A Riemannian metric on a complex manifold M is, by definition, a complex tensor field of type $\binom{0}{2}$ satisfying (CRM1)–(CRM3).

Let g be a Riemannian metric on a complex manifold M and let $(U, z^1, \ldots, z^m)$ be a complex chart on M. From (CRM1) we obtain the following equalities:

$$g_{ij} = g_{ji}, \qquad g_{\bar{\imath}\bar{\jmath}} = g_{\bar{\jmath}\bar{\imath}}, \qquad g_{i\bar{\jmath}} = g_{\bar{\jmath}i},$$

while (CRM3) implies

$$\overline{g_{ij}} = g_{\bar{\imath}\bar{\jmath}}, \qquad \overline{g_{i\bar{\jmath}}} = g_{\bar{\imath}j}.$$

A Riemannian metric g on a complex manifold M is called *hermitian* if

$$g(J^M X, J^M Y) = g(X, Y)$$

for all $X, Y \in \mathfrak{X}^{\mathbb{C}}(M)$. Let g be a hermitian metric on a complex manifold M. It can easily be shown that $g(X, Y) = 0$ if both vector fields X and Y are either of type (1,0) or of type (0,1). Therefore, for any complex chart $(U, z^1, \ldots, z^m)$ on M one obtains the equalities

$$g_{ij} = 0, \qquad g_{\bar{\imath}\bar{\jmath}} = 0,$$

which imply the following local representation of g:

$$g|_U = g_{i\bar{\jmath}} dz^i \otimes d\bar{z}^{\bar{\jmath}} + g_{\bar{\imath}j} d\bar{z}^{\bar{\imath}} \otimes dz^j. \tag{3.140}$$

Let g be a hermitian metric on a complex manifold M. The 2-form Φ defined by

$$\Phi(X, Y) \equiv g(X, J^M Y),$$

is called the *fundamental form* associated with g. Using a complex chart $(U, z^1, \ldots, z^m)$, one has

$$\Phi|_U = -2\sqrt{-1} g_{i\bar{\jmath}} dz^i \wedge d\bar{z}^{\bar{\jmath}}. \tag{3.141}$$

A hermitian metric g on M is called a *Kähler metric* if the fundamental form Φ associated with g is closed, i.e.,

$$d\Phi = 0.$$

A complex manifold endowed with a Kähler metric is said to be a *Kähler manifold*. The fundamental form Φ associated with a hermitian metric g can be locally represented as

$$\Phi = -2\sqrt{-1}\partial\bar{\partial}K,$$

where K is a real valued function. The function K entering the last relation is called a *Kähler potential* of the hermitian metric g.

Note also that the expression for the Ricci curvature tensor R in the case under consideration has the form

$$R = 2\sqrt{-1} d^{(1,0)} d^{(0,1)} \ln \mathcal{G}, \tag{3.142}$$

where $\mathcal{G}$ is the determinant of the matrix $(g_{i\bar{\jmath}})$, see Kobayashi & Nomizu (1963).

EXAMPLE 3.12 Let V be a complex linear space and let $P(V)$ be the projective space of V. Recall that $P(V)$ is the set of one-dimensional subspaces of V. Denote by pr the canonical projection from the set $V^\times \equiv V - \{0\}$ onto $P(V)$. Suppose that V is endowed with a positive definite hermitian scalar product $(\cdot, \cdot)$, and define a real valued function F on $V^\times$ by

$$F(v) \equiv \ln(v, v).$$

There exists a unique 2-form Φ^{FS} on $P(V)$ satisfying the relation

$$\mathrm{pr}^* \Phi^{FS} = -2\sqrt{-1} d^{(1,0)} d^{(0,1)} F. \qquad (3.143)$$

This form determines a Kähler metric on $P(V)$, called the *Fubini-Study metric.*

3.4.2 Verma modules and flag manifolds

Let G be a complex connected semisimple Lie group of rank r and let $\mathfrak{g}$ be a Lie algebra of G. Fix a Cartan subalgebra $\mathfrak{h}$ of $\mathfrak{g}$ and a base Π of the corresponding system of roots. As above, denote by h_i and $x_{\pm i}$ the Cartan and Chevalley generators of $\mathfrak{g}$ associated with the base Π. Let $L(\lambda)$ be the Verma module of the highest weight λ and let v_λ be the corresponding highest weight vector. By definition, one has

$$h v_\lambda = \langle \lambda, h \rangle v_\lambda, \qquad h \in \mathfrak{h}, \qquad (3.144)$$

$$x v_\lambda = 0, \qquad x \in \mathfrak{n}_+. \qquad (3.145)$$

Since $L(\lambda)$ is a finite-dimensional $\mathfrak{g}$-module, then $\lambda_i \equiv \langle \lambda, h_i \rangle$, $i = 1, \ldots, r$, are nonnegative integers. The conditions (3.144)–(3.145) are equivalent to

$$h_i v_\lambda = \lambda_i v_\lambda, \qquad i = 1, \ldots, r, \qquad (3.146)$$

$$x_{+i} v_\lambda = 0, \qquad i = 1, \ldots, r. \qquad (3.147)$$

Using the relations

$$[x_{+j}, x_{-i}] = \delta_{ji} h_i$$

and (3.147), one obtains

$$x_{+j}(x_{-i} v_\lambda) = \delta_{ji} \lambda_i v_\lambda. \qquad (3.148)$$

If for some i one has $\lambda_i = 0$, then (3.148) implies

$$x_{+j}(x_{-i} v_\lambda) = 0.$$

Hence, either $x_{-i} v_\lambda = 0$ or $x_{-i} v_\lambda$ is a singular vector. Suppose that the vector $x_{-i} v_\lambda$ is a singular vector. Any singular vector in $L(\lambda)$ should be proportional to v_λ. The vector $x_{-i} v_\lambda$ is of weight $\lambda - \pi_i$ and, for this reason, it cannot be proportional to v_λ. Therefore, $x_{-i} v_\lambda = 0$. On the other hand, if for some i one has $\lambda_i \neq 0$, it follows from (3.148) that

$$x_{+i}(x_{-i} v_\lambda) = \lambda_i v_\lambda \neq 0,$$

and, therefore, $x_{-i} v_\lambda \neq 0$.

Thus, if $\lambda_i = 0$, then $x_{-i}v_\lambda = 0$; and if $\lambda_i = 0$, then $x_{-i}v_\lambda \neq 0$.

Let ρ_λ be the representation of G which is obtained by 'integration' of the representation of $\mathfrak{g}$ in $L(\lambda)$. Consider the projective space $P(L(\lambda))$ and denote by pr_λ the canonical projection from $L(\lambda)^\times \equiv L(\lambda) - \{0\}$ onto $P(L(\lambda))$. The representation ρ_λ, as any representation, defines the left action of G in $P(L(\lambda))$ satisfying the condition

$$a \cdot \mathrm{pr}_\lambda(v) = \mathrm{pr}_\lambda(\rho_\lambda(a)v) \tag{3.149}$$

for any $a \in G$ and $v \in L(\lambda)$. Denote by p_λ the point of $P(L(\lambda))$ corresponding to the vector v_λ. As follows from (3.144) and (3.145), the Lie algebra $\mathfrak{g}_{p_\lambda}$ of the isotropy subgroup G_{p_λ} contains the Borel subalgebra $\mathfrak{b}_+$. Hence, $\mathfrak{g}_{p_\lambda}$ is a parabolic subalgebra of $\mathfrak{g}$. Furthermore, the above discussion implies that this Lie algebra contains the vectors x_{-i} for all i such that $\lambda_i = 0$ and does not contain the vectors x_{-i} for all i such that $\lambda_i \neq 0$. Taking into account the general structure of parabolic subalgebras described in subsection 3.1.2, we conclude that

$$\mathfrak{g}_{p_\lambda} = \mathfrak{p}_{+i_1,\ldots,i_k},$$

where the positive integers $i_1, \ldots, i_k$ make up the set of all $i \in \{1, \ldots, r\}$ for which $\lambda_i = 0$. Since a parabolic subgroup of a connected Lie group is connected, then one concludes that

$$G_{p_\lambda} = P_{+i_1,\ldots,i_k},$$

and the orbit of p_λ is a submanifold of $P(L(\lambda))$ diffeomorphic to the flag manifold $F_{-i_1,\ldots,i_k}$. Define the embedding $\iota_{-i_1,\ldots,i_k}$ of the flag manifold $F_{-i_1,\ldots,i_k}$ into the projective space $P(L(\lambda))$ by

$$\iota_{-i_1,\ldots,i_k}(aP_{-i_1,\ldots,i_k}) \equiv a \cdot p_\lambda. \tag{3.150}$$

With this embedding, the Fubini–Study metric on $\mathbb{P}(L(\lambda))$ induces a Kähler metric on $F_{-i_1,\ldots,i_k}$; hence, the flag manifold $F_{-i_1,\ldots,i_k}$ is a Kähler manifold.

3.4.3 Generalised Plücker relations

Let φ_- be a holomorphic mapping from a complex manifold M of dimension 1 to the flag manifold $F_- = G/B_+$. Using the language of algebraic geometry, we call φ_- a holomorphic curve in F_-. For any parabolic subgroup $P_{+i_1,\dots,i_k}$ there is the natural projection $\pi_{-i_1,\dots,i_k}$ from F_- onto the flag manifold $F_{-i_1,\dots,i_k}$, defined by

$$\pi_{-i_1,\dots,i_k}(aB_+) \equiv aP_{+i_1,\dots,i_k}. \qquad (3.151)$$

Using this projection we can define a holomorphic curve $\varphi_{-i_1,\dots,i_k}$ in $F_{-i_1,\dots,i_k}$ as

$$\varphi_{-i_1,\dots,i_k} \equiv \pi_{-i_1,\dots,i_k} \circ \varphi_-.$$

Thus, any holomorphic curve φ_- in F_- generates a family of holomorphic curves $\varphi_{-i_1,\dots,i_k}$ labelled by the parabolic subgroups containing the Borel subgroup B_+.

Consider the flag manifold $F_{-i_1,\dots,i_k}$ as the orbit in the projectivised representation space $P(L(\lambda))$. The mapping $\varphi_{-i_1,\dots,i_k}$ is not, in general, an embedding. Nevertheless, the Kähler metric on $F_{-i_1,\dots,i_k}$ specifies a symmetric tensor field of type $\binom{0}{2}$ on M. This tensor field is called a *pseudo-metric* on M; we will denote it by g_λ. The fundamental form Φ_λ associated with this pseudo-metric is given by

$$\Phi_\lambda = (\varphi_-^* \circ \pi_{-i_1\dots,i_k}^* \circ \iota_{-i_1,\dots,i_k}^*)\Phi_\lambda^{FS}, \qquad (3.152)$$

where Φ_i^{FS} is the fundamental form associated with the Fubini–Study metric on $P(L(\lambda))$.

Proposition 3.25 *Any holomorphic local lift $\widetilde{\varphi}_-$ of the mapping φ_- to G leads to the following local representation of the fundamental form Φ_λ:*

$$\Phi_\lambda = -2\sqrt{-1}d^{(1,0)}d^{(0,1)}\ln(\widetilde{\varphi}_- v_\lambda, \widetilde{\varphi}_- v_\lambda).$$

Proof By definition, the mapping $\widetilde{\varphi}_-$ satisfies the relation

$$\varphi_- = \pi_- \circ \widetilde{\varphi}_-. \qquad (3.153)$$

Define the mapping $\tau_\lambda : G \to L(\lambda)^\times$ by

$$\tau_\lambda(a) \equiv \rho_\lambda(a)v_\lambda, \qquad a \in G,$$

and prove the following equality:

$$\mathrm{pr}_\lambda \circ \tau_\lambda = \iota_{-i_1,\dots,i_k} \circ \pi_{-i_1,\dots,i_k} \circ \pi_-. \qquad (3.154)$$

Indeed, using (3.149), for any $a \in G$ we obtain

$$(\mathrm{pr}_\lambda \circ \tau_\lambda)(a) = \mathrm{pr}_\lambda(\rho_\lambda(a)v_\lambda) = g \cdot p_\lambda.$$

On the other hand, it follows from (3.151) and (3.150) that

$$(\iota_{-i_1,\dots,i_k} \circ \pi_{-i_1,\dots,i_k} \circ \pi_-)(a) = \iota_{-i_1,\dots,i_k}(aP_{+i_1,\dots,i_k}) = a \cdot p_\lambda.$$

Hence, equality (3.154) is true. Relations (3.153) and (3.154) result in

$$\iota_{-i_1,\dots,i_k} \circ \pi_{-i_1,\dots,i_k} \circ \varphi_- = \mathrm{pr}_\lambda \circ \tau_\lambda \circ \widetilde{\varphi}_-.$$

Taking (3.152) and (3.143) into account, we obtain

$$\Phi_\lambda = (\widetilde{\varphi}_-^* \circ \tau_\lambda * \circ \, \mathrm{pr}_\lambda^*)\Phi_\lambda^{FS} = -2\sqrt{-1} d^{(1,0)} d^{(0,1)} (\widetilde{\varphi}_-^* \circ \tau_\lambda^* \circ F). \quad (3.155)$$

The assertion of the proposition is the direct consequence of (3.155). □

Proposition 3.25 shows that the function

$$K_\lambda \equiv \ln(\widetilde{\varphi}_- v_\lambda, \widetilde{\varphi}_- v_\lambda)$$

is a Kähler potential of the pseudo-metric g_λ on M, having Φ_λ as its fundamental form. Choosing different lifts $\widetilde{\varphi}_-$, we obtain different Kähler potentials.

Let us suppose now that $\varphi_{-*p}(\partial_{-p}) \in \mathcal{M}'_{-p}$ for any $p \in M$. Here the subset $\mathcal{M}'_{-p}$ of $T_{\varphi_-(p)}(F_-)$ is defined in section 3.2.5. The most convenient, for our purposes, choice of the lift $\widetilde{\varphi}_-$ can be constructed as follows. Suppose that the hermitian bilinear form $(\cdot\,,\cdot)$ in V_λ is chosen in such a way that the corresponding representation of the group G is σ-unitary with σ being the Chevalley conjugation. Denote by Σ the antiholomorphic automorphism of G corresponding to σ and define the mapping $\Sigma_- : F_- \to F_+ = G/B_-$ by

$$\Sigma_-(aB_+) \equiv \Sigma(a)B_-.$$

Now introduce the mapping $\varphi_+ : M \to F_+ = G/B_-$ given by

$$\varphi_+ \equiv \Sigma_- \circ \varphi_-.$$

The mapping φ_+ is antiholomorphic and, as follows from theorem 3.5, there exists the mapping φ such that

$$\varphi_\pm = \pi_\pm \circ \varphi.$$

The mapping φ, by construction, satisfies the reality condition. Construct for φ a local decomposition of form (3.77). With the help of (3.82), determine the mapping γ_- corresponding to the mapping μ_-. Finally choose the mapping $\widetilde{\varphi}_-$ as

$$\widetilde{\varphi}_- \equiv \mu_- \gamma_-.$$

It is clear that $\widetilde{\varphi}_-$ is a holomorphic local lift of φ_- to G. Thus, taking the σ-unitarity of the considered representation of the group

G into account, we obtain the following expression for the Kähler potential K_λ:

$$K_\lambda = \ln(\mu_-\gamma_-v_\lambda, \mu_-\gamma_-v_\lambda) = \ln(v_\lambda, (\mu_-\gamma_-)^\dagger(\mu_-\gamma_-)v_\lambda).$$

Further, propositions 3.23 and 3.24 imply that the above expression can be rewritten as

$$K_\lambda = \ln(v_\lambda, (\mu_+\gamma_+)^{-1}(\mu_-\gamma_-)v_\lambda),$$

where μ_- is the mapping arising in the decomposition of φ of form (3.76) and γ_- is the mapping determined from (3.82). Using theorem 3.4 we now conclude that

$$K_\lambda = \ln(v_\lambda, \gamma v_\lambda),$$

where γ is the mapping satisfying the abelian Toda equations (3.105). Using parametrisation (3.106), we obtain

$$K_\lambda = \sum_{i=1}^{r} f_i\lambda_i.$$

Here the Toda fields f_i satisfy equations (3.107).

For the fundamental representations L_i, $i = 1, \ldots, r$, denote the corresponding pseudo-metrics by g_i and the associated fundamental forms by Φ_i. In this case the Kähler potential K_i of the pseudo-metric g_i coincides with the Toda field f_i.

Theorem 3.7 *Under the conditions described above, the Ricci curvature tensors R_i of the pseudo-metrics g_i on M are connected with the corresponding fundamental forms Φ_i by the relations*

$$R_i = \sum_{j=1}^{r} k_{ij}\Phi_j. \tag{3.156}$$

Proof Since the Toda field f_i is the Kähler potential of g_i then

$$\Phi_i = -2\sqrt{-1}d^{(1,0)}d^{(0,1)}f_i. \tag{3.157}$$

Comparing (3.140) and (3.141) and using equations (3.107), for the pseudo-metrics g_i we find the expression

$$\begin{aligned} g_i &= \partial_-\partial_+f_i(dz^-\otimes dz^+ + dz^+\otimes dz^-) \\ &= 2k_i\exp[-(kf)_i](dz^-\otimes dz^+ + dz^+\otimes dz^-). \end{aligned}$$

Taking (3.142) into account and using again equations (3.107), we have

$$R_i = -2\sqrt{-1}\sum_{j=1}^{r} k_{ij} d^{(1,0)} d^{(0,1)} f_j;$$

and, taking account of (3.157), we arrive at (3.156). □

Relations (3.156) are called the *generalised infinitesimal Plücker relations*. The validity of these relations was first conjectured in Givental' (1989), and proved in Positsel'skii (1991). Our proof is based on the special choice of the Kähler potentials of the pseudometrics, and in this sense is more close to the proof of the standard Plücker formulas, see Griffiths & Harris (1978). Note also, that the relation between the abelian Toda fields and the Kähler potentials in question has been established for the A_r series in Gervais & Matsuo (1993), and for the other classical series (B_r, C_r, D_r) in Gervais & Saveliev (1996), using explicit calculations in a local coordinate parametrisation of the flag manifolds. The discussion above follows the lines of Razumov & Saveliev (1994) and is valid for an arbitrary simple Lie algebra $\mathfrak{g}$. The corresponding generalised global Plücker formulas are also valid (F. E. Burstall, personal communication).

4
Toda-type systems and their explicit solutions

4.1 General remarks

Roughly speaking, the Toda-type systems (3.102)–(3.104), in the form given in section 3.2.5, represent systems of elliptic partial differential equations. In physical applications one more often deals with systems of hyperbolic partial differential equations. For this reason we start the present chapter with a brief review of the procedure for obtaining the Toda-type systems and constructing their general solutions for the case of hyperbolic systems.

We begin by introducing the notion of a chiral manifold, see Gervais & Matsuo (1993). Let M be a two-dimensional manifold. Suppose that there exists an atlas $\{(U_\alpha; z^-_\alpha, z^+_\alpha)\}_{\alpha \in \mathcal{A}}$ of M, such that

$$\partial z^-_\alpha / \partial z^+_\beta = 0, \qquad \partial z^+_\alpha / \partial z^-_\beta = 0$$

for all $\alpha, \beta \in \mathcal{A}$, and $\bigcup_{\alpha \in \mathcal{A}} = M$. In this case we say that the atlas $\{(U_\alpha; z^-_\alpha, z^+_\alpha)\}_{\alpha \in \mathcal{A}}$ endows M with the structure of a *chiral manifold*. Here, any chart $(U; z^-, z^+)$ is called a *chiral chart* on M. A smooth function f on M is called *(anti)chiral* if

$$\partial f / \partial z^+ = 0 \qquad (\partial f / \partial z^- = 0)$$

for any chiral chart on M. A mapping φ from a chiral manifold M to a manifold N is called *(anti)chiral* if, with respect to any chiral chart on M and any chart on N, it is described by (anti)chiral functions.

Now let M be a simply connected chiral manifold, G a complex matrix semisimple Lie group, and $\mathfrak{g}$ the Lie algebra of G. A connection on the trivial fibre bundle $M \times G \to M$ is described by a $\mathfrak{g}$-valued 1-form ω. The form ω corresponding to a flat connection satisfies the zero curvature condition (3.40). For any such connection we can point out a mapping $\varphi : M \to G$, such that

$$\omega = \varphi^{-1} d\varphi. \tag{4.1}$$

On the other hand, any mapping $\varphi : M \to G$ generates, via this relation, a flat connection which is denoted ${}^{\varphi}\omega$.

Let ω be a flat connection on $M \times G \to M$. Choose a chiral chart $(U; z^-, z^+)$ on M and write

$$\omega = \omega_- dz^- + \omega_+ dz^+.$$

In terms of the $\mathfrak{g}$-valued functions $\omega_\pm$, the zero curvature condition (3.40) takes the form (3.45), which can be considered as a system of nonlinear partial differential equations. To obtain a nontrivial system we should impose some restrictions on the form ω.

Suppose that the Lie algebra $\mathfrak{g}$ is endowed with a $\mathbb{Z}$-gradation. The first condition we impose on the form ω is the requirement that ω_- takes values in $\widetilde{\mathfrak{b}}_-$, and ω_+ takes values in $\widetilde{\mathfrak{b}}_+$, where $\widetilde{\mathfrak{b}}_\pm$ are the parabolic subalgebras of $\mathfrak{g}$ defined by (3.19). We call this requirement the general grading condition. A mapping $\varphi : M \to G$ is said to satisfy the general grading condition if the corresponding connection ${}^{\varphi}\omega$ satisfies the general grading condition.

Now fix two positive integers $l_\pm$ and define the subspace $\widetilde{\mathfrak{m}}_\pm$ of $\mathfrak{g}$ by (3.54). We say that the connection ω satisfies the specified grading condition if ω_- takes values in $\widetilde{\mathfrak{m}}_- \oplus \widetilde{\mathfrak{h}}$ and ω_+ takes values in $\widetilde{\mathfrak{h}} \oplus \widetilde{\mathfrak{m}}_+$. Here the subalgebra $\widetilde{\mathfrak{h}}$ is defined by (3.20). A mapping φ is said to satisfy the specified grading condition if the corresponding connection ${}^{\varphi}\omega$ satisfies this condition.

Denote by $\widetilde{N}_\pm$ the connected Lie subgroups of G corresponding to the nilpotent subalgebras $\widetilde{\mathfrak{n}}_\pm$ defined by (3.18). Further, let $\widetilde{H}$ be a connected Lie subgroup of G corresponding to the subalgebra $\widetilde{\mathfrak{h}}$. It can be shown that any mapping $\varphi : M \to G$ has the following local modified Gauss decompositions:

$$\varphi = \mu_+ \nu_- \eta_- = \mu_- \nu_+ \eta_+, \tag{4.2}$$

where the mappings $\nu_\pm$ take values in the Lie groups $\widetilde{N}_\pm$, the mappings $\eta_\pm$ take values in the Lie group $\widetilde{H}$, and the mappings $\mu_\pm$ take values in subsets $a_\pm \widetilde{N}_\pm$ for some elements $a_\pm \in G$. The latter condition is equivalent to the requirement that $\mu_-^{-1}\partial_-\mu_-$ takes values in $\widetilde{\mathfrak{n}}_-$, while $\mu_+^{-1}\partial_+\mu_+$ takes values in $\widetilde{\mathfrak{n}}_+$.

A mapping $\varphi : M \to G$ satisfies the specified grading condition if and only if for any local modified Gauss decompositions (4.2) the mapping μ_- is chiral and $\mu_-^{-1}\partial_-\mu_-$ takes values in $\widetilde{\mathfrak{m}}_-$; while the mapping μ_+ is antichiral and $\mu_+^{-1}\partial_+\mu_+$ takes values in $\widetilde{\mathfrak{m}}_+$. In this case one can write

$$\mu_\pm^{-1}\partial_\pm\mu_\pm = \sum_{m=1}^{l_\pm} \lambda_{\pm m}, \tag{4.3}$$

where the mappings $\lambda_{\pm m}$ take values in $\mathfrak{g}_{\pm m}$. Here the mappings λ_{-m} are chiral, and the mappings λ_{+m} are antichiral.

Now choose some fixed elements $x_\pm$ belonging to the subspaces $\mathfrak{g}_{\pm l_\pm}$. The adjoint representation of G generates the action of the subgroup $\widetilde{H}$ in the grading subspaces $\mathfrak{g}_{\pm m}$. Denote the orbits of the elements $x_\pm$ by $\mathcal{O}_\pm$. Consider a mapping φ satisfying the specified grading condition, and suppose that the mappings $\lambda_{\pm l_\pm}$ entering representation (4.3) take values in $\mathcal{O}_\pm$. We say in such a case that the mapping φ is admissible. It is clear that there are local mappings $\gamma_\pm : M \to \widetilde{H}$, such that

$$\lambda_{\pm l_\pm} = \gamma_\pm x_\pm \gamma_\pm^{-1}.$$

Now one can prove that there exists a local $\widetilde{H}$-gauge transformation which brings the connection ${}^\varphi\omega$ to the connection ω with

$$\omega_- = \sum_{m=1}^{l_- -1} \upsilon_{-m} + x_- + \gamma^{-1}\partial_-\gamma, \tag{4.4}$$

$$\omega_+ = \gamma^{-1}\left(\sum_{m=1}^{l_+ -1} \upsilon_{+m} + x_+\right)\gamma. \tag{4.5}$$

Here the mapping $\gamma : M \to \widetilde{H}$ is given by

$$\gamma = \gamma_+^{-1}\eta_-\eta_+^{-1}\gamma_- \equiv \gamma_+^{-1}\eta\gamma_-, \tag{4.6}$$

and the mappings $\upsilon_{\pm m} : M \to \mathfrak{g}_{\pm m}$ are defined as

$$\sum_{m=1}^{l_- -1} \upsilon_{-m} \equiv \gamma_-^{-1}\eta^{-1}(\nu_-^{-1}\partial_-\nu_-)\eta\gamma_-, \tag{4.7}$$

$$\sum_{m=1}^{l_+ -1} \upsilon_{+m} \equiv \gamma_+^{-1}\eta(\nu_+^{-1}\partial_+\nu_+)\eta^{-1}\gamma_+. \tag{4.8}$$

We restrict ourselves by the choice $l_- = l_+ \equiv l$; asymmetric systems corresponding $l_- \neq l_+$ can be considered in the same way.

In this case the zero curvature condition for the connection with the components given by (4.4) and (4.5) leads to the equations

$$\partial_+ v_{-m} = [x_-, \gamma^{-1} v_{+(l-m)} \gamma] + \sum_{n=1}^{l-m-1} [v_{-(m+n)}, \gamma^{-1} v_{+n} \gamma], \quad (4.9)$$

$$\partial_+(\gamma^{-1}\partial_-\gamma) = [x_-, \gamma^{-1} x_+ \gamma] + \sum_{m=1}^{l-1} [v_{-m}, \gamma^{-1} v_{+m} \gamma], \quad (4.10)$$

$$\partial_- v_{+m} = [x_+, \gamma v_{-(l-m)} \gamma^{-1}] + \sum_{n=1}^{l-m-1} [v_{+(m+n)}, \gamma v_{-n} \gamma^{-1}]. \quad (4.11)$$

Denote by $\widetilde{H}_\pm$ the isotropy subgroups of the elements $x_\pm$. System (4.9)–(4.11) possesses the symmetry transformations of the form

$$\gamma' = \xi_+^{-1} \gamma \xi_-, \qquad v'_{\pm m} = \xi_\pm^{-1} v_{\pm m} \xi_\pm, \quad (4.12)$$

where $\xi_- : M \to \widetilde{H}_-$ is a chiral mapping, and $\xi_+ : M \to \widetilde{H}_+$ is an antichiral one. Call a system of partial differential equations of form (4.9)–(4.11) a Toda-type system. In this chapter we illustrate the general system (4.9)–(4.11) by considering concrete examples for $l = 1, 2$. It will also be shown how the general formulas from the preceding chapter for the solution of system (4.9)–(4.11) work for the equations under consideration.

Recall that to find the general solution to equations (4.9)–(4.11) we choose some mappings $\gamma_\pm : M \to \widetilde{H}$ and $\lambda_{\pm m} : M \to \mathfrak{g}_{\pm m}$, $m = 1, \ldots, l-1$. Here the mappings γ_- and λ_{-m} are chiral, while the mappings γ_+ and λ_{+m} are antichiral. Then we define the mappings $\lambda_\pm$ by

$$\lambda_\pm \equiv \gamma_\pm x_\pm \gamma_\pm^{-1} + \sum_{m=1}^{l-1} \lambda_{\pm m}.$$

The next step is to find the mappings $\mu_\pm$. They are obtained by the integration of the equations

$$\partial_\pm \mu_\pm = \mu_\pm \lambda_\pm. \quad (4.13)$$

One can easily check that the general solution to these equations is represented by the series of nested integrals

$$\mu_\pm(z^\pm) = a_\pm \Bigg[\sum_{n=0}^{\infty} \int_{c^\pm}^{z^\pm} dy_1^\pm \int_{c^\pm}^{y_1^\pm} dy_2^\pm \dots \int_{c^\pm}^{y_{n-1}^\pm} dy_n^\pm \times \lambda_\pm(y_n^\pm) \cdots \lambda_\pm(y_1^\pm) \Bigg], \quad (4.14)$$

where $a_\pm$ are some elements of G, and $c^\pm$ are some real numbers. It is clear that for such a solution we have

$$\mu_\pm(c^\pm) = a_\pm.$$

Moreover, the mappings $\mu_\pm$ take values in $a_\pm \widetilde{N}_\pm$. In fact, as was discussed in section 3.3.1, to obtain almost all solutions it suffice to take the elements $a_\pm$ belonging to the subgroups $\widetilde{N}_\pm$.

Further, we use the Gauss decomposition

$$\mu_+^{-1} \mu_- = \nu_- \eta \nu_+^{-1} \quad (4.15)$$

to obtain the mapping η. Then the mapping γ is calculated with the help of (4.6). Finally, the mappings $\upsilon_{\pm m}$, $m = 1, \dots, l-1$, are determined using (4.7) and (4.8), where the mappings $\nu_\pm$ can be found from (4.15).

Exercises

4.1 Derive the generalisation of equations (4.9)–(4.11) for the case where $l_+ \neq l_-$.

4.2 Abelian Toda systems

For the case where $l = 1$ equations (4.9)–(4.11) take the form

$$\partial_+(\gamma^{-1} \partial_- \gamma) = [x_-, \gamma^{-1} x_+ \gamma], \quad (4.16)$$

which are called the Toda equations. Consider the principal embedding of the Lie algebra $\mathfrak{sl}(2, \mathbb{C})$ into $\mathfrak{g}$, and the $\mathbb{Z}$-gradation defined by the grading operator $q = h/2$, where h is the Cartan generator of the $\mathfrak{sl}(2, \mathbb{C})$-subalgebra. Recall that for such a gradation we do not use tildes in notations. The subgroup H is generated here by a Cartan subalgebra $\mathfrak{h}$ of $\mathfrak{g}$ and, therefore, it is an abelian group. The discussion of abelian Toda systems given in section 3.2 is based on a local parametrisation of the mapping

$\gamma : M \to H$. Any such parametrisation leads to a loss of some global information. Therefore, at the beginning of this section we consider the transition to a global parametrisation of the mapping γ.

First note that, for the case under consideration, a convenient local parametrisation of the mapping γ is

$$\gamma = \exp\left(-\sum_{i=1}^{r} f_i h_i\right). \tag{4.17}$$

Using this parametrisation, we obtain from (4.16) the equations

$$\partial_+ \partial_- f_i = 2k_i \exp\left(\sum_{j=1}^{r} k_{ij} f_j\right).$$

Introducing the notation

$$\beta_i = \exp(-f_i),$$

we arrive at the following system of partial differential equations:

$$\partial_+(\beta_i^{-1}\partial_-\beta_i) = -2k_i \prod_{j=1}^{r} \beta_j^{-k_{ij}}, \qquad i = 1, \ldots, r. \tag{4.18}$$

Consider now classical Lie groups and investigate, to what extent the introduced parametrisation is a global one.

4.2.1 Lie group $\mathrm{SL}(r+1, \mathbb{C})$

Let us begin with the Lie group $\mathrm{SL}(r+1, \mathbb{C})$. Using the formulas given in section 1.3.1, one sees that the Cartan matrix k in this case has the form

$$k = \begin{pmatrix} 2 & -1 & 0 & \cdots & 0 & 0 & 0 \\ -1 & 2 & -1 & \cdots & 0 & 0 & 0 \\ 0 & -1 & 2 & \cdots & 0 & 0 & 0 \\ \vdots & \vdots & \vdots & \ddots & \vdots & \vdots & \vdots \\ 0 & 0 & 0 & \cdots & 2 & -1 & 0 \\ 0 & 0 & 0 & \cdots & -1 & 2 & -1 \\ 0 & 0 & 0 & \cdots & 0 & -1 & 2 \end{pmatrix}.$$

After some calculations one obtains the following expression for the inverse of k:

$$k^{-1} = \frac{1}{r+1} \begin{pmatrix} r & r-1 & r-2 & \cdots & 3 & 2 & 1 \\ r-1 & 2(r-1) & 2(r-2) & \cdots & 6 & 4 & 2 \\ r-2 & 2(r-2) & 3(r-2) & \cdots & 9 & 6 & 3 \\ \vdots & \vdots & \vdots & \ddots & \vdots & \vdots & \vdots \\ 3 & 6 & 9 & \cdots & 3(r-2) & 2(r-2) & r-2 \\ 2 & 4 & 6 & \cdots & 2(r-2) & 2(r-1) & r-1 \\ 1 & 2 & 3 & \cdots & r-2 & r-1 & r \end{pmatrix}.$$

Using this expression we obtain

$$2k_i = 2\sum_{j=1}^{r}(k^{-1})_{ij} = i(r-i+1).$$

The group H consists of all complex nondegenerate diagonal matrices from $\mathrm{SL}(r+1,\mathbb{C})$, and we have the following parametrisation of the mapping γ:

$$\gamma = \begin{pmatrix} \beta_1 & 0 & \ldots & 0 & 0 \\ 0 & \beta_1^{-1}\beta_2 & \ldots & 0 & 0 \\ \vdots & \vdots & \ddots & \vdots & \vdots \\ 0 & 0 & \ldots & \beta_{r-1}^{-1}\beta_r & 0 \\ 0 & 0 & \ldots & 0 & \beta_r^{-1} \end{pmatrix}$$

It is clear that the parametrisation of γ via the functions β_i is global. Note that in this case equations (4.16) have the form

$$\partial_+(\beta_1^{-1}\partial_-\beta_1) = -r\beta_1^{-2}\beta_2,$$
$$\partial_+(\beta_i^{-1}\partial_-\beta_i) = -i(r-i+1)\beta_{i-1}\beta_i^{-2}\beta_{i+1}, \quad 1 < i < r,$$
$$\partial_+(\beta_r^{-1}\partial_-\beta_r) = -r\beta_{r-1}\beta_r^{-2}.$$

The simplest abelian Toda system, the Liouville equation, and its general solution were considered in example 3.11. Let us consider a more difficult case of the abelian Toda system based on the Lie group $\mathrm{SL}(3,\mathbb{C})$. As was discussed above, a global parametrisation of the mapping γ can be chosen here as

$$\gamma = \begin{pmatrix} \beta_1 & 0 & 0 \\ 0 & \beta_1^{-1}\beta_2 & 0 \\ 0 & 0 & \beta_2^{-1} \end{pmatrix}, \tag{4.19}$$

where β_1 and β_2 are functions taking values in $\mathbb{C}^\times$. The elements $x_\pm$ are given by

$$x_+ = \begin{pmatrix} 0 & \sqrt{2} & 0 \\ 0 & 0 & \sqrt{2} \\ 0 & 0 & 0 \end{pmatrix}, \qquad x_- = \begin{pmatrix} 0 & 0 & 0 \\ \sqrt{2} & 0 & 0 \\ 0 & \sqrt{2} & 0 \end{pmatrix}.$$

The corresponding abelian Toda equations have the form

$$\partial_+(\beta_1^{-1}\partial_-\beta_1) = -2\beta_1^{-2}\beta_2, \tag{4.20}$$

$$\partial_+(\beta_2^{-1}\partial_-\beta_2) = -2\beta_1\beta_2^{-2}. \tag{4.21}$$

Parametrise the mappings $\mu_\pm$ as follows:

$$\mu_+ = \begin{pmatrix} 1 & \mu_{+12} & \mu_{+13} \\ 0 & 1 & \mu_{+23} \\ 0 & 0 & 1 \end{pmatrix}, \qquad \mu_- = \begin{pmatrix} 1 & 0 & 0 \\ \mu_{-21} & 1 & 0 \\ \mu_{-31} & \mu_{-32} & 1 \end{pmatrix}.$$

Then, for the mappings $\mu_\pm^{-1}$ we obtain the expressions

$$\mu_+^{-1} = \begin{pmatrix} 1 & -\mu_{+12} & -\mu_{+13} + \mu_{+12}\mu_{+23} \\ 0 & 1 & -\mu_{+23} \\ 0 & 0 & 1 \end{pmatrix},$$

$$\mu_-^{-1} = \begin{pmatrix} 1 & 0 & 0 \\ -\mu_{-21} & 1 & 0 \\ -\mu_{-31} + \mu_{-32}\mu_{-21} & -\mu_{-32} & 1 \end{pmatrix}.$$

Hence, one can write

$$\mu_+^{-1}\partial_+\mu_+ = \begin{pmatrix} 0 & \partial_+\mu_{+12} & \partial_+\mu_{+13} - \mu_{+12}\partial_+\mu_{+23} \\ 0 & 0 & \partial_+\mu_{+23} \\ 0 & 0 & 0 \end{pmatrix},$$

$$\mu_-^{-1}\partial_-\mu_- = \begin{pmatrix} 0 & 0 & 0 \\ \partial_-\mu_{-21} & 0 & 0 \\ \partial_-\mu_{-31} - \mu_{-32}\partial_-\mu_{-21} & \partial_-\mu_{-32} & 0 \end{pmatrix}.$$

Representing the mappings $\gamma_\pm$ in the form

$$\gamma_\pm = \begin{pmatrix} \beta_{\pm 1} & 0 & 0 \\ 0 & \beta_{\pm 1}^{-1}\beta_{\pm 2} & 0 \\ 0 & 0 & \beta_{\pm 2}^{-1} \end{pmatrix}, \tag{4.22}$$

we obtain

$$\lambda_+ = \gamma_+ x_+ \gamma_+^{-1} = \begin{pmatrix} 0 & \sqrt{2}\beta_{+1}^2 \beta_{+2}^{-1} & 0 \\ 0 & 0 & \sqrt{2}\beta_{+1}^{-1}\beta_{+2}^2 \\ 0 & 0 & 0 \end{pmatrix},$$

$$\lambda_- = \gamma_- x_- \gamma_-^{-1} = \begin{pmatrix} 0 & 0 & 0 \\ \sqrt{2}\beta_{-1}^{-2}\beta_{-2} & 0 & 0 \\ 0 & \sqrt{2}\beta_{-1}\beta_{-2}^{-2} & 0 \end{pmatrix}.$$

Using the above relations we see that the matrix equations (4.13) are equivalent to the system of equations

$$\partial_+ \mu_{+12} = \sqrt{2}\beta_{+1}^2 \beta_{+2}^{-1}, \quad \partial_+ \mu_{+23} = \sqrt{2}\beta_{+1}^{-1}\beta_{+2}^2,$$
$$\partial_+ \mu_{+13} - \mu_{+12}\partial_+ \mu_{+23} = 0,$$
$$\partial_- \mu_{-21} = \sqrt{2}\beta_{-1}^{-2}\beta_{-2}, \quad \partial_- \mu_{-32} = \sqrt{2}\beta_{-1}\beta_{-2}^{-2},$$
$$\partial_- \mu_{-31} - \mu_{-32}\partial_- \mu_{-21} = 0.$$

The general solution of this system is

$$\mu_{+12}(z^+) = m_{+12} + \sqrt{2}\int_{c^+}^{z^+} dy_1^+ \beta_{+1}^2(y_1^+)\beta_{+2}^{-1}(y_1^+),$$

$$\mu_{+23}(z^+) = m_{+23} + \sqrt{2}\int_{c^+}^{z^+} dy_1^+ \beta_{+1}^{-1}(y_1^+)\beta_{+2}^2(y_1^+),$$

$$\mu_{+13}(z^+) = m_{+13} + m_{+12}\sqrt{2}\int_{c^+}^{z^+} dy_1^+ \beta_{+1}^{-1}(y_1^+)\beta_{+2}^2(y_1^+)$$
$$+ 2\int_{c^+}^{z^+} dy_2^+ \int_{c^+}^{y_2^+} dy_1^+ \beta_{+1}^2(y_1^+)\beta_{+2}^{-1}(y_1^+)\beta_{+1}^{-1}(y_2^+)\beta_{+2}^2(y_2^+),$$

$$\mu_{-21}(z^-) = m_{-21} + \sqrt{2}\int_{c^-}^{z^-} dy_1^- \beta_{-1}^{-2}(y_1^-)\beta_{-2}(y_1^-),$$

$$\mu_{-32}(z^-) = m_{-32} + \sqrt{2}\int_{c^-}^{z^-} dy_1^- \beta_{-1}(y_1^-)\beta_{-2}^{-2}(y_1^-),$$

$$\mu_{-31}(z^-) = m_{-31} + m_{-32}\sqrt{2}\int_{c^-}^{z^-} dy_1^- \beta_{-1}(y_1^-)\beta_{-2}^{-2}(y_1^-)$$
$$+ 2\int_{c^-}^{z^-} dy_2^- \int_{c^-}^{y_2^-} dy_1^- \beta_{-1}(y_1^-)\beta_{-2}^{-2}(y_1^+)\beta_{-1}^{-2}(y_2^-)\beta_{-2}(y_2^-),$$

where $m_{\pm ij}$ are arbitrary complex numbers.

Now we should find the mapping η entering the Gauss decomposition (4.15) of the mapping $\mu_+^{-1}\mu_-$. Here we use the following Gauss decomposition of a general element a of the Lie group $\mathrm{SL}(3,\mathbb{C})$:

$$a = n_- h n_+^{-1},$$

where $n_\pm \in N_\pm$ and $h \in H$. As follows from the formulas of section 3.2.4 the explicit form of the nonzero matrix elements of the matrices n_-, h and n_+ is

$$n_{-21} = \frac{a_{21}}{a_{11}}, \quad n_{-31} = \frac{a_{31}}{a_{11}}, \quad n_{-32} = \frac{a_{11}a_{32} - a_{12}a_{31}}{a_{11}a_{22} - a_{21}a_{12}},$$

$$h_{11} = a_{11}, \quad h_{22} = \frac{a_{11}a_{22} - a_{21}a_{12}}{a_{11}}, \quad h_{33} = \frac{1}{a_{11}a_{22} - a_{21}a_{12}},$$

$$n_{+12} = -\frac{a_{12}}{a_{11}}, \qquad n_{+13} = \frac{a_{12}a_{23} - a_{13}a_{22}}{a_{11}a_{22} - a_{21}a_{12}},$$

$$n_{+23} = -\frac{a_{11}a_{23} - a_{13}a_{21}}{a_{11}a_{22} - a_{21}a_{12}}.$$

Actually, here we need only the expressions for the matrix elements of h. The expressions for the matrix elements of $n_\pm$ will be used later for the construction of the general solution of a higher grading generalisation of the system under consideration.

Having introduced the notation $\kappa \equiv \mu_+^{-1}\mu_-$, we find for the matrix elements of the mapping κ the expressions

$$\kappa_{11} = 1 - \mu_{+12}\mu_{-21} - \mu_{+13}\mu_{-31} + \mu_{+12}\mu_{+23}\mu_{-31},$$
$$\kappa_{12} = -\mu_{+12} - \mu_{+13}\mu_{-32} + \mu_{+12}\mu_{+23}\mu_{-32},$$
$$\kappa_{13} = -\mu_{+13} + \mu_{+12}\mu_{+23}, \quad \kappa_{21} = \mu_{-21} - \mu_{+23}\mu_{-31},$$
$$\kappa_{22} = 1 - \mu_{+23}\mu_{-32}, \quad \kappa_{23} = -\mu_{+23},$$
$$\kappa_{31} = \mu_{-31}, \quad \kappa_{32} = \mu_{-32}, \quad \kappa_{33} = 1.$$

After some calculations we obtain

$$\eta_{11} = 1 - \mu_{+12}\mu_{-21} - \mu_{+13}\mu_{-31} + \mu_{+12}\mu_{+23}\mu_{-31},$$
$$\eta_{22} = \eta_{11}^{-1}(1 - \mu_{+13}\mu_{-31} - \mu_{+23}\mu_{-32} + \mu_{+13}\mu_{-32}\mu_{-21}),$$
$$\eta_{33} = \eta_{11}^{-1}\eta_{22}^{-1}.$$

From relations (4.6), (4.19) and (4.22) it follows that

$$\beta_1 = \beta_{+1}^{-1}\beta_{-1}\eta_{11}, \qquad \beta_2 = \beta_{+2}^{-1}\beta_{-2}\eta_{11}\eta_{22}.$$

Thus, we finally come to the following expression for the general solution of system (4.20), (4.21):

$$\beta_1 = \beta_{+1}^{-1}\beta_{-1}(1 - \mu_{+12}\mu_{-21} - \mu_{+13}\mu_{-31} + \mu_{+12}\mu_{+23}\mu_{-31}), \quad (4.23)$$
$$\beta_2 = \beta_{+2}^{-1}\beta_{-2}(1 - \mu_{+13}\mu_{-31} - \mu_{+23}\mu_{-32} + \mu_{+13}\mu_{-32}\mu_{-21}); \quad (4.24)$$

which is parametrised by two chiral, β_{-1}, β_{-2}; and two antichiral, β_{+1}, β_{+2} functions taking values in $\mathbb{C}^\times$.

4.2.2 Lie group $\widetilde{\mathrm{SO}}(2r+1,\mathbb{C})$

The situation is more complicated for the case of the Lie group $\widetilde{\mathrm{SO}}(2r+1,\mathbb{C})$. The Cartan matrix has now the form

$$k = \begin{pmatrix} 2 & -1 & 0 & \cdots & 0 & 0 & 0 \\ -1 & 2 & -1 & \cdots & 0 & 0 & 0 \\ 0 & -1 & 2 & \cdots & 0 & 0 & 0 \\ \vdots & \vdots & \vdots & \ddots & \vdots & \vdots & \vdots \\ 0 & 0 & 0 & \cdots & 2 & -1 & 0 \\ 0 & 0 & 0 & \cdots & -1 & 2 & -2 \\ 0 & 0 & 0 & \cdots & 0 & -1 & 2 \end{pmatrix}$$

After some algebra, one arrives at the following expression for the inverse matrix:

$$k^{-1} = \frac{1}{2}\begin{pmatrix} 2 & 2 & 2 & \cdots & 2 & 2 & 2 \\ 2 & 4 & 4 & \cdots & 4 & 4 & 4 \\ 2 & 4 & 6 & \cdots & 6 & 6 & 6 \\ \vdots & \vdots & \vdots & \ddots & \vdots & \vdots & \vdots \\ 2 & 4 & 6 & \cdots & 2(r-2) & 2(r-2) & 2(r-2) \\ 2 & 4 & 6 & \cdots & 2(r-2) & 2(r-1) & 2(r-1) \\ 1 & 2 & 3 & \cdots & r-2 & r-1 & r \end{pmatrix}.$$

This expression allows one to show that

$$2k_i = i(2r-i+1), \quad 1 \le i < r, \qquad 2k_r = r(r+1)/2.$$

The group H consists of all diagonal matrices from $\widetilde{\mathrm{SO}}(2r+1,\mathbb{C})$, and the mapping γ has the form

$$\gamma = \begin{pmatrix} \alpha & 0 & 0 \\ 0 & 1 & 0 \\ 0 & 0 & (\alpha^T)^{-1} \end{pmatrix},$$

where α is a mapping taking values in the Lie group $\mathrm{D}(r,\mathbb{C})$ of all complex diagonal $r \times r$ matrices. Using the explicit expressions for the Cartan generators h_i given in table 1.4, we obtain that, in terms of the functions β_i, the mapping α can be written as

$$\alpha = \begin{pmatrix} \beta_1 & 0 & \dots & 0 & 0 \\ 0 & \beta_1^{-1}\beta_2 & \dots & 0 & 0 \\ \vdots & \vdots & \ddots & \vdots & \vdots \\ 0 & 0 & \dots & \beta_{r-2}^{-1}\beta_{r-1} & 0 \\ 0 & 0 & \dots & 0 & \beta_{r-1}^{-1}\beta_r^2 \end{pmatrix}.$$

From this expression we conclude that the function β_r enters the parametrisation of the mapping γ only in the form β_r^2. Therefore, to obtain a global parametrisation of γ one must use the function β_r^2 as a basic object. Introducing the functions

$$\delta_i \equiv \beta_i, \quad 1 \le i < r, \qquad \delta_r \equiv \beta_r^2,$$

we obtain the equations

$$\begin{aligned} &\partial_+(\delta_1^{-1}\partial_-\delta_1) = -2r\delta_1^{-2}\delta_2, \\ &\partial_+(\delta_i^{-1}\partial_-\delta_i) = -i(2r-i+1)\delta_{i-1}\delta_i^{-2}\delta_{i+1}, \quad 1 < i < r, \\ &\partial_+(\delta_r^{-1}\partial_-\delta_r) = -r(r+1)\delta_{r-1}\delta_r^{-1}; \end{aligned}$$

which describe the abelian Toda system associated with the Lie group $\widetilde{\mathrm{SO}}(2r+1,\mathbb{C})$.

4.2.3 Lie group $\widetilde{\mathrm{Sp}}(r,\mathbb{C})$

We proceed to the case of the Lie group $\widetilde{\mathrm{Sp}}(r,\mathbb{C})$. The Cartan matrix in this case is the transpose of the Cartan matrix for $\widetilde{\mathrm{SO}}(2r+1,\mathbb{C})$. The inverse matrices are certainly connected by the transposition. After some calculations we obtain

$$2k_i = i(2r-i), \quad 1 \le i < r, \qquad 2k_r = r^2.$$

The group H consists of all diagonal matrices from $\widetilde{\mathrm{Sp}}(r,\mathbb{C})$; hence, the mapping γ can be represented as

$$\gamma = \begin{pmatrix} \alpha & 0 \\ 0 & (\alpha^T)^{-1} \end{pmatrix}, \tag{4.25}$$

where α is a mapping taking values in the Lie group $\mathrm{D}(r,\mathbb{C})$. Using the explicit expressions for the Cartan generators h_i given in table 1.4, we obtain that, in terms of the functions β_i, the mapping α

has the form

$$\alpha = \begin{pmatrix} \beta_1 & 0 & \dots & 0 & 0 \\ 0 & \beta_1^{-1}\beta_2 & \dots & 0 & 0 \\ \vdots & \vdots & \ddots & \vdots & \vdots \\ 0 & 0 & \dots & \beta_{r-2}^{-1}\beta_{r-1} & 0 \\ 0 & 0 & \dots & 0 & \beta_{r-1}^{-1}\beta_r \end{pmatrix}.$$

Therefore, in this case the functions β_i provide a global parametrisation of the mapping γ. The corresponding abelian Toda system is

$$\begin{aligned} &\partial_+(\beta_1^{-1}\partial_-\beta_1) = -(2r-1)\beta_1^{-2}\beta_2, \\ &\partial_+(\beta_i^{-1}\partial_-\beta_i) = -i(2r-i)\beta_{i-1}\beta_i^{-2}\beta_{i+1}, \quad 1 < i < r, \\ &\partial_+(\beta_r^{-1}\partial_-\beta_r) = -r^2\beta_{r-1}^2\beta_r^{-2}. \end{aligned}$$

4.2.4 Lie group $\widetilde{\mathrm{SO}}(2r,\mathbb{C})$

As in the last example take the Lie group $\widetilde{\mathrm{SO}}(2r,\mathbb{C})$. It follows from the formulas of section 1.3.4 that the explicit form of the Cartan matrix for this case is

$$k = \begin{pmatrix} 2 & -1 & 0 & \cdots & 0 & 0 & 0 \\ -1 & 2 & -1 & \cdots & 0 & 0 & 0 \\ 0 & -1 & 2 & \cdots & 0 & 0 & 0 \\ \vdots & \vdots & \vdots & \ddots & \vdots & \vdots & \vdots \\ 0 & 0 & 0 & \cdots & 2 & -1 & -1 \\ 0 & 0 & 0 & \cdots & -1 & 2 & 0 \\ 0 & 0 & 0 & \cdots & -1 & 0 & 2 \end{pmatrix},$$

and for the inverse of k one has the expression

$$k^{-1} = \frac{1}{4}\begin{pmatrix} 4 & 4 & 4 & \cdots & 4 & 2 & 2 \\ 4 & 8 & 8 & \cdots & 8 & 4 & 4 \\ 4 & 8 & 12 & \cdots & 12 & 6 & 6 \\ \vdots & \vdots & \vdots & \ddots & \vdots & \vdots & \vdots \\ 4 & 8 & 12 & \cdots & 4(r-2) & 2(r-2) & 2(r-2) \\ 2 & 4 & 6 & \cdots & 2(r-2) & r & r-2 \\ 2 & 4 & 6 & \cdots & 2(r-2) & r-2 & r \end{pmatrix}.$$

It is not difficult to see that

$$2k_i = i(2r-i-1), \quad 1 \le i < r-1, \qquad 2k_{r-1} = 2k_r = r(r-1)/2.$$

Here the group H is the same as for the case of the Lie group $\widetilde{\mathrm{Sp}}(r, \mathbb{C})$, and the mapping γ has the form of (4.25), where the mapping α, in terms of the functions β_i, is

$$\alpha = \begin{pmatrix} \beta_1 & 0 & \dots & 0 & 0 \\ 0 & \beta_1^{-1}\beta_2 & \dots & 0 & 0 \\ \vdots & \vdots & \ddots & \vdots & \vdots \\ 0 & 0 & \dots & \beta_{r-2}^{-1}\beta_{r-1}\beta_r & 0 \\ 0 & 0 & \dots & 0 & \beta_{r-1}^{-1}\beta_r \end{pmatrix}.$$

From this expression we conclude that a global parametrisation of the mapping γ can be realised with the functions

$$\delta_i \equiv \beta_i, \quad 1 \le i < r-2, \qquad \delta_{r-1} \equiv \beta_{r-1}\beta_r, \qquad \delta_r \equiv \beta_r^2,$$

and the corresponding Toda system is

$$\begin{aligned} &\partial_+(\delta_1^{-1}\partial_-\delta_1) = -(2r-2)\delta_1^{-2}\delta_2, \\ &\partial_+(\delta_i^{-1}\partial_-\delta_i) = -i(2r-i-1)\delta_{i-1}\delta_i^{-2}\delta_{i+1}, \quad 1 < i < r, \\ &\partial_+(\delta_{r-1}^{-1}\partial_-\delta_{r-1}) = -\frac{r(r-1)}{2}(\delta_{r-2}\delta_{r-1}^{-2}\delta_r + \delta_{r-2}\delta_r^{-1}), \\ &\partial_+(\delta_r^{-1}\partial_-\delta_r) = -r(r-1)\delta_{r-2}\delta_r^{-1}. \end{aligned}$$

Exercises

4.2 Due to the symmetry of the root system or the Dynkin diagram for $\mathfrak{sl}(r, \mathbb{C})$, one can perform the reductions, sometimes called foldings, $\mathfrak{sl}(2r+1, \mathbb{C}) \to \mathfrak{so}(2r+1, \mathbb{C})$ and $\mathfrak{sl}(2r, \mathbb{C}) \to \mathfrak{sp}(r, \mathbb{C})$. Quite naturally this symmetry is manifested in the Cartan matrix entering the corresponding abelian Toda system. Obtain the abelian Toda systems associated with the algebras $\mathfrak{so}(2r+1, \mathbb{C})$ and $\mathfrak{sp}(r, \mathbb{C})$ from those for the complex special linear algebra using the relevant equations for of the Toda fields f_i, see (4.17).

4.3 Using the folding $\mathfrak{so}(7, \mathbb{C}) \to G_2$, obtain the abelian Toda equations for the case of the algebra G_2.

4.4 Show that the general solution of equations (4.20) and (4.21) can be described by the chiral and antichiral functions $f_a^{\mp}, 0 \le a \le 2$, entering the decomposition $\beta_1 = \sum_{a=0}^{2} f_a^+ f_a^-$ and submitted to the condition $\det \partial_\pm^a f_b^\pm = \pm 2\sqrt{2}$. Represent the solution of this condition as nested integrals of two arbitrary (anti)chiral functions.

4.5 Starting from representation (4.14) with the boundary condition $a_\pm = e$ and relation (3.126) with $|u\rangle = |v\rangle = |i\rangle$, where $|i\rangle$ is the highest weight vector of the ith fundamental representation, find the solution for the abelian Toda system (4.18) in the form of finite sums of nested integrals. Use a Verma basis.

4.3 Nonabelian Toda systems

4.3.1 Lie group $\widetilde{\mathrm{Sp}}(r, \mathbb{C})$

Let us begin our discussion of nonabelian Toda-type systems with an example based on the complex symplectic group $\widetilde{\mathrm{Sp}}(r, \mathbb{C})$. Endow the corresponding Lie algebra $\widetilde{\mathfrak{sp}}(r, \mathbb{C})$ with a $\mathbb{Z}$-gradation associated with the $\mathfrak{sl}(2, \mathbb{C})$-subalgebra constructed as follows. Define the Cartan generator h as the element with the characteristic

0 0 ... 0 2

Using the explicit form of the inverse of the Cartan matrix for the Lie algebra $\mathfrak{sp}(r, \mathbb{C})$, given in the previous section, and the relation

$$h = \sum_{i,j=1}^{r} (k^{-1})_{ij} n_j h_i, \tag{4.26}$$

where n_j are the labels entering the characteristic of h, we obtain

$$h = \sum_{i=1}^{n} i h_i.$$

The explicit form of the element h is

$$h = \begin{pmatrix} I_r & 0 \\ 0 & -I_r \end{pmatrix}.$$

Introduce a $\mathbb{Z}$-gradation of $\widetilde{\mathfrak{sp}}(r, \mathbb{C})$ choosing as the grading operator the element $h/2$. We obtain three grading subspaces, $\mathfrak{g}_0$ and $\mathfrak{g}_{\pm 1}$. The subspace $\mathfrak{g}_0 \equiv \widetilde{\mathfrak{h}}$ is formed by $2r \times 2r$ matrices a of the block form

$$a = \begin{pmatrix} x & 0 \\ 0 & -x^T \end{pmatrix},$$

where x is an arbitrary complex $r \times r$ matrix. The subspace $\mathfrak{g}_{+1} = \widetilde{\mathfrak{n}}_+$ is composed of the matrices a of the block form

$$a = \begin{pmatrix} 0 & y \\ 0 & 0 \end{pmatrix},$$

with y being an arbitrary complex $r \times r$ matrix which satisfies the relation $y^T = y$. Finally, the subspace $\mathfrak{g}_{-1} = \widetilde{\mathfrak{n}}_-$ consists of the matrices

$$a = \begin{pmatrix} 0 & 0 \\ z & 0 \end{pmatrix},$$

where the complex $r \times r$ matrix z satisfies the condition $z^T = z$. Consider now the corresponding subgroups of $\widetilde{\mathrm{Sp}}(r, \mathbb{C})$. The subgroup $\widetilde{H}$ is formed by $2r \times 2r$ matrices a of the block form

$$a = \begin{pmatrix} X & 0 \\ 0 & (X^T)^{-1} \end{pmatrix},$$

where X is an arbitrary complex nondegenerate $r \times r$ matrix. The subgroup $\widetilde{N}_+$ is composed of the matrices a of the block form

$$a = \begin{pmatrix} I_r & Y \\ 0 & I_r \end{pmatrix},$$

with Y being an arbitrary complex $r \times r$ matrix which satisfies the relation $Y^T = Y$. Finally, the subspace $\widetilde{N}_-$ consists of the matrices

$$a = \begin{pmatrix} I_r & 0 \\ Z & I_r \end{pmatrix},$$

where the complex $r \times r$ matrix Z satisfies the condition $Z^T = Z$.

One can see that the Chevalley generators of an $\mathfrak{sl}(2, \mathbb{C})$-subalgebra corresponding to the Cartan generator h can be chosen in the form

$$x_+ = \begin{pmatrix} 0 & I_r \\ 0 & 0 \end{pmatrix}, \qquad x_- = \begin{pmatrix} 0 & 0 \\ I_r & 0 \end{pmatrix}.$$

Parametrise the mapping γ as

$$\gamma = \begin{pmatrix} \beta & 0 \\ 0 & (\beta^T)^{-1} \end{pmatrix},$$

where the mapping β takes values in the Lie group $\mathrm{GL}(r, \mathbb{C})$. Using this parametrisation we obtain the following matrix equation:

$$\partial_+(\beta^{-1}\partial_-\beta) = -\beta^{-1}\beta^{T-1}. \tag{4.27}$$

Now construct the general solution of this equation.

Write the mappings $\mu_\pm$ in the form

$$\mu_+ = \begin{pmatrix} I_r & \mu_{+12} \\ 0 & I_r \end{pmatrix}, \qquad \mu_- = \begin{pmatrix} I_r & 0 \\ \mu_{-21} & I_r \end{pmatrix},$$

where the mappings μ_{+12} and μ_{-21} satisfy the relations $\mu_{+12}^T = \mu_{+12}$ and $\mu_{-21}^T = \mu_{-21}$. Parametrising the mappings $\gamma_\pm$ as

$$\gamma_\pm = \begin{pmatrix} \beta_\pm & 0 \\ 0 & (\beta_\pm^T)^{-1} \end{pmatrix},$$

we find that, in our case, equations (4.13) are equivalent to the equations

$$\partial_+ \mu_{+12} = \beta_+ \beta_+^T, \qquad \partial_- \mu_{-21} = (\beta_-^T)^{-1} \beta_-^{-1}.$$

The general solution of these equations is given by

$$\mu_{+12}(z^+) = m_{+12} + \int_{c^+}^{z^+} dy^+ \beta_+(y^+) \beta_+^T(y^+),$$

$$\mu_{-21}(z^-) = m_{-21} + \int_{c^-}^{z^-} dy^- \beta_-^{T-1}(y^-) \beta_-^{-1}(y^-),$$

where m_{+12} and m_{-21} are constant complex $r \times r$ matrices satisfying the relations $m_{+12}^T = m_{+12}$ and $m_{-21}^T = m_{-21}$. For the mapping $\kappa \equiv \mu_+^{-1} \mu_-$ we have the representation

$$\kappa = \begin{pmatrix} I_r - \mu_{+12}\mu_{-21} & -\mu_{+12} \\ \mu_{-21} & I_r \end{pmatrix}.$$

Here it is convenient to consider all the matrices arising in our construction as 2×2 matrices over the associative algebra $\mathrm{Mat}(r, \mathbb{C})$. The Gauss decomposition for such matrices is given in section 3.2.4. Using the formulas obtained there, we obtain the following representation for the general solution of equation (4.27):

$$\beta = \beta_+^{-1}(I_r - \mu_{+12}\mu_{-21})\beta_-,$$

where β_- and β_+ are arbitrary chiral and antichiral mappings taking values in $\mathrm{GL}(r, \mathbb{C})$. Note also that the subgroups $\widetilde{H}_\pm$ in the case under consideration are isomorphic to the Lie group $\widetilde{\mathrm{O}}(r, \mathbb{C})$, and the symmetry transformations (4.12) look as follows:

$$\beta' = \varepsilon_+^{-1} \beta \varepsilon_-,$$

where the mappings $\varepsilon_\pm$ satisfy the relation

$$\varepsilon_\pm^T \varepsilon_\pm = I_r;$$

the mapping ε_- is chiral and the mapping ε_+ is antichiral.

4.3.2 Lie group $\widetilde{\mathrm{SO}}(2r+1,\mathbb{C})$

Consider now some nonabelian Toda systems associated with the complex special orthogonal group $\widetilde{\mathrm{SO}}(2r+1,\mathbb{C})$. Introduce a $\mathbb{Z}$-gradation of the Lie algebra $\tilde{\mathfrak{o}}(2r+1,\mathbb{C})$ connected with the $\mathfrak{sl}(2,\mathbb{C})$-subalgebra having as its Cartan generator h the element with the characteristic

$$\overbrace{2 \;\cdots\; 2}^{n} \;\; \overbrace{0 \;\cdots\; 0 \;\; 0}^{r-n}$$

Here n is a positive integer, such that $0 < n < r$. Using the explicit form of the inverse of the Cartan matrix for the Lie algebra $\mathfrak{o}(2r+1,\mathbb{C})$, given in section 4.2.2, and relation (4.26), we obtain

$$h = \sum_{i=1}^{n} i(2n-i+1)h_i + \frac{n(n+1)}{2}\left(2\sum_{i=n+1}^{r-1} h_i + h_r\right).$$

One can verify that the corresponding Chevalley generators of the $\mathfrak{sl}(2,\mathbb{C})$-subalgebra can be chosen in the form

$$x_\pm = \sum_{i=1}^{n-1} \sqrt{i(2n-i+1)}\,x_{\pm i} + \sqrt{\frac{n(n+1)}{2}}\,x_{\pm n,\ldots,r}.$$

Here and in what follows we use the notation

$$\begin{aligned} x_{+i_k,\ldots,i_3,i_2,i_1} &\equiv [x_{+i_k}, \ldots [x_{+i_3}, [x_{+i_2}, x_{+i_1}]] \ldots], \\ x_{-i_k,\ldots,i_3,i_2,i_1} &\equiv [\ldots [[x_{-i_1}, x_{-i_2}], x_{-i_3}] \ldots, x_{-i_k}]. \end{aligned}$$

It is convenient to write the matrices $x_\pm$ in the following block form:

$$x_+ = \begin{pmatrix} a_+ & b_+ & 0 \\ 0 & 0 & -b_+^T \\ 0 & 0 & -a_+^T \end{pmatrix}, \qquad x_- = \begin{pmatrix} a_- & 0 & 0 \\ b_- & 0 & 0 \\ 0 & -b_-^T & -a_-^T \end{pmatrix},$$

where $a_\pm$ are $n\times n$ matrices, b_+ is an $n\times(2(r-n)+1)$ matrix, and b_- is a $(2(r-n)+1)\times k$ matrix. The explicit form of the matrix a_+ is

$$a_+ = \begin{pmatrix} 0 & \sqrt{1\cdot 2n} & \cdots & 0 & 0 \\ 0 & 0 & \cdots & 0 & 0 \\ \vdots & \vdots & \ddots & \vdots & \vdots \\ 0 & 0 & \cdots & 0 & \sqrt{(n-1)(n+2)} \\ 0 & 0 & \cdots & 0 & 0 \end{pmatrix},$$

while for the matrix b_+ we have

$$b_+ = \begin{pmatrix} 0 & \cdots & 0 & 0 & 0 & \cdots & 0 \\ \vdots & \ddots & \vdots & \vdots & \vdots & \ddots & \vdots \\ 0 & \cdots & 0 & 0 & 0 & \cdots & 0 \\ 0 & \cdots & 0 & \sqrt{n(n+1)} & 0 & \cdots & 0 \end{pmatrix}.$$

The matrices a_- and b_- are the transposes of a_+ and b_+,

$$a_- = (a_+)^t, \qquad b_- = (b_+)^t.$$

The matrix valued function γ can be also written in a block form

$$\gamma = \begin{pmatrix} \alpha & 0 & 0 \\ 0 & \beta & 0 \\ 0 & 0 & (\alpha^T)^{-1} \end{pmatrix}, \tag{4.28}$$

where the mapping α takes values in $\mathrm{D}(n, \mathbb{C})$, and the mapping β takes values in $\widetilde{\mathrm{SO}}(2(r-n)+1, \mathbb{C})$ which can be written as

$$\beta^T \beta = I_{2(r-n)+1}.$$

Using the above formulas we obtain the relation

$$\gamma^{-1} x_+ \gamma = \begin{pmatrix} \alpha^{-1} a_+ \alpha & \alpha^{-1} b_+ \beta & 0 \\ 0 & 0 & -(\alpha^{-1} b_+ \beta)^T \\ 0 & 0 & -(\alpha^{-1} a_+ \alpha)^T \end{pmatrix},$$

which allows to write equations (4.16) as

$$\partial_+(\alpha^{-1}\partial_-\alpha) = [a_-, \alpha^{-1} a_+ \alpha] - \alpha^{-1} b_+ \beta b_-\,,$$
$$\partial_+(\beta^{-1}\partial_-\beta) = b_- \alpha^{-1} b_+ \beta - (b_- \alpha^{-1} b_+ \beta)^T.$$

To be more concrete, consider the Lie group $\widetilde{\mathrm{SO}}(7, \mathbb{C})$ and put $n = 2$. In this case the $\mathrm{SL}(2, \mathbb{C})$ subgroup is generated by the elements

$$h = 4h_1 + 6h_2 + 3h_3, \qquad x_\pm = 2x_{\pm 1} + \sqrt{3} x_{\pm 2,3},$$

and the grading operator is

$$q = 2h_1 + 3h_2 + \tfrac{3}{2} h_3. \tag{4.29}$$

The grading subspaces have the form

$$\mathfrak{g}_0 = \mathfrak{g}^{-\alpha_3} \oplus \mathfrak{h} \oplus \mathfrak{g}^{+\alpha_3}, \tag{4.30}$$
$$\mathfrak{g}_{\pm 1} = \mathfrak{g}^{\pm\alpha_1} \oplus \mathfrak{g}^{\pm\alpha_2} \oplus \mathfrak{g}^{\pm(\alpha_2+\alpha_3)} \oplus \mathfrak{g}^{\pm(\alpha_2+2\alpha_3)}, \tag{4.31}$$
$$\mathfrak{g}_{\pm 2} = \mathfrak{g}^{\pm(\alpha_1+\alpha_2)} \oplus \mathfrak{g}^{\pm(\alpha_1+\alpha_2+\alpha_3)} \oplus \mathfrak{g}^{\pm(\alpha_1+\alpha_2+2\alpha_3)}, \tag{4.32}$$
$$\mathfrak{g}_{\pm 3} = \mathfrak{g}^{\pm(\alpha_1+2\alpha_2+2\alpha_3)}. \tag{4.33}$$

Parametrise the mapping γ as

$$\gamma = e^{f_+ x_{+3}} e^{f_- x_{-3}} e^{f_1 h_1 + f_2 h_2 + f_3 h_3}.$$

Such a parametrisation leads to a mapping of form of (4.28) with

$$\alpha = \begin{pmatrix} e^{f_1} & 0 \\ 0 & e^{-f_1+f_2} \end{pmatrix},$$

$$\beta = \begin{pmatrix} e^{-f_2+2f_3}(1+f_-f_+)^2 & \sqrt{2}f_+(1+f_-f_+) & -e^{f_2-2f_3}f_+^2 \\ \sqrt{2}e^{-f_2+2f_3}f_-(1+f_-f_+) & 1+2f_-f_+ & -\sqrt{2}e^{f_2-2f_3}f_+ \\ -e^{-f_2+2f_3}f_-^2 & -\sqrt{2}f_- & e^{f_2-2f_3} \end{pmatrix}.$$

Direct calculations give

$$\begin{aligned} \gamma^{-1}\partial_-\gamma = {} & \partial_- f_1\, h_1 + \partial_- f_2\, h_2 + (\partial_- f_3 + f_-\partial_- f_+)\, h_3 \\ & + e^{f_2-2f_3}\partial_- f_+\, x_{+3} + e^{-f_2+2f_3}(\partial_- f_- - f_-^2\partial_- f_+)\, x_{-3}. \end{aligned}$$

Further, one can see that

$$\begin{aligned} [x_-, \gamma^{-1}x_+\gamma] = {} & -4e^{-2f_1+f_2}h_1 - 3e^{f_1-f_2}(1+2f_-f_+)\,(2h_2+h_3) \\ & + 6e^{f_1-2f_3}f_+\, x_{+3} + 6e^{f_1-2f_2+2f_3}f_-(1+f_-f_+)\, x_{-3}. \end{aligned}$$

Using the above relations, we obtain the following system of equations:

$$\begin{aligned} & \partial_+\partial_- f_1 = -4e^{-2f_1+f_2}, \\ & \partial_+\partial_- f_2 = -6e^{f_1-f_2}(1+2f_-f_+), \\ & \partial_+(\partial_- f_3 + f_-\partial_- f_+) = -3e^{f_1-f_2}(1+2f_-f_+), \\ & \partial_+(e^{f_2-2f_3}\partial_- f_+) = 6e^{f_1-2f_3}f_+, \\ & \partial_+(e^{-f_2+2f_3}(\partial_- f_- - f_-^2\partial_- f_+)) = 6e^{f_1-2f_2+2f_3}f_-(1+f_-f_+). \end{aligned}$$

Note that in some physical applications, in particular, in relation to black holes and relativistic string models, a constrained version of these equations arises, namely,

$$\begin{aligned} & \partial_+\partial_- v_1 = 2e^{v_1} - e^{v_2} - e^{v_3}, \\ & \partial_+\partial_- v_2 = -e^{v_1} + 2e^{v_2} - 2\frac{\sinh\frac{v_2-v_3}{4}}{\cosh^3\frac{v_2-v_3}{4}}\partial_+ v_4 \partial_- v_4, \\ & \partial_+\partial_- v_3 = -e^{v_1} + 2e^{v_2} + 2\frac{\sinh\frac{v_2-v_3}{4}}{\cosh^3\frac{v_2-v_3}{4}}\partial_+ v_4 \partial_- v_4, \\ & \partial_+\left(\tanh^2\frac{v_2-v_3}{4}\partial_- v_4\right) + \partial_-\left(\tanh^2\frac{v_2-v_3}{4}\partial_+ v_4\right) = 0; \end{aligned}$$

for more detail see Gervais & Saveliev (1992); Barbashov, Nesterenko & Chervyakov (1982) and Leznov & Saveliev (1992). Here

$$
\begin{aligned}
v_1 &= -2f_1 + f_2, \\
v_2 &= f_1 - f_2 + 2\text{Arsinh}\,(f_- f_+)^{1/2}, \\
v_3 &= f_1 - f_2 - 2\text{Arsinh}\,(f_- f_+)^{1/2}, \\
\partial_+ v_4 &= -(1 + f_- f_+)\partial_+[\ln\,(f_- e^{-f_2+2f_3})] + \partial_+[\ln\, f_- f_+]/2, \\
\partial_- v_4 &= \frac{1 + f_- f_+}{1 + 2f_- f_+}\partial_-[\ln\,(f_+ e^{f_2-2f_3})] - \partial_-[\ln\, f_- f_+]/2.
\end{aligned}
$$

We suggest that the reader perform such a reduction as an exercise.

Exercises

4.6 Prove the compatibility condition $\partial_+(\partial_- v_4) = \partial_-(\partial_+ v_4)$, and obtain the above given constrained system.

4.4 Higher grading systems

In general, system (4.9)–(4.11) describes the Toda-type fields coupled to matter fields parametrising the mappings $v_{\pm m}$; see Gervais & Saveliev (1995). For the case of affine Lie algebras this system was studied in Ferreira *et al.* (1996). Here we conventionally call the fields parametrising the mapping γ the Toda fields, while the fields entering a parametrisation of the mappings $v_\pm$ are called the matter fields. The reason for this becomes clear from the observation that, using a relevant specialisation of the Inönü–Wigner contraction, one can bring to zero the back reaction to the Toda fields for some or all matter fields. In particular, we can define the subalgebra $\mathfrak{k}_0$ of $\mathfrak{g}$ as

$$\mathfrak{k}_0 \equiv \mathfrak{g}_o \oplus \mathfrak{g}_{\pm l} \oplus \mathfrak{g}_{\pm 2l} \oplus \dots,$$

and take as $\mathfrak{k}_1$ the direct sum of all the remaining grading subspaces. Performing the Inönü–Wigner contraction now, as is described in section 1.1.11, we arrive at the case where equation (4.10) does not contain the mappings $v_{\pm m}$. As a result, we obtain the equation which looks similar to the equation describing some standard Toda system, but with a different meaning for the elements $x_\pm$ which belong here to the subspaces $\mathfrak{g}_{\pm l}$. Note that this type of system has been discussed in Ferreira, Miramontes & Guillén (1995). Evidently, there are many other meaningful possibilities for obtaining contracted systems.

In this section we discuss the equations corresponding to the case where $l = 2$ when a system of type (4.9)–(4.11) is rewritten in the form

$$\partial_+(\gamma^{-1}\partial_-\gamma) = [x_-, \gamma^{-1}x_+\gamma] + [v_-, \gamma^{-1}v_+\gamma], \qquad (4.34)$$

$$\partial_+ v_- = [x_-, \gamma^{-1}v_+\gamma], \qquad (4.35)$$

$$\partial_- v_+ = [x_+, \gamma v_- \gamma^{-1}], \qquad (4.36)$$

where we denote $v_{\pm 1}$ simply by $v_\pm$. Note that this system looks very similar to a system based on a semi-integral gradation of the Lie algebra $\mathfrak{g}$; see Leznov (1985) and Fehér *et al.* (1992); and to the supersymmetric Toda system associated with a superalgebra, see Leites, Saveliev & Serganova (1986), where, however, the mappings $v_\pm$ are parametrised by odd functions. Some special abelian cases of system (4.34)–(4.36) were also considered in Chao & Hou (1994); moreover, in Chao & Hou (1995) these authors have studied an asymmetric, as they called a heterotic Toda system, corresponding to the Lie algebra $\mathfrak{sl}(r+1, \mathbb{C})$ endowed with the principal gradation, and a choice when $l_- = 1, l_+ = 2$.

4.4.1 Lie group SL$(r+1, \mathbb{C})$

Consider the Lie group SL$(r+1, \mathbb{C})$ and endow the corresponding Lie algebra $\mathfrak{sl}(r+1, \mathbb{C})$ with the principal gradation. Parametrise the mapping γ as in (4.17) and the mappings $v_\pm$ by

$$v_\pm = \sum_{i=1}^{r} q_{\pm i} x_{\pm i}.$$

The general form of the elements $x_\pm \in \mathfrak{g}_{\pm 2}$ is

$$x_\pm = \sum_{i=1}^{r-1} c_i x_{\pm i, i+1}$$

with some constants c_i. One can easily see that the corresponding equations look as follows:

$$\partial_+(\beta_i^{-1}\partial_-\beta_i) = -c_i^2 \prod_{j=1}^{r} \beta_j^{-k_{ij}-k_{i+1,j}}$$

$$-c_{i-1}^2 \prod_{j=1}^{r} \beta_j^{-k_{i-1,j}-k_{i,j}} - \prod_{j=1}^{r} \beta_j^{-k_{ij}} q_{-i} q_{+i},$$

$$\partial_\pm q_{\mp i} = \pm(c_{i-1} \prod_{j=1}^{r} \beta_j^{-k_{i-1,j}} q_{\pm(i-1)} - c_i \prod_{j=1}^{r} \beta_j^{-k_{i+1,j}} q_{\pm(i+1)}),$$

where $c_i \equiv 0$ for $i = 0, r$; and $q_{\pm i} \equiv 0$ for $i = 0, r+1$. Introducing the functions

$$\delta_i \equiv \prod_{j=1}^{r} \beta_j^{-k_{ij}},$$

we arrive at the equivalent system of equations

$$\partial_+(\delta_i^{-1} \partial_- \delta_i) = \sum_{j=1}^{r} k_{ij} \delta_j (c_j^2 \delta_{j+1} + c_{j-1}^2 \delta_{j-1} + q_{-j} q_{+j}),$$

$$\partial_\pm q_{\mp i} = \pm(c_{i-1} \delta_{i-1} q_{\pm(i-1)} - c_i \delta_{i+1} q_{\pm(i+1)});$$

compare this with the equations given in Gervais & Saveliev (1995).

For the case where $r = 2$ there is actually only one possibility of choosing the elements $x_\pm$, namely,

$$x_+ = \begin{pmatrix} 0 & 0 & 1 \\ 0 & 0 & 0 \\ 0 & 0 & 0 \end{pmatrix}, \qquad x_- = \begin{pmatrix} 0 & 0 & 0 \\ 0 & 0 & 0 \\ 1 & 0 & 0 \end{pmatrix}.$$

This choice gives us the following equations:

$$\partial_+(\beta_1^{-1} \partial_- \beta_1) = -(\beta_1 \beta_2)^{-1} - \beta_1^{-2} \beta_2 q_{-1} q_{+1},$$

$$\partial_+(\beta_2^{-1} \partial_- \beta_2) = -(\beta_1 \beta_2)^{-1} - \beta_1 \beta_2^{-2} q_{-2} q_{+2},$$

$$\partial_+ q_{-1} = -\beta_1 \beta_2^{-2} q_{+2}, \qquad \partial_+ q_{-2} = \beta_1^{-2} \beta_2 q_{+1},$$

$$\partial_- q_{+1} = \beta_1 \beta_2^{-2} q_{-2}, \qquad \partial_- q_{+2} = -\beta_1^{-2} \beta_2 q_{-1}.$$

The subgroups $\widetilde{H}_\pm$ are isomorphic to the Lie group $\mathrm{GL}(1, \mathbb{C})$ and the mappings $\xi_\pm$, entering symmetry transformations (4.12), can be parametrised as follows:

$$\xi_\pm = \begin{pmatrix} \varepsilon_\pm & 0 & 0 \\ 0 & \varepsilon_\pm^{-2} & 0 \\ 0 & 0 & \varepsilon_\pm \end{pmatrix},$$

where ε_- is a chiral function, ε_+ is an antichiral one, and both of them take values in $\mathbb{C}^\times$. Using such a parametrisation, we see that in our case symmetry transformations (4.12) are given by the relations

$$\beta_1' = \varepsilon_+^{-1}\varepsilon_-\beta_1, \qquad \beta_2' = \varepsilon_+\varepsilon_-^{-1}\beta_2,$$
$$q_{-1}' = \varepsilon_-^3 q_{-1}, \qquad q_{-2}' = \varepsilon_-^{-3} q_{-2},$$
$$q_{+1}' = \varepsilon_+^{-3} q_{+1}, \qquad q_{+2}' = \varepsilon_+^3 q_{+2}.$$

We now proceed to the construction of the general solution of the system under consideration. Using (4.22) we obtain the expressions

$$\lambda_+ = \begin{pmatrix} 0 & \zeta_{+1} & \beta_{+1}\beta_{+2} \\ 0 & 0 & \zeta_{+2} \\ 0 & 0 & 0 \end{pmatrix}, \quad \lambda_+ = \begin{pmatrix} 0 & 0 & 0 \\ \zeta_{-1} & 0 & 0 \\ \beta_{-1}^{-1}\beta_{-2}^{-2} & \zeta_{-2} & 0 \end{pmatrix},$$

where ζ_{-1}, ζ_{-2} are arbitrary chiral functions, while ζ_{+1}, ζ_{+2} are arbitrary antichiral functions. The system of equations determining the mappings $\mu_\pm$ looks in our case as follows:

$$\partial_+\mu_{+12} = \zeta_{+1}, \quad \partial_+\mu_{+23} = \zeta_{+2},$$
$$\partial_+\mu_{+13} - \mu_{+12}\partial_+\mu_{+23} = \beta_{+1}\beta_{+2},$$
$$\partial_-\mu_{-21} = \zeta_{-1}, \quad \partial_-\mu_{-32} = \zeta_{-2},$$
$$\partial_-\mu_{-31} - \mu_{-32}\partial_-\mu_{-21} = \beta_{-1}^{-1}\beta_{-2}^{-1}.$$

The general solution of this system is

$$\mu_{+12}(z^+) = m_{+12} + \int_{c^+}^{z^+} dy_1^+ \zeta_{+1}(y_1^+),$$

$$\mu_{+23}(z^+) = m_{+23} + \int_{c^+}^{z^+} dy_1^+ \zeta_{+2}(y_1^+),$$

$$\mu_{+13}(z^+) = m_{+13} + m_{+12}\int_{c^+}^{z^+} dy_1^+ \zeta_{+2}(y_1^+)$$
$$+ \int_{c^+}^{z^+} dy_1^+ \beta_{+1}(y_1^+)\beta_{+2}(y_2^+) + \int_{c^+}^{z^+} dy_2^+ \int_{c^+}^{y_2^+} dy_1^+ \zeta_{+1}(y_1^+)\zeta_{+2}(y_2^+),$$

$$\mu_{-21}(z^-) = m_{-21} + \int_{c^-}^{z^-} dy_1^- \zeta_{-1}(y_1^-),$$

$$\mu_{-32}(z^-) = m_{-32} + \int_{c^-}^{z^-} dy_1^- \zeta_{-2}(y_1^-),$$

$$\mu_{-31}(z^-) = m_{-31} + m_{-32}\int_{c^-}^{z^-} dy_1^- \zeta_{-1}(y_1^-)$$

$$+\int_{c^-}^{z^-} dy_1^- \beta_{-1}^{-1}(y_1^-)\beta_{-2}^{-1}(y_1^-) + \int_{c^-}^{z^-} dy_2^- \int_{c^-}^{y_2^-} dy_1^- \zeta_{-2}(y_1^-)\zeta_{-1}(y_2^-),$$

where $m_{\pm ij}$ are arbitrary complex numbers. It is clear that we have for the functions β_1, β_2 expressions (4.23) and (4.24), where the matrix elements of the mappings $\mu_\pm$ are given above.

To find the expressions for the functions $q_{\pm 1}$ and $q_{\pm 2}$ we use the analogues of relations (3.95) and (3.92), which in our case look as follows:

$$\nu_+^{-1}\partial_+\nu_+ = (\eta^{-1}\nu_-^{-1}\lambda_+\nu_-\eta)_{\tilde{\mathfrak{n}}_+},$$
$$\nu_-^{-1}\partial_-\nu_- = (\eta\nu_+^{-1}\lambda_-\nu_+\eta^{-1})_{\tilde{\mathfrak{n}}_-}.$$

Substituting these equalities into (4.8) and (4.7), one obtains

$$\upsilon_+ = (\gamma_+^{-1}\nu_-^{-1}\lambda_+\nu_-\gamma_+)_{\tilde{\mathfrak{n}}_+},$$
$$\upsilon_- = (\gamma_-^{-1}\nu_+^{-1}\lambda_-\nu_+\gamma_-)_{\tilde{\mathfrak{n}}_-}.$$

Now, taking into account the explicit formulas for the Gauss decomposition (4.4) given in section 4.1.2, we come to the expressions

$$\begin{aligned}
q_{+1} &= \beta_{+1}^{-2}\beta_{+2}\zeta_{+1} \\
&\quad + \beta_{+1}^{-1}\beta_{+2}^{2}\frac{\mu_{-32} + \mu_{+12}\mu_{-31} - \mu_{+12}\mu_{-32}\mu_{-21}}{1 - \mu_{+13}\mu_{-31} - \mu_{+23}\mu_{-32} + \mu_{+13}\mu_{-32}\mu_{-21}}, \\
q_{+2} &= \beta_{+1}\beta_{+2}^{-2}\zeta_{+2} \\
&\quad - \beta_{+1}^{2}\beta_{+2}^{-1}\frac{\mu_{-21} + \mu_{+23}\mu_{-31}}{1 - \mu_{+12}\mu_{-21} - \mu_{+13}\mu_{-31} + \mu_{+12}\mu_{+23}\mu_{-31}}, \\
q_{-1} &= \beta_{-1}^{2}\beta_{-2}^{-1}\zeta_{-1} \\
&\quad - \beta_{-1}\beta_{-2}^{2}\frac{\mu_{+23} - \mu_{+13}\mu_{-21}}{1 - \mu_{+13}\mu_{-31} - \mu_{+23}\mu_{-32} + \mu_{+13}\mu_{-32}\mu_{-21}}, \\
q_{-2} &= \beta_{-1}^{-1}\beta_{-2}^{2}\zeta_{-2} \\
&\quad + \beta_{-1}^{2}\beta_{-2}\frac{\mu_{+12} + \mu_{+13}\mu_{-32} - \mu_{+12}\mu_{+23}\mu_{-32}}{1 - \mu_{+12}\mu_{-21} - \mu_{+13}\mu_{-31} + \mu_{+12}\mu_{+23}\mu_{-31}},
\end{aligned}$$

where, as above, the functions $\mu_{\pm ij}$ are determined by four chiral functions $\beta_{-1,-2}$ and $\zeta_{-1,-2}$, and by four antichiral functions $\beta_{+1,+2}$ and $\zeta_{+1,+2}$.

4.4.2 Lie group $\widetilde{\mathrm{SO}}(7,\mathbb{C})$

Now consider the Lie group $\widetilde{\mathrm{SO}}(7,\mathbb{C})$, and provide the corresponding Lie algebra $\widetilde{\mathfrak{o}}(7,\mathbb{C})$ with the gradation defined by the grading operator (4.29). The corresponding grading subspaces are given by (4.30)–(4.33). The subgroup $\widetilde{H}$ is isomorphic to $\mathrm{D}(2,\mathbb{C})\times\widetilde{\mathrm{SO}}(3,\mathbb{C})$. Introduce for it the following local parametrisation:

$$\gamma = e^{ah_3} e^{b(x_{-3}+x_{+3})} e^{dh_3} e^{a_2h_2} e^{h_1a_1},$$

where a, b, d, a_1 and a_2 are complex functions. The mapping γ is here of form (4.28) with

$$\alpha = \begin{pmatrix} e^{a_1} & 0 \\ 0 & e^{-a_1+a_2} \end{pmatrix},$$

$$\beta = \begin{pmatrix} e^{-a_2+2(a+d)}\cosh^2 b & \frac{1}{\sqrt{2}}e^{2a}\sinh 2b & -e^{a_2+2(a-d)}\sinh^2 b \\ \frac{1}{\sqrt{2}}e^{-a_2+2d}\sinh 2b & \cosh 2b & -\frac{1}{\sqrt{2}}e^{a_2-2d}\sinh 2b \\ -e^{-a_2-2(a-d)}\sinh^2 b & -\frac{1}{\sqrt{2}}e^{-2a}\sinh 2b & e^{a_2-2(a+d)}\cosh^2 b \end{pmatrix}.$$

After some lengthy but simple calculations, we obtain

$$\begin{aligned} \gamma^{-1}\partial_-\gamma = \partial_-a_1\, h_1 &+ \partial_-a_2\, h_2 + (\cosh 2b\, \partial_-a + \partial_-d)\, h_3 \\ &+ e^{a_2-2d}(\sinh 2b\, \partial_-a + \partial_-b)\, x_{+3} \\ &+ e^{-a_2+2d}(-\sinh 2b\, \partial_-a + \partial_-b)\, x_{-3}. \end{aligned}$$

Now we should choose the elements $x_\pm \in \mathfrak{g}_{\pm 2}$. Seemingly, the simplest, though rather nontrivial, possibility arises when one takes

$$x_\pm \equiv x_{\pm 1,3,2,3}.$$

For such a choice of $x_\pm$ we obtain

$$\begin{aligned} &[x_-, \gamma^{-1}x_+\gamma] \\ &\quad = -4e^{a_1+a_2-2(a+d)}\cosh^2 b(h_1+h_2+h_3) + 2e^{-2a-a_1}\sinh 2b\, x_{-3}. \end{aligned}$$

Introduce now an appropriate parametrisation of the mappings $v_\pm$. Convenient bases in the subspaces $\mathfrak{g}_{\pm 1}$ are formed by the elements

$$e_{\pm 0} \equiv x_{\pm 1}, \quad e_{\pm 1} \equiv \tfrac{1}{2}x_{\pm 3,2,3}, \quad e_{\pm 2} \equiv \tfrac{1}{\sqrt{2}}x_{\pm 2,3}, \quad e_{\pm 3} \equiv x_{\pm 2}.$$

Using these bases, we parametrise the mappings $v_\pm$ as

$$v_\pm \equiv \sum_{i=0}^{3} q_{\pm i} e_{\pm i}.$$

Define the functions $\widetilde{q}_{+i}$ by

$$\gamma^{-1} v_+ \gamma \equiv \sum_{i=0}^{3} \widetilde{q}_{+i} e_{+i}.$$

The explicit forms of the functions $\widetilde{q}_{+i}$ are

$$\begin{aligned}
\widetilde{q}_{+0} &= e^{-2a_1+a_2} q_{+0}, \\
\widetilde{q}_{+1} &= e^{a_1-2(a+d)} \cosh^2 b \, q_{+1} \\
&\quad - \tfrac{1}{\sqrt{2}} e^{a_1-2d} \sinh 2b \, q_{+2} - e^{a_1+2(a-d)} \sinh^2 b \, q_{+3}, \\
\widetilde{q}_{+2} &= -\tfrac{1}{\sqrt{2}} e^{-2a+a_1-a_2} \sinh 2b \, q_{+1} \\
&\quad + e^{a_1-a_2} \cosh 2b \, q_{+2} + \tfrac{1}{\sqrt{2}} e^{2a+a_1-a_2} \sinh 2b \, q_{+3}, \\
\widetilde{q}_{+3} &= -e^{a_1-2a_2+2(-a+d)} \sinh^2 b \, q_{+1} \\
&\quad + \tfrac{1}{\sqrt{2}} e^{a_1-2a_2+2d} \sinh 2b \, q_{+2} + e^{a_1-2a_2+2(a+d)} \cosh^2 b \, q_{+3}.
\end{aligned}$$

We will also need the expressions for the functions $\widetilde{q}_{-i}$ defined as

$$\gamma v_- \gamma^{-1} \equiv \sum_{i=0}^{3} \widetilde{q}_{-i} e_{-i}.$$

Using the parametrised form of the mapping γ, we come to the representation

$$\begin{aligned}
\widetilde{q}_{-0} &= e^{-2a_1+a_2} q_{-0}, \\
\widetilde{q}_{-1} &= e^{a_1-2(a+d)} \cosh^2 b \, q_{-1} \\
&\quad - \tfrac{1}{\sqrt{2}} e^{a_1-a_2-2a} \sinh 2b \, q_{-2} - e^{a_1-2a_2+2(-a+d)} \sinh^2 b \, q_{-3}, \\
\widetilde{q}_{-2} &= -\tfrac{1}{\sqrt{2}} e^{a_1-2d} \sinh 2b \, q_{-1} \\
&\quad + e^{a_1-a_2} \cosh 2b \, q_{-2} + \tfrac{1}{\sqrt{2}} e^{a_1-2a_2+2d} \sinh 2b \, q_{-3}, \\
\widetilde{q}_{-3} &= -e^{a_1+2(a-d)} \sinh^2 b \, q_{-1} \\
&\quad + \tfrac{1}{\sqrt{2}} e^{a_1-a_2+2a} \sinh 2b \, q_{-2} + e^{a_1-2a_2+2(a+d)} \cosh^2 b \, q_{-3}.
\end{aligned}$$

Now one can obtain the following relation:

$$\begin{aligned}
[v_-, \gamma^{-1} v_+ \gamma] &= -q_{-0} \widetilde{q}_{+0} \, h_1 \\
&\quad - (q_{-1}\widetilde{q}_{+1} + q_{-2}\widetilde{q}_{+2} + q_{-3}\widetilde{q}_{+3}) \, h_2 - (q_{-1}\widetilde{q}_{+1} + \tfrac{1}{2} q_{-2}\widetilde{q}_{+2}) \, h_3 \\
&\qquad - \tfrac{1}{\sqrt{2}} (q_{-2}\widetilde{q}_{+1} - q_{-3}\widetilde{q}_{+2}) \, x_{+3} - \tfrac{1}{\sqrt{2}} (q_{-1}\widetilde{q}_{+2} - q_{-2}\widetilde{q}_{+3}) \, x_{-3}.
\end{aligned}$$

With these formulas we come in the case under consideration to the following equations for the Toda fields:

$$\begin{aligned}
\partial_+ \partial_- a_1 &= -4 e^{-a_1+a_2-2(a+d)} \cosh^2 b - q_{-0} \widetilde{q}_{+0}, \\
\partial_+ \partial_- a_2 &= -4 e^{-a_1+a_2-2(a+d)} \cosh^2 b - q_{-1}\widetilde{q}_{+1} - q_{-2}\widetilde{q}_{+2} - q_{-3}\widetilde{q}_{+3},
\end{aligned}$$

$$
\begin{aligned}
\partial_+(\partial_- d + \cosh 2b\, \partial_- a) &= -4e^{-a_1+a_2-2(a+d)} \cosh^2 b \\
&\quad - \tfrac{1}{2}(q_{-1}\widetilde{q}_{+1} + q_{-2}\widetilde{q}_{+2}), \\
\partial_+(e^{a_2-2d}(\partial_- b + \sinh 2b\, \partial_- a)) &= \tfrac{1}{\sqrt{2}}(q_{-3}\widetilde{q}_{+2} - q_{-2}\widetilde{q}_{+1}), \\
\partial_+(e^{-a_2+2d}(\partial_- b - \sinh 2b\, \partial_- a)) &= 2e^{-2a-a_1} \sinh 2b \\
&\quad + \tfrac{1}{\sqrt{2}}(q_{-2}\widetilde{q}_{+3} - q_{-1}\widetilde{q}_{+2});
\end{aligned}
$$

see Gervais & Saveliev (1995). Now, using the relations

$$
\begin{aligned}
[x_-, \gamma^{-1} v_+ \gamma] &= -2\widetilde{q}_{+1} e_{-0} + 2\widetilde{q}_{+0} e_{-1}, \\
[x_+, \gamma v_- \gamma^{-1}] &= 2\widetilde{q}_{-1} e_{+0} - 2\widetilde{q}_{-0} e_{+1},
\end{aligned}
$$

we obtain the equations for the matter fields

$$
\begin{aligned}
\partial_\pm q_{\mp 0} &= \mp 2\widetilde{q}_{\pm 1}, \qquad & \partial_\pm q_{\mp 1} &= \pm 2\widetilde{q}_{\pm 0}, \\
\partial_\pm q_{\mp 2} &= 0, \qquad & \partial_\pm q_{\mp 3} &= 0.
\end{aligned}
$$

Let us give an example of a contracted form of the above considered system. Here we use the notation of section 1.1.11. To perform the Inönü–Wigner contraction of the algebra $\widetilde{\mathfrak{so}}(7, \mathbb{C})$ with the chosen gradation, define

$$
\begin{aligned}
\mathfrak{k}_0 &\equiv \mathfrak{g}_{-2} \oplus \mathfrak{g}_0 \oplus \mathfrak{g}_{+2}, \\
\mathfrak{k}_1 &\equiv \mathfrak{g}_{-3} \oplus \mathfrak{g}_{-1} \oplus \mathfrak{g}_{+1} \oplus \mathfrak{g}_{+3}.
\end{aligned}
$$

After the contraction we arrive at the algebra $\mathfrak{g}'$ which is not already simple; it is the semi-direct sum of the Lie algebra $\mathfrak{gl}(1, \mathbb{C}) \times \widetilde{\mathfrak{o}}(5, \mathbb{C})$ and the ten-dimensional commutative subalgebra with a basis formed by the elements $x_{\pm 1}, x_{\pm 2}, x_{\pm 3,2}, x_{\pm 3,2,3}, x_{\pm 2,1,3,2,3}$. Here $\mathfrak{gl}(1, \mathbb{C})$ is generated by the element $2h_2 + h_3$, while the subalgebra $\widetilde{\mathfrak{o}}(5, \mathbb{C})$ is spanned by the elements $x_{\pm 1,2}$, $x_{\pm 1,3,2}$, $x_{\pm 1,3,2,3}$, $x_{\pm 3}$. The resulting system has the form

$$
\begin{aligned}
\partial_+\partial_- a_1 &= -4e^{-a_1+a_2-2(a+d)} \cosh^2 b, \\
\partial_+\partial_- a_2 &= -4e^{-a_1+a_2-2(a+d)} \cosh^2 b, \\
\partial_+(\partial_- d + \cosh 2b\, \partial_- a) &= -4e^{-a_1+a_2-2(a+d)} \cosh^2 b, \\
\partial_+(e^{a_2-2d}(\partial_- b + \sinh 2b\, \partial_- a)) &= 0, \\
\partial_+(e^{-a_2+2d}(\partial_- b - \sinh 2b\, \partial_- a)) &= 2e^{-2a-a_1} \sinh 2b, \\
\partial_+ q_{-0} &= -2e^{a_1-2d}(e^{-2a} \cosh^2 b\, q_{+1} \\
&\quad - \tfrac{1}{\sqrt{2}} \sinh 2b\, q_{+2} - e^{2a} \sinh^2 b\, q_{+3}), \\
\partial_- q_{+0} &= 2e^{a_1-a_2-2a}(e^{a_2-2d} \cosh^2 b\, q_{-1} \\
&\quad - \tfrac{1}{\sqrt{2}} \sinh 2b\, q_{-2} - e^{-a_2+2d} \sinh^2 b\, q_{-3}),
\end{aligned}
$$

$$\partial_+ q_{-1} = 2e^{-2a_1+a_2} q_{+0}, \qquad \partial_- q_{+1} = -2e^{-2a_1+a_2} q_{-0},$$
$$\partial_\pm q_{\mp 2} = 0, \qquad \partial_\pm q_{\mp 3} = 0;$$

compare with those in Gervais & Saveliev (1995). In accordance with the general integration scheme discussed above, this contracted system, as well as the initial one, can be solved in an explicit way.

Exercises

4.7 Using a substitution, analogous to those for the functions $v_a, 1 \leq a \leq 4$, given in the end of section 4.3.2, obtain the constrained version for nonabelian systems considered in the present section.

References

Ablowitz, M.J. & Segur, H. (1981). *Solitons and Inverse Scattering Transform.* Philadelphia: SIAM.

Barbashov, B. M., Nesterenko, V. V. & Chervyakov, A. M. (1982). General solutions of nonlinear equations in the geometric theory of the relativistic string. *Communications in Mathematical Physics*, **84**, 471–9.

Bourbaki, N. (1975). *Éléments de Mathematique, Groupes et Algèbres de Lie. Chs VII-VIII.* Paris: Hermann.

Bourbaki, N. (1982). *Éléments de Mathematique, Groupes et Algèbres de Lie. Ch. IX.* Paris: Masson.

Burstall, F. E. & Rawnsley, J. H. (1990). *Twistor Theory for Riemannian Symmetric Spaces.* New York: Springer.

Calogero, F. & Degasperis, A. (1982). *Spectral Transform and Solitons.* Amsterdam: North-Holland.

Chao, L. & Hou, B.-Y. (1994). On the solutions of two-extended principal conformal Toda theory. *Annals of Physics*, **230**, 1–20.

Chao, L. & Hou, B.-Y. (1995). Heterotic Toda fields. *Nuclear Physics*, **B436**, 638–58.

Delduc, F., Ragoucy, E. & Sorba, P. (1992). Towards a classification of W algebras arising from non-abelian Toda theories. *Physics Letters*, **B279**, 319–25.

Dickey, L. A. (1991). *Soliton Equations and Hamiltonian Systems.* Singapore: World Scientific.

Dirac, P. A. M. (1958). *The Principles of Quantum Mechanics.* Oxford: Clarendon Press.

Dorfman, I. (1993). *Dirac Structures and Integrability of Nonlinear Evolution Equations.* New York: Wiley.

Dubrovin, B. A., Fomenko, A. T. & Novikov, S. P. (1985). *Modern Geometry: Methods and Applications, 2: The geometry and topology of manifolds.* New York: Springer.

Dubrovin, B. A., Fomenko, A. T. & Novikov, S. P. (1992). *Modern Geometry: Methods and Applications, 1: The Geometry of Surfaces, Transformation Groups, and Fields.* New York: Springer.

Dynkin, E. B. (1957a). Semi-simple subalgebras of semi-simple Lie algebras. *American Mathematical Society Translation Series 2*, **6**, 111–244.

Faddeev, L. D. & Takhtadjan, L. A. (1987). *Hamiltonian Methods in the Theory of Solitons.* Berlin: Springer.

Fehér, L., O'Raifeartaigh, L., Ruelle, P., Tsutsui, I. & Wipf, A. (1992). On Hamiltonian reductions of the Wess-Zumino-Novikov-Witten theories. *Physics Reports*, **222**, 1–64.

Ferreira, L. A., Gervais, J.-L., Guillén, J. S. & Saveliev, M. V. (1996). Affine Toda systems coupled to matter fields. *Nuclear Physics*, **B470**, 236–90.

Ferreira, L. A., Miramontes, J.L. & Guillén, J. S. (1995). Solitons, tau-functions and Hamiltonian reduction for non-abelian Toda theories. *Nuclear Physics*, **B449**, 631–79.

Fushchich, W. I., Serov, N. I. & Shtelen, W. M. (1989). *Symmetry Analysis and Exact Solutions of Nonlinear Equations of Mathematical Physics.* Kiev: Naukova Dumka.

Gelfand, I. M. & Retakh, V. S. (1991). Determinants of matrices over noncommutative rings. *Functional Analysis and its Applications*, **25**, 91–102.

Gelfand, I. M. & Retakh, V. S. (1992). A theory of noncommutative determinants and characteristic functions of graphs. *Functional Analysis and its Applications*, **26**, 1–20.

Gervais, J.-L. & Matsuo, Y. (1993). Classical A_n-W-geometry. *Communications in Mathematical Physics*, **152**, 317–68.

Gervais, J.-L. & Saveliev, M. V. (1992). Black holes from non-abelian Toda theories. *Physics Letters*, **B286**, 271–78.

Gervais, J.-L. & Saveliev, M. V. (1995). Higher grading generalisations of the Toda systems. *Nuclear Physics*, **B453**, 449–76.

Gervais, J.-L. & Saveliev, M. V. (1996). W-geometry of the Toda systems associated with non-exceptional simple Lie algebras. *Communications in Mathematical Physics*, **180**, 265–96.

Givental', A. B. (1989). Plücker formulae and Cartan matrices. *Russian Mathematical Surveys*, **44:3**, 193–4.

Gorbatsevich, V. V., Onishchik, A. L. & Vinberg, E. B. (1994). *Lie Groups and Lie Algebras. III.* New York: American Mathematical Society.

Goto, M. & Grosshans, F. (1978). *Semisimple Lie Algebras.* New York: Marcel Dekker.

Griffiths, P. & Harris, J. (1978). *Principles of Algebraic Geometry.* New York: Interscience.

Harish-Chandra (1953). Representations of semi-simple Lie groups on a Banach space, I. *Transactions of the American Mathematical Society*, **75**, 185–243.

Helgason, S. (1978). *Differential Geometry, Lie Groups, and Symmetric Spaces.* New York: Academic Press.

Humphreys, J. E. (1972). *Introduction to Lie Algebras and Representation Theory.* New York: Springer.

Ibragimov, N. H. (1987). *Transformation Groups Applied to Mathematical Physics.* Dordrecht: D. Reidel.

Inönü, E. & Wigner, E. P. (1953). On the contraction of groups and their representations. *Proceedings of the National Academy of Science of the USA*, **39**, 510–24.

Jimbo, M. & Miwa, T. (1995). *Algebraic Analysis of Solvable Lattice Models.* Providence: American Mathematical Society.

Kac, V. G. (1990). *Infinite Dimensional Lie Algebras.* Cambridge University Press.

Kelley, J. L. (1957). *General Topology.* New York: D. Van Nostrand.

Kirillov, A. A. (1976). *Elements of the Theory of Representations.* New York: Springer.

Kobayashi, S. & Nomizu, K. (1963). *Foundations of Differential Geometry, vol. I.* New York: Wiley.

Kobayashi, S. & Nomizu, K. (1969). *Foundations of Differential Geometry, vol. II.* New York: Wiley.

Kosniowski, C. (1980). *A First Course in Algebraic Topology.* Cambridge University Press.

Kostant, B. (1959). The principal three-dimensional subgroup and the Betti-numbers of a complex simple Lie group. *American Journal of Mathematics*, **81**, 973–1032.

Kostrikin, A. I. & Manin, Yu. I. (1989). *Linear Algebra and Geometry.* New York: Gordon and Breach.

Krasil'shchik, I. S., Lychagin, V. V. & Vinogradov, A. M. (1986). *Geometry of Jet Spaces, and Nonlinear Partial Differential Equations.* New York: Gordon and Breach.

Leites, D., Saveliev, M. & Serganova, V. (1986). Embeddings of Lie superalgebras $\mathfrak{osp}(m|n)$ into simple Lie superalgebras and integrable dynamical systems. In *Group-Theoretical Methods in Physics, 1*, eds. V. V. Dodonov, V. I. Manko & M. A. Markov, pp. 255–97. Utrecht: VNU Science Press.

Leznov, A. N. (1985). The internal symmetry group and methods of field theory for integrating exactly soluble dynamic systems. In *Group Theoretical Methods in Physics*, eds. M. A. Markov, V. I. Man'ko & A. E. Shabad, pp. 443–57. New York: Harwood Academic Publishers.

Leznov, A. N. & Saveliev, M. V. (1989). Exactly and completely integrable nonlinear dynamical systems. *Acta Applicandae Mathematicae*, **16**, 1–74.

Leznov, A. N. & Saveliev, M. V. (1992). *Group Theoretical Methods for Integration of Nonlinear Dynamical Systems.* Basel: Birkhauser.

Leznov, A. N., Smirnov, V. G. & Shabat, A. B. (1982). Internal symmetry group and integrability conditions for two-dimensional dynamical systems. *Theoretical and Mathematical Physics*, **51**, 10–21.

Lorente, M. & Gruber, B. (1972). Classification of semisimple subalgebras of simple Lie algebras. *Journal of Mathematical Physics*, **13**, 1639–63.

Manakov, S. V., Novikov, S. P., Pitaevsky, L. P. & Zakharov, V. E. (1984). *Theory of Solitons: The Method of the Inverse Scattering Problem.* New York: Plenum Press.

Marchenko, V. A. (1988). *Nonlinear Equations and Operator Algebras.* Dordrecht: D. Reidel.

Najmark, M. A. & Stern, A. I. (1982). *Theory of group representations.* Berlin: Springer.

Narasimhan, R. (1968). *Analysis on Real and Complex Manifolds.* Amsterdam: North-Holland.

Olver, P. J. (1986). *Application of Lie Groups to Differential Equations.* New York: Springer.

Ovsiannikov, L. V. (1982). *Group Analysis of Differential Equations.* New York: Academic Press.

Positsel'skii, L. E. (1991). Local Plücker formulas for a semisimple Lie group. *Functional Analysis and its Applications*, **25**, 291–2.

Razumov, A. V. & Saveliev, M. V. (1994). Differential geometry of Toda systems. *Communications in Analysis and Geometry*, **2**, 461–511.

Rudin, W. (1964). *Principles of Mathematical Analysis.* New York: McGraw Hill.

Saletan, E. I. (1961). Contraction of Lie groups. *Journal of Mathematical Physics*, **2**, 1–21.

Saveliev, M. V. & Vershik, A. M. (1990). A new class of infinite-dimensional Lie algebras (continuum Lie algebras) and associated nonlinear systems. In *Differential Geometric Methods in Mathematical Physics*, eds. V. Bruzzo, C. Bartocci & R. Cianci, pp. 162–70. New York: Springer.

Serre, J.-P. (1966). *Algèbres de Lie Semisimple Complexes.* New York: Benjamin.

Warner, F. W. (1983). *Foundations of Differentiable Manifolds and Lie*

Groups. New York: Springer.
Zhelobenko, D. P. (1994). *Representations of Reductive Lie Algebras*. Moscow: Nauka, in Russian.

Index

Printed in the United States
By Bookmasters